Library of Congress Cataloging-in-Publication Data

Names: Adler, Alexander, editor.
Title: The Lamiaceae family: an overview / [edited by] Alexander Adler.
Other titles: Plant science research and practices.
Identifiers: LCCN 2019058205 (print) | LCCN 2019058206 (ebook) | ISBN 9781536170788 (hardcover) | ISBN 9781536171181 (adobe pdf) Subjects: LCSH: Lamiaceae.
Classification: LCC QK495.L25 L36 2020 (print) | LCC QK495.L25 (ebook) | DDC 583/.96--dc23
LC record available at https://lccn.loc.gov/2019058205
LC ebook record available at https://lccn.loc.gov/2019058206

Published by Nova Science Publishers, Inc. † New York

CONTENTS

Preface **vii**

Chapter 1 Chemical Composition, Bioactivity and Commercial Potential of Some Himalayan Lamiaceae Species **1**
Chandra Shekhar Mathela and Rajendra Chandra Padalia

Chapter 2 Lamiaceae as a Source of Natural Antioxidants: Different Procedures for the Extraction of Polyphenolic Compounds from *Origanum Vulgare* and *O. Heracleoticum* **95**
Vanja M. Tadić, Snežana Brašanac Vukanović, Ana Žugić, Vesna Pešić Vukašinović and Nada Blagojević

Chapter 3 Liposome Encapsulation of *Origanum compactum* Essential Oil as the Active Ingredient: Assuring the Drug Effect **141**
Merima Ibišević, Vanja M. Tadić, Snežana Savić, Jelena Đoković and Ivana Nešić

Chapter 4	Impact of Green Synthesised Silver Nanoparticles on *In Vitro*-Grown *Lavandula officinalis* Plantlets *Pınar Nartop and Züleyha Günbeldek*	**181**
Chapter 5	Variation of Secondary Metabolites in *Lamiaceae* Herbs Water Extracts *Hasib Ahmad and Yoichi Matsubara*	**211**
Chapter 6	Homoploid Hybridization and Its Role in Emergence and Diversity of the Genus *Origanum* L. (Lamiaceae) *Tuncay Dirmenci, Turan Arabacı, Taner Özcan, Türker Yazıcı, Esra Martin, Sevcan Çelenk, Mikail Açar and Devrim Üzel*	**241**
Chapter 7	Aromatic Species of the Lamiaceae Family from Ecuador *Eduardo Valarezo, Jandry Rosales, Nixon Cumbicus and Miguel Angel Meneses*	**297**
Index		**327**

PREFACE

The Lamiaceae Family: An Overview provides information on the ethnobotanical, chemical diversity, pharmacological activities and commercial importance of over 100 Himalayan Lamiaceae species belonging to Nepeta, Ocimum, Mentha, Elsholtzia, Ajuga, Origanum, Thymus, Hyssopus, Calamintha, Plectranthus, Coleus, Perilla, Pogostemon, Salvia, Stachys, Teucrium and Scutellaria.

Next, the authors provide systematic and thorough information on the phenolic composition of two Origanum species extracts obtained using different solvents and/or methods shown to possess antioxidant properties that might be correlated to their traditionally established usage.

In order to protect essential oils from degradation, to improve their stability and bioavailability, different methods of their encapsulation in various colloidal systems such as microcapsules, microspheres, nanoemulsions and liposomes are introduced. In addition, literature data revealed that the incorporation of synthetic as well as herbal drugs into phospholipids carriers such as liposomes tended to improve their efficacy.

Following this, in vitro-grown plantlets of Lavandula officinalis were subjected to green-synthesized silver nanoparticles at 1, 2 and 4 mg/L concentrations. Growth parameters such as elongation, shoot and root formation and biomass accumulations are investigated to understand the effects of silver nanoparticles on micropropagation.

Additionally, the variation of major secondary metabolites in water extracts of 13 different Lamiaceae herbs are investigated. After raising for 8 weeks in the greenhouse, the herbs were harvested and cryopreserved using liquid nitrogen. Afterward, sample extracts (20% w/v) were prepared from the leaf using ultrapure water and analyzed using UPLC-MS/MS.

The authors presents some evidence about homoploid hybridization of Origanum genus put forward by Ietswaart. If habitat, ecological and geographic isolations occur between the hybrids and their parents, these hybrids can be speciated via homoploid hybridization in the future.

In conclusion, an overview of aromatic species belonging to the family Lamiaceae is presented, and the chemical composition of the volatile fraction of these existing species in Ecuador is also reported.

Chapter 1 - The Lamiaceae (formerly Labiateae, mint family) is a cosmopolitan family with ~200 genera and ~7000 species distributed worldwide and are known as the valuable source of diverse groups of secondary metabolites viz. terpenoids, alkaloids, glycosides, anthocyanins, flavonoids, lignans, tannins, phenolics, carotenoids, steroids and saponins used in flavours, medicine, nutrition and health-care products. In modern age, natural products have played a pivotal role in pharmaceutical, cosmeseuticals, agricultural, food-flavour and allied industries due to enormous advancement in isolation, characterization and synthesis of bioactive lead molecules and pharmaceutical intermediates. The demand of plants and their natural product isolates is on rise due to bio-safety and global acceptance. Aromatic plants are source of essential oils and aroma constituents considered as valuable natural products for flavors, fragrances, cosmetics, pharmaceuticals and aromatherapy. Himalaya is bestowed with enormous plant biodiversity and considered an ideal habitat for various plants of medicinal and commercial importance. It is believed that a large number of the plants described in "*Ayurvedic Materia Medica*" are distributed in the Himalayan region. Among enormous Himalayan plant biodiversity, Lamiaceae species have dominance in providing large number of molecules of bioactivity and flavour interest and have been in focus in search for diverse structural diversity and biological activity for commercial applications. Chemical investigations on Himalayan

Lamiaceae plants revealed large variations in phytoconstituents owing to different endogenous and exogenous factors. The existence of chemotypic variations with phytochemical diversity and distinct bio-efficacy within species could also be accredited to genetic and climatic and environmental conditions. The objective of this chapter, therefore, is to provide information on the ethnobotanical, chemical diversity, pharmacological activities and commercial importance of over 100 Himalayan Lamiaceae species belonging to *Nepeta*, *Ocimum*, *Mentha*, *Elsholtzia*, *Ajuga*, *Origanum*, *Thymus*, *Hyssopus*, *Calamintha*, *Plectranthus*, *Coleus*, *Perilla*, *Pogostemon*, *Salvia*, *Stachys*, *Teucrium* and *Scutellaria*.

Chapter 2 - Different species belonging to *Oregano* genera have been traditionally used in folk medicine to alleviate conditions such as asthma, bronchitis, cough, diarrhea, indigestion, stomachache, menstrual disorders, general infections, inflammation-related illnesses and diabetes. Besides, they have been widely used in agriculture, cosmetic and food industries. For centuries, they have been recognized as culinary herbs, flavoring substances in food products, alcoholic beverages and perfumery for their spicy fragrance. The benefits of *Oregano* species on human health have been attributed to their phytochemical composition. The most important pharmacologically active phytochemicals found in species of this genus have been categorized into two main groups, namely essential oils and phenolic compounds.

Most of the studies dealing with *Oregano* species have been focused on essential oils, frequently ignoring more polar fractions represented by polyphenolic compounds. Therefore, the authors used different extraction methods and solvents varying in polarity to establish suitable conditions for obtaining extracts rich in polyphenolic compounds presumingly possessing traditionally established effects. Hence, our research focused on the correlation between the investigated antioxidant properties and chemical profile of different types of extracts obtained from *Origanum vulgare* L. (*O. vulgare* ssp. *viride* (Boiss.) Hayek) and *O. heracleoticum* L. (*O. vulgare* L. ssp. *hirtum*) belonging to Lamiaceae family.

It is known that the extraction protocol might affect the qualitative and quantitative composition of plant extracts. Therefore, it is important to define the extraction conditions, including among the others the type of solvent and temperature, in order to achieve the satisfactory efficacy of the process, as well as to ensure the extraction of the compounds of interest avoiding their chemical modification. In this context, the aim of the current study was to establish correlation of chemical composition and antioxidant activity of the extracts obtained from the abovementioned species applying infusion, maceration and Soxhlet extraction (OVEI, OVEM and OVES, for *O. vulgare*, and OHEI, OHEM and OHES for *O. heracleoticum*, respectively). Taking into account that water herbal infusions are most often used traditional remedies, this type of conventional extraction was our first choice in the performed investigation. However, bearing in mind that industrial scale production usually uses ethanol as extraction vehicle, Soxhlet extraction with this solvent was employed as a model method of an exhaustive extraction of the plant material. But, as the Soxhlet extraction requires high temperature that might be disadvantageous for thermolabile substances, maceration was employed, as well, as the procedure performed at ambient temperature.

This research provided systematic and thorough information on the phenolic composition of two *Origanum* species extracts obtained using different solvents and/or methods shown to possess antioxidant properties that might be correlated to their traditionally established usage.

Chapter 3 - *Origanum compactum* Benth, Lamiaceae, is an endemic Moroccan medicinal herb. Although known as culinary spice in the country of origin Morocco, *O. compactum* has been traditionally used in the treatment of many diseases such as colitis, bronchopulmonary, gastric acidity and gastrointestinal diseases. Numerous studies showed that the *O. compactum* extracts, essential oil and its main components, thymol and carvacrol possess broad spectrum of pharmacological and therapeutic activities such as antibacterial, antifungal, antioxidant and anticancer.

Essential oils generally in pure form are not suitable for administration, because of their instability and susceptibility to the presence of extrinsic factors, like oxygen, light and temperature. In order to protect essential oils from degradation, to improve their stability and bioavailability, different methods of their encapsulation in various colloidal systems such as microcapsules, microspheres, nanoemulsions and liposomes were introduced. In addition, literature data revealed that the incorporation of synthetic as well as herbal drugs into phospholipids carriers such as liposomes tended to improve their efficacy, with the possibility to modulate the pharmacokinetics and/or tissue distribution of the drug in a beneficial way. In order to provide a starting active ingredient that could be incorporated in various pharmaceutical forms, the aim of our investigation was to prepare liposomal dispersion with encapsulated *O. compactum* essential oil with defined chemical and antibacterial properties. Liposomal dispersion was prepared by dissolving Phosal IP 40 (Lipoid, Germany), defined amounts of *O. compactum* essential oil and *Aqua ad injectabilia* applying homogenization. Further, in order to optimize homogenization, the obtained dispersion was homogenized in a high pressure homogenizer. Chemical composition was determined applying GC/MS and GC/FID methods, while the antibacterial activity was evaluated by agar diffusion and microdilution methods on standardized bacterial strains from ATCC collection *Staphylococcus aureus* ATCC 25923, *Enterococcus faecalis* ATCC 51299, *Escherichia coli* ATCC 25922, as well as on clinical strains collected from patients: *Staphylococcus sp.*, *E. faecalis*, *E. coli and Streptococcus agalactiae*.

Results indicated that *O. compactum* essential oil was rich in thymol and carvacrol, as main components. It possessed wide spectrum of activities against all bacterial strains tested, although in different degrees. The inhibition zones varied from 20.3 ± 0.57 mm to 47.1 ± 1.15 mm, and the values for MIC ranged from 0.195 mg/ml to 0.781 mg/ml for tested bacteria. For determining the degree of liposome encapsulation, GC/MS and GC/FID were applied and it was confirmed that more than 75% of thymol and more than 65% of carvacrol were encapsulated.

The obtained results of microbiological analysis indicated that *O. compactum* essential oil had potent antibacterial properties, what might place it as a potential replacement for conventional antimicrobial products. Undesirable effects can be diminished by effective incorporation into phospholipid carriers such as liposomes.

Chapter 4 - Metal nanoparticles have been continuously used in many applications in nanotechnology. Silver nanoparticles are the most used ones amongst the other metal nanoparticles in nanobiotechnology. They have strong antiseptic, antibacterial, antifungal and antiviral effects depending on their sizes and shapes. Moreover, they are eco-friendly elements which have relatively low toxicity towards the environment and human. These nanoparticles can also be used in plant biotechnology applications. Our previous studies showed that green synthesized silver nanoparticles can be used safely and effectively as a surface sterilization agent for *Lamiaceae* seeds. In this study, *in vitro*-grown plantlets of *Lavandula officinalis* were subjected to green-synthesized silver nanoparticles at 1, 2 and 4 mg/L concentrations. Growth parameters such as elongation, shoot and root formation and biomass accumulations were investigated to understand the effects of silver nanoparticles on micropropagation of *L. officinalis*. The results showed that silver nanoparticles enhanced single shoot formations and average node length in *L. officinalis* node cultures.

Chapter 5 - *Lamiaceae*, also known as Mint family is well known for including many medicinally important plants used in flavouring, preservative and especially medicinal purpose due to having curative and preventive properties. They have also found their use in combating crop pest and diseases owing to containing a complex mixture of secondary metabolites. However, information about the variation of these compounds in different *Lamiaceae* herbs especially in water extracts is a less ventured area. As such, in the present study, the variation of major secondary metabolites in water extracts of 13 different *Lamiaceae* herbs was investigated. After raising for 8 weeks in the greenhouse, the herbs were harvested and cryopreserved using liquid nitrogen. Afterward, sample extracts (20% w/v) were prepared from the leaf using ultrapure water and analyzed using UPLC-MS/MS. The top three peaks observed in the chromatogram of each herb and their corresponding m/z values were cross-checked with the mass bank and the identified compounds were considered to be the major components of the extract. About six flavonoids (apigenin, camphenol, luteolin, epigallocatechin, oenin and limocitrin) and two compounds involved in flavonoid synthesis (shikimic acid and coumaric acid) were identified. Furthermore, some other phenolic compounds like geniposide, rosmarininc acid and camphene were also identified. Among the 13 herbs, presence of apigenin in 9, kaempferol in 7, geniposide in 6, shikimic acid in 5, luteolin in 5 and coumaric acid in 2 were determined. Oenin and camphene were only found in hyssop whereas epigallocatechin was only found in rosemary. Rosmarinic acid and limocitrin were also found individually only in catnip and thyme respectively. From these results, it can be stated that the herbs under *Lamiaceae* family contain a variety of secondary metabolites which undoubtedly plays an important role in their characteristics as well as potential use.

Chapter 6 - Hybrid specimens can be occurred as same ploidy level (homoploidy) or not (polyploidy) with its parents. Homoploid hybrids are originated without any change in their ploidy level (2n) and polyploid hybrids are produced with the change in ploidy level (3n, 4n etc.). According to recent studies, the frequency of natural hybridization (homoploid and polyploid) has been estimated to 25% of all flowering plants. Recent studies show that the genus *Origanum* has homoploid hybrids. Hybridization is a common phenomenon among the species of *Origanum* in overlapping distribution habitats (natural or cultivated areas). Ietswaart suggested that the hybridization process is a significant factor in the speciation of *Origanum*, and *Origanum* is probably caused by the hybridization of the old tribe Saturejeae (now tribe Mentheae) genera. Based on the hypothesis of Ietswaart, we can discuss the emergence of the genus *Origanum*, its intrageneric speciation and its hybrids having the possibility for speciation via hybridization using morphological and cytological evidence. During the field studies, it was seen that many populations of wide-distributed species were overlapping, and some *Origanum* putative hybrid specimens were also observed in these overlapping areas with morphological differences from their parents. All studied *Origanum* species and hybrids have 2n = 30 chromosome numbers to define homoploidy. Palynological characters were revealed, and intermediate pollen structure (size, P/E ratio etc.) was seen in the hybrids between their parents. In addition, DNA extractions were done from all the hybrids and their parents. nrITS DNA was sequenced. Hybrid specimens have some heterozygous DNA loci in comparison to their parents. This study presents some evidence about homoploid the hybridization of the genus *Origanum* put forward by Ietswaart. If habitat, ecological and geographic isolations occur between the hybrids and their parents, these hybrids can be speciation via homoploid hybridization in the future.

Chapter 7 - Family Lamiaceae is amongst the largest families of flowering plants with about 250 genera and over 7,000 species distributed around the world. It is considered as the important source of essential oils. In Ecuador 219 species belonging to 27 genera of this family have been registered, of which about 29 species are endemic. In this chapter an over view of aromatic species belonging to the family Lamiaceae is presented, the chemical composition of the volatile fraction of these existing species in Ecuador is also reported. The plant material was collected in southern Ecuador and botanical specimens was identified by Herbario UTPL. The essential oils were obtained by hydrodistillation and then analysed by gas chromatography/flame ionization detector (GC/FID) and gas chromatography/mass spectrometry (GC/MS). Nine species were studied, including two species of the genus Clinopodium, two of the genus Hyptis, two of the genus Lepechinia, one of the genus Minthostachys, one of the genus Origanum and one of the genus Scutellaria. These species grow in the Andean forests, moors and dry inter-Andean valleys, over 1000 m a.s.l., less frequently in the dry forests of the Coast, Galapagos and Amazon. The principal constituents in essential oil of *Clinopodium nubigenum* was carvacrol acetate (38.1%), in *Clinopodium taxifolium* was Z-caryophyllene (24.1%) and in *Hyptis suaveolens* was eucalyptol (39.0%). Aromadendrene (24.6%) and □-3-carene (19.9%) were the principal constituents in essential oils of *Lepechinia paniculata* and *Lepechinia radula*, respectively. *Minthostachys mollis* essential oil reported (Z)-dihidro carvone (13.2%). In the essential oil of *Origanum vulgare* was identified γ-terpinene with a relative abundance of 20.8%. Germacrene D was principal constituents in essential oils of *Scutellaria volubilis* with 20.4% and in *Hyptis pectinata* with 34.1%.

In: The Lamiaceae Family: An Overview ISBN: 978-1-53617-078-8
Editor: Alexander Adler

Chapter 1

CHEMICAL COMPOSITION, BIOACTIVITY AND COMMERCIAL POTENTIAL OF SOME HIMALAYAN LAMIACEAE SPECIES

Chandra Shekhar Mathela[1]
and Rajendra Chandra Padalia[2]
[1]Department of Chemistry, Kumaun University,
Nainital, Uttarakhand, India
[2]CSIR-Central Institute of Medicinal and Aromatic Plants (CIMAP),
Research Center, Pantnagar, Uttarakhand, India

ABSTRACT

The Lamiaceae (formerly Labiateae, mint family) is a cosmopolitan family with ~200 genera and ~7000 species distributed worldwide and are known as the valuable source of diverse groups of secondary metabolites viz. terpenoids, alkaloids, glycosides, anthocyanins, flavonoids, lignans, tannins, phenolics, carotenoids, steroids and saponins used in flavours, medicine, nutrition and health-care products. In modern age, natural products have played a pivotal role in pharmaceutical, cosmeseuticals, agricultural, food-flavour and allied industries due to enormous advancement in isolation, characterization and synthesis of bioactive lead molecules and pharmaceutical intermediates. The demand of plants and their natural product isolates is on rise due to bio-safety and global acceptance. Aromatic plants are source of essential oils and aroma constituents considered as valuable natural products for flavors, fragrances, cosmetics, pharmaceuticals and aromatherapy. Himalaya is bestowed with enormous plant biodiversity and considered an ideal habitat for various plants of medicinal and commercial importance. It is believed that a large number of the plants described in "*Ayurvedic Materia Medica*" are distributed in the Himalayan region. Among enormous Himalayan plant biodiversity, Lamiaceae species have dominance in providing large number of molecules of bioactivity and flavour interest and have been in focus in search for diverse structural diversity and biological activity for commercial applications. Chemical investigations on Himalayan Lamiaceae plants revealed large variations in phytoconstituents owing to different endogenous and exogenous factors. The existence of chemotypic variations with phytochemical diversity and distinct bio-efficacy within species could also be accredited to genetic and climatic and environmental conditions. The objective of this chapter, therefore, is to provide information on the ethnobotanical, chemical diversity, pharmacological activities and commercial importance of over 100 Himalayan Lamiaceae species belonging to *Nepeta*, *Ocimum*, *Mentha*, *Elsholtzia*, *Ajuga*, *Origanum*, *Thymus*, *Hyssopus*, *Calamintha*, *Plectranthus*, *Coleus*, *Perilla*, *Pogostemon*, *Salvia*, *Stachys*, *Teucrium* and *Scutellaria*.

Keywords: Lamiaceae, medicinal and aromatic plants, bioactivity, Nepeta, Elsholtzia, Ajuga

INTRODUCTION

Plants were considered as the valuable reservoir of diverse natural products useful as biofuels, medicine, nutrition and other health-care products. Medicinal and aromatic plants are nature's gift to mankind as source of numerous structurally diverse bioactive molecules useful for industrial products, and are used as pharmaceutical intermediates, and chemical entities for synthetic modern medicines, nutraceuticals, food supplements, perfumery and cosmetic industries (Mathela and Kumar, 2018; Tripathi and Pandey, 2017). Of now, plants are used not only in traditional system of medicine, but are also used a source of numerous bioactive lead molecules or prototype for synthesis of various pharmacologically active compounds for drugs development in pharmaceutical industry (Mathela, 2017). Medicinal and aromatic plants are, therefore, receiving considerable attention all over the world for safe and cost-effective, preventive and curative therapies. WHO report indicates 80% of the world's population use the medicinal and aromatic plants through traditional systems for healthcare. The international market for medicinal and aromatic plants related trade is estimated ~US$ 62 billion, which is expected to increase upto ~US$ 5 trillion by the year 2050 (Mittal and Singh, 2011). Medicinal and aromatic plants synthesise and preserve a variety of secondary metabolites viz. terpenoids, alkaloids, glycosides, anthocyanins, flavonoids, lignans, tannins, phenolics, caretenoids, steroids and saponins and various other classes of constituents useful for human being for diverse applications (Ramakrishna and Ravishankar, 2011). The function and efficacy of medicinal and aromatic plants are based upon the reservoirs of these secondary metabolites. Various bioactive constituents belonging to structurally diverse secondary metabolites were found to possess antioxidant, antiseptic, anti-inflammatory, anti-asthmatic, anticancer, antispasmodic, antipyretic, diuretic, diaphoretic, sedative, antimicrobial, analgesic, fungicidal, herbicidal, insecticidal and insect repellent etc. A large number of compounds in use as medicine today are derivatives of these naturally occurring secondary metabolites or bioactive constituents (Mathela, 2017; Padalia, 2012; Sharma and Canoo, 2013). Moreover, the aromatic plants are source of volatile secondary metabolites (essential oils) as well which are used in preparation of various industrial commodities. In recent times, the demand of natural essential oils and their isolates for the food flavor, food preservation, cosmetic, and aromatherapy industries is on rise because of their relative bio-safety value, global acceptance, and amenability for diverse functional applications (Padalia et al. 2018).

The Himalayan region is characterised by diverse geographic features and covers a broad range of ecological habitats and enormous biodiversity. The Himalaya, spread over Northern part (East to West) is capped with snow-capped peaks, is characterized by the biological diversity (Figure 1). The name 'Himalaya' originated from Sanskrit literature from hima (snow) and alaya (abode) means 'abode of snow' that attracted pilgrim, philosophers, tourists, and scientists of all over the world for its scenic and ethnic magnificence and immense biodiversity. Based on geology, altitude, geography, etc., the Himalaya is categorized it in three broad zones: the northern or Tibetan zone, the Central or Himalayan zone and the outer or sub Himalayan zone; and based on altitude into the Greater Himalaya averaging ≥6100 m; the lesser Himalaya between 3600-4000 m; and the outer Himalaya (Siwalik ranges) between 800- 1500 m (Ahmad, 2011; Nandy et al. 2006; Valdiya, 2001). Indian Himalayan region sustain entire array of climatic conditions from the tropical to the deep arctic conditions, from the wettest spots of Arunachal Pradesh to the cold deserts of Ladakh. Hence it supports repository of plant diversity ranging from humid tropical evergreen to moist mixed deciduous Sal forests, marshes, swamps, mixed deciduous forests, subtropical pine forests, broad leaved temperate forests, temperate conifer forests, sub alpine and alpine vegetation and alpine meadows. The rich biodiversity of medicinal and aromatic plants of this Indian Himalayan region were utilised by native communities for traditional medicine from ancient times. Uttarakhand situated centrally in the long stretch of the Himalaya and is characterized by diverse geographical features ranging from snow-capped mountain in the North to tropical forests in the South. It has been divided into two regions- the western region (Garhwal division) and the eastern region (Kumaon division). These includes 13 district viz. six district in Kumaun (Almora, Nainital, Udham Singh Nagar, Champawat, Pithoragarh, Bageshwar) and seven district in Garhwal (Chamoli, Dehradun, Haridwar, Pauri Garhwal, Tehri Garhwal, Rudraprayag, Uttarkashi) (Figure 2). Uttarakhand Himalaya is also renowned for its forest resources and biodiversity. It has almost all major climatic zones from the valley regions to the highly elevated alpine meadows. The Valley of Flowers National Park, Nanda Devi National Park, Govind National Park and Gangotri National Park and tropical rainforests are the storehouse of various endemic and potential medicinal and aromatic plants used by local inhabitants for variety of application and in folk medicine (Kumari et al. 2009; Nandi et al. 2006; Samal et al. 2004). Therefore, our research programs mainly focused and targeted on various medicinal and aromatic plants of Uttarakhand Himalaya which are being used in folk medicines. Among enormous plant biodiversity in Himalayan region, a few families viz. Lamiaceae, Asteraceae, Apiaceae, Lauraceae, Poaceae, Rutaceae, Myrtaceae, Cupressaceae, Zingiberaceae and Piperaceae were found promising for various bioactive molecules and essential oils for commercial applications (Sharma and Canoo, 2013). Among them, Lamiaceae is one of the family focused by various researchers for diverse chemical diversity and biological functions. The Lamiaceae (formerly Labiateae, mint family) is a cosmopolitan family with ~200 genera and ~7000 species distributed worldwide (Heywoo, 1978; Raja, 2012). The original family name is Labiatae, originated due to its floral characteristics with petals fused into an upper lip and a lower lip of the flowering plants belonging to this family (Uritu et al. 2018). Leaves, often aromatic, are whorled or decussate emerging oppositely as pairs in right angles to the previous one. Various member of this family are commercially important and are cultivated worldwide for ornamental, culinary, medicine, food-flavor, fragrance and cosmetics (Lee et al. 2011). They are also being used in various traditional systems of medicines. The most well known plants of this family cultivated for medicinal and essential oil production are basil, mints, rosemary, thyme, Origanum, lavender, sage, Nepeta, Hyssopus, lemon balm, Elsholtzia, savory, Perilla, marjoram etc (Harley et al. 2004; Stanko et al. 2016). The plants belonging to this family consist of different class of pharmacologically active secondary metabolites with a wide spectrum of bioactivity for drug discovery as well as cosmetic, food, and pesticides industries (Hussein, 2018; Khaled-Khodja et al. 2014). Chemical investigations on Himalayan Lamiaceae plants revealed interesting variations due to genetic and climatic/environmental conditions. The focus of this chapter is to provide information on the ethnobotanical uses, chemical diversity and pharmacological activities and commercial aspects of over 100 Himalayan Lamiaceae species belonging to genus Nepeta, Ocimum, Mentha, Elsholtzia, Ajuga, Origanum, Thymus, Hyssopus,

Calamintha, Coleus, Perilla, Plectranthus, Pogostemon, Salvia, Stachys, Teucrium, and Scutellaria.

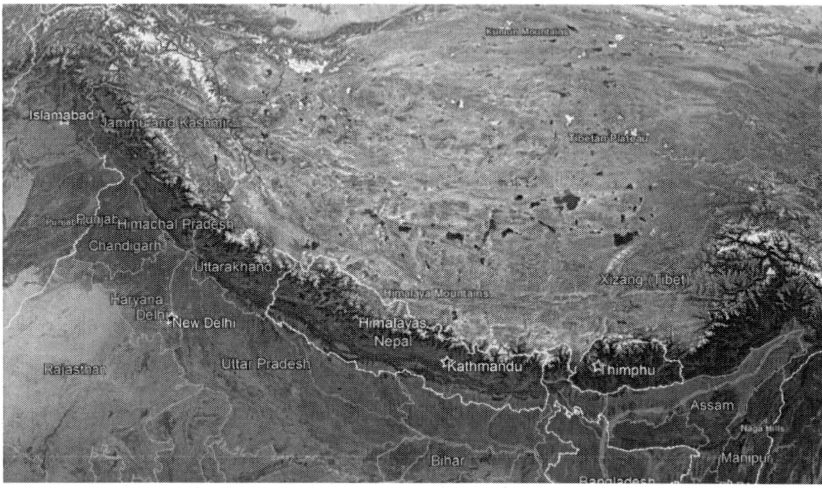

Figure 1. Google Earth map of the Himalayan regions (Source: Google Earth Pro).

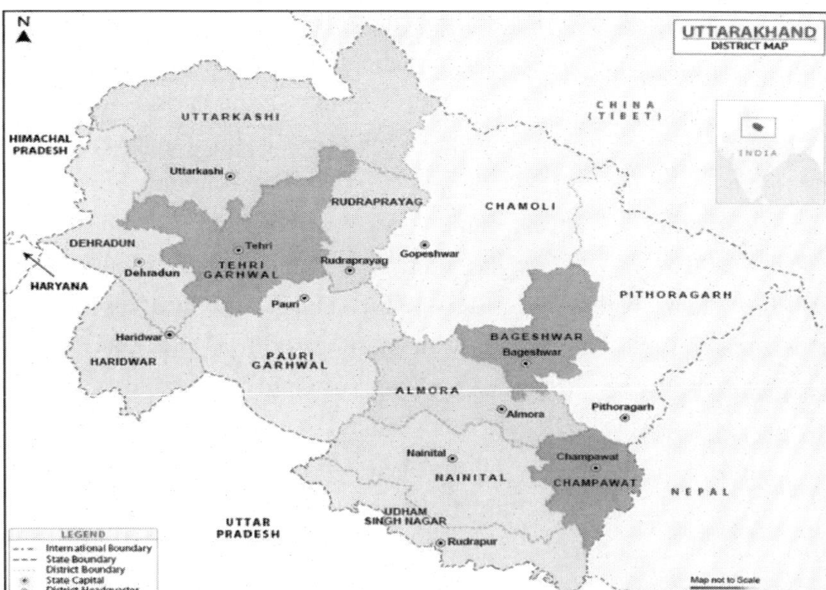

Figure 2. District and region wise map of Uttarakhand (Source: MAPs of India@ www.mapsofindia.com).

Chemical Diversity in Himalayan Lamiaceae Species

The literature reports on Himalayan Lamiaceae species showed huge chemical diversity in the chemical composition some of which are listed below.

2.1. *Ajuga* spp.

The genus *Ajuga* consists of ca. 100-300 species distributed in the temperate and warm temperate regions of Asia, Europe, Australia, North America and Africa. Most members of this genus are flowering annual and perennial herbaceous in habitat and used extensively in traditional system of medicine worldwide (Khaled-Khodja et al. 2014; Qing et al. 2017). *A. parviflora* Benth., *A. bracteosa* Wall Benth., and *A. macrosperma* Wall ex Benth. are the most common and abundantly distributed members of this genus in Himalayan region (Babu, 1977; Beauchamp et al. 1996). The genus *Ajuga* attracted attention since the report in 1976 which showed that *A. remota* grown in Kenya is not attacked by African armyworms, which later resulted the isolation and identification of three neo-clerodane class of diterpenoids, ajugarins-I, ajugarins-II and ajugarins-III having strong antifeedant activity (Beauchamp et al. 1996; Castro et al. 2008; Kubo et al. 1976). This was followed by detailed studies globally on different *Ajuga* spp. for their bioactive metabolites viz. clerodane, neo-clerodane and abietanes diterpenoids, triterpenoids, phytoecdysteroids, iridoids, sesquiterpenoids, flavonoids, polyketides and alkaloids (Camps and Coll, 1993; Chen et al. 2018; Coll and Tandron, 2008; Ono et al. 2011; Qing et al. 2017). These metabolites isolated from different *Ajuga* spp. were found to possess broad spectrum of biological and pharmacological activities (Qing et al. 2017). In our works on Himalayan *Ajuga* spp., we have isolated three neo-clerodane diterpenoids (deoxyajugarin-I, ajugarin-I, chlorohydrin and 3β-acetoxyclerodinin C) along with ajugarin-I and II, ajugamarin F4, dihydroclerodin-I, clerodinins C, clerodinins D, 15-α-ethoxy--14-hydroajugapitin and 15-β-ethoxy-14-hydroajugapitin from *Ajuga parviflora* naturally occurring in different parts of Uttarakhand (Beauchamp et al. 1996). Later on, three new phytoecdysteroids, ajugacetalsterones C, ajugacetalsterones, and breviflorasterone were also isolated and identified on the basis of extensive NMR studies from the roots of *Ajuga macrosperma* var. *brevifolia* (Castro et al. 2008). In another study, sterols viz. dehydroclerosterol, clerosterol, campestrol, and sitosterol were also identified in *A. bracteosa*, *A. macrosperma* and *A. parviflora* from Himalayan region (Singh et al. 2014). Phytoecdysteroids, clerodane and non-clerodane diterpenoids isolated from *Ajuga* spp. were found to possess physiological activity which might be responsible for the activity of these plants in folk medicines.

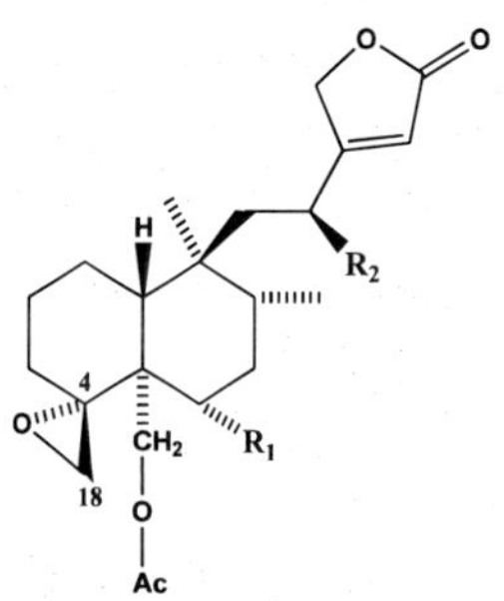

Ajugarin-I (R_1=AcO & R_2 = H)
Ajugarin-II (R_1= OH & R_2 = H)
Deoxyajugarin-I (R_1=AcO & R_2 = H; with C-4,18 deoxy)
Ajugarin-I Chlorohydrin (R_1=AcO & R_2 = H; with C-4α-OH, C-18-Cl)
Ajugarmarin F4 (R_1= AcO & R_2 = $EtCHMeCO_2$)

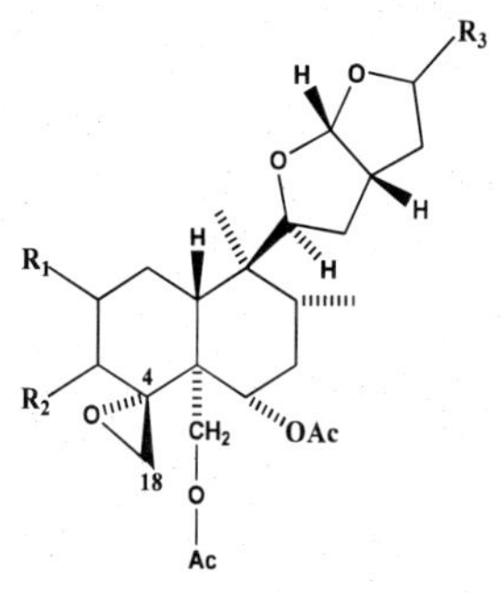

Dihydroclerodin-I (R_1=H & R_2 = H & R_3 = H)
Clerodinin C (R_1=H & R_2 = H & R_3 = β-OEt)
Clerodinin D (R_1=H & R_2 = H & R_3 = α-OEt)
3β-Acetoxyclerodinin C (R_1=H & R_2 = β-AcO & R_3 = β-OEt)
15β-Ethoxy-14-hydroajugapitin (R_1=α-OH & R_2 = β-$EtCHMeCO_2$ & R_3 = β-OEt)
15α-Ethoxy-14-hydroajugapitin (R_1=α-OH & R_2 = β-$EtCHMeCO_2$ & R_3 = α-OEt)

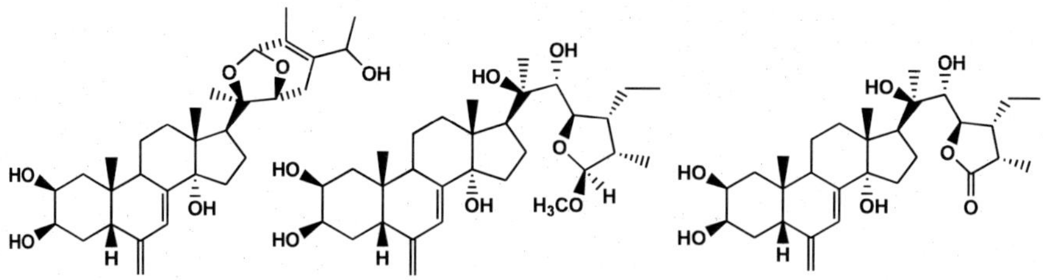

Ajugacetalsterone C **Ajugacetalsterone D** **Breviflorasterone**

Figure 3. Chemical markers of *Ajuga parviflora* and *A. macrosperma* var. *brevifolia*.

2.2. *Calamintha* spp.

The plants of this genus are medium to large size erect herbaceous perennials which is widely distributed in shade and damp forest areas, at an elevation of 1700 - 2500 m in the hill region. The genus *Calamintha* has been considered taxonomically very close to few other Lamiaceae genera such as *Clinopodium, Micromeria, Satureja* and *Acinos*, resulting into existence of several synonyms. Consequently, these have undergone various taxonomic revisions from time to time (Bozovic and Ragno, 2017; Demirci et al. 2011). Most of the *Calamintha* spp. are aromatic with pleasant mint-like smell and used as spices in local cuisines as well used in folk medicine mainly as stimulant, digestive, tonic, antiseptic etc (Chevallier, 2001; Mancini et al. 2013). The essential oil compositions of *Calamintha* spp. from different origin are dominated by presence of p-menthane-type monoterpenoids which constitute the major part of their oil composition. Various chemotypes were reportes based upon the dominance of the C-3 oxygenated p-menthanes (pulegone, menthone, isomenthone, piperitone, piperitone oxides) and C-6 oxygenated p-menthane (carvone) in *Calamintha* spp. (Baldovini et al. 2000; Bozovic and Ragno, 2017). The *C. umbrosa* (M.Bieb.) Rchb. is an important member of this genus. The essential oil composition of *C. umbrosa* was studied from Uttarakhand Himalaya which revealed the distribution of two distinct chemotypes with distinct composition. *cis*-Piperitone oxide (63.05%) and piperitenone oxide (17.98%) rich chemotype was identified from Kedarnath region of Uttarakhand (Shah et al. 1992). The other chemotype identified of *C. umbrosa* was dominated by β-caryophyllene (11.66%), germacrene D (9.81%), spathulenol (9.44%), β-pinene (7.47%), terpinolene (6.86%), and α-cubenol (5.01%) as major constituents (Kumar and Mathela, 2017).

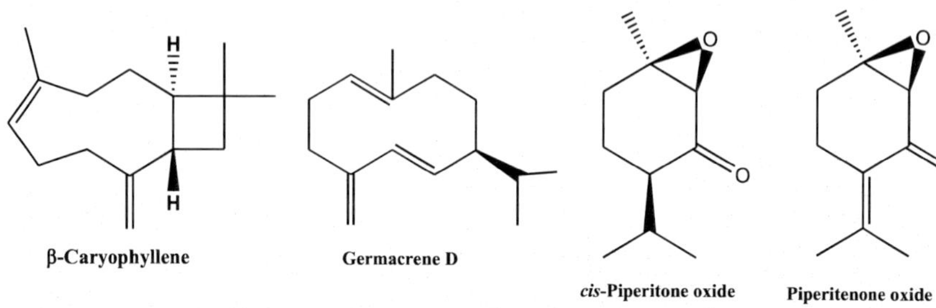

Figure 4. Chemical markers of *Calamintha umbrosa.*

2.3. *Elsholtzia* spp.

The genus *Elsholtzia* is represented by about three dozen species widely distributed in temperate and sub-temperate regions of East Asia, Africa, North America, and Europe. Most members of the genus *Elsholtzia* are mostly aromatic herbs or undershrub in habitat and are extensive part of folk medicines (Guo et al. 2012; Chauhan et al. 2019). In India, *Elsholtzia* is represented by 10 species/varities viz. *E. strobilifera* Benth., *E. eriostachya* var. *pusilla* Benth., *E. eriostachya* Benth., *E. densa* Benth., *E. pilosa* Benth., *E. polystachya* Benth., *E. flava* Benth., *E. cristata* Willd., *E. incisa* Benth., and *E. incisa* var. *major* Benth. widely distributed in sub-temperate to temperate Himalaya regions (Mathela et al. 2009; Naithani, 1985; Stracheyi, 1975) and extensively studied. *Elsholtzia* spp. owing to their diverse biological and pharmacological properties has been in focus for their phytochemical investigations worldwide. Numerous classes of constituents, such as flavonids, coumarins, glycosides, furans, phytosterol, terpenoids, phenylpropanoids etc. with a wide spectrum of pharmacological activities were found as main chemical constituents in this genus (Guo et al. 2012; Liu et al. 2007). Most members of this genus are aromatic and possess diverse volatile constituents used in food-flavor, fragrance and cosmetics products (Chauhan et al. 2019; Liu et al. 2007). Previous studies on essential oil composition of Himalayan *Elsholtzia* spp. showed existence of several chemotypes with dominance of acylfurans viz. rosefuran, rosefuran epoxide, perillene, elsholtziaketone, dehydroelsholtzia ketone which are the characteristic constituents besides monoterpenoids viz. 1,8-cineole, citral, pinocarvone, geranyl acetate, piperitone, piperitenone, γ-terpinene and (*E*)-β-ocimene (Chauhan et al. 2019; Mathela et al. 2009). The essential oils of *E. eriostachya* and *E. eriostachya* var. *pusilla* collected from high altitude region were dominated by rosefuran epoxide (85.18%) and citral (91.70%), respectively as their characteristic constituents (Mathela et al. 2009; Pant et al. 1992). The uniqueness of the two oils was the existence of almost exclusive content of rosefuran epoxide and citral, two commercially important chemicals. The essential oil of *E. pilosa* was characterised by 1,8-cineole (34.37-49.94%) as major constituent (Bestmann et al. 1988; Mathela et al. 2009). In *E. polystachya* collected from different location of Uttarakhand, three distinct chemotypes represented by rosefuran (48.15%)/1,8-cineole (19.02%); perillene (61.21%); and 1,8-cineole (37.30%)/(*E*)-β-ocimene (18.22%) (Mathela at al. 1992; 2009). In *E. flava* as well three distinct chemotypes characterized by rosefuran (45.05%), estragol (40.51%) and piperitenone (30.81%)/carvacrol (4.84%), respectively were identified (Bestmann et al. 1997; Mathela et al. 2009). The inflorescense and leaf essential oils of *E. densa* collected from Chopta, Rudraprayag were characterised by citral (52.2%) and pinocarvone (51.9%), respectively as marker constituents (Chauhan et al. 2019). In another study, rosefuran epoxide (86.0%) was reported as the major constituent of the essential oil *E. densa* from Malari region (Melkani et al. 1994). Moreover, dehydroelsholtzia ketone (86.6%) and pinocarvone (51.88%) rich essential oil compositions were characterized for *E. cristata*, and *E. strobilifera* respectively, from Uttarakhand (Bisht et al. 1985; Mathela et al. 2009). The essential oil from *E. incisa* growing locally was characterized by 1,8-cineole (41.1%), and γ-terpinene (19.1%) as major constituents (Bestmann et al. 1995). Other reports on *Elsholtzia* spp. from different geographical regions in hills also revealed the existence of various chemotypes with dominance of citral, linalool, geranyl acetate, rosefuran, dehydroelsholtzia ketone, elsholtzia ketone, 1,8-cineole as marker constituents in Himalayan *Elsholtzia* pp. (Bestmann et al. 1992; Chauhan et al. 2019; Thappa et al. 1999).

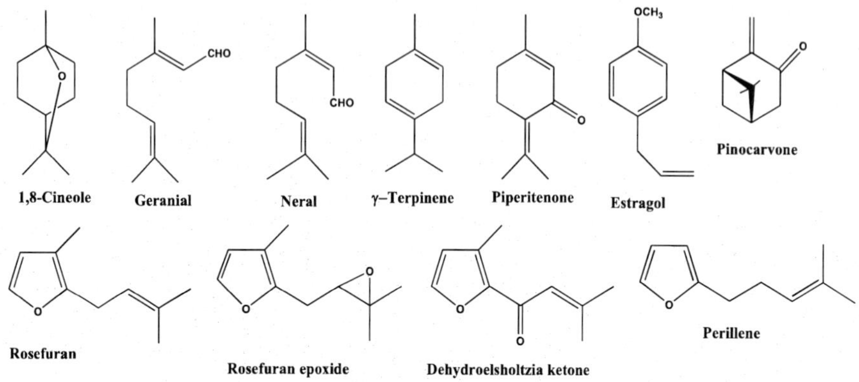

Figure 5. Chemical markers of *Elsholtzia* spp.

2.4. *Hyssopus* spp.

The genus *Hyssopus* is comprised of about one dozen species that are aromatic perennial herbs or shrubs. Some of these are cultivated in the East Mediterranean to central Asia (Fathiazad et al. 2011). *Hyssopus officinalis* L. popularly known as 'Hyssop' is one of the important member of this genus distributed in temperate regions of Asia, Europe and America. It is an evergreen perennial shrub, 50-120 cm in height with smaller linear leaves with purplish-blue flower. In India, it is widely distributed in western Himalayan region at an altitude of 8000-11000 ft (Chopra et al. 2009; Pandey et al. 2014). *H. officinalis* is cultivated for its minty flavor, condiment and spices in food industries as well culinary and traditional medicinal use (Dragland et al. 2003; Fathiazad and Hamedeyazdan, 2011; Pandey et al. 2014; Shah, 1991). Phenolic acids, polyphenols including flavanoids and their glycosides, tannins and essential oil were reported as the main constituents (Fathiazad and Hamedeyazdan, 2011; Srivastava et al. 2018). The essential oil of *H. officinalis* has been extensively studied for its volatile constituents for use in cosmetic, fragrance food and pharmaceutical products. The main constituents characterised in the essential oil of *H. officinalis* are pinocamphone, isopinocamphone, pinenes, myrtenyl acetate, pinocarvone, 1,8-cineole and linalool (Pandey et al. 2014). The essential oil of *H. officinalis*, collected from Malari was found to be dominated by isopinocamphone (43.8%), 1,8-cineole (13.0%), β-pinene (12.8%) and pinocarvone (11.0%) as major constituents (Sah et al. 2008). The essential oil of *H. officinalis* cultivated in the north Indian plains was found to possess pinocamphone (49.1%), β-pinene (18.4%) and isopinocamphone (9.7%) (Garg et al. 1999). Pinocarvone (32.1%) and isopinocamphone (30.7%) were identified as major constituents of the essential oil *H. officinalis* grown in Garhwal region as well (Mohan et al. 2012). The hydro-distilled essential oil from the leaf, flower and stem of *H. officinalis* grown in Chamoli were found to be characterised by *cis*-pinocamphone (49.7–57.7%), pinocarvone (5.5–24.9%), β-pinene (5.7–9.3%), and 1,8-cineole (2.9–8.0%) as major constituents (Pandey et al. 2014). In another study, isopinocamphone (38.1%), pinocarvone (20.3%), 1-8-cineole (12.2%), and β-pinene (10.2%) were reported as the major constituents of the essential oil of *H. officinalis* (Shah, 1991). However, pinocamphone (53.54%), β-pinene (9.91%) and limonene (7.19%) were reported as the characteristic constituents of *H. officinalis* grown in Kashmir (Khan et al. 2012). The essential oil composition of *H. officinalis* from Himalayan region is comparable with ISO standard for hyssop oil of trade that specifies the highest content for: isopinocamphone (25-45%), pinocamphone (8.0-25%) and β-pinene (7.0-20.0%) (Moro et al. 2011).

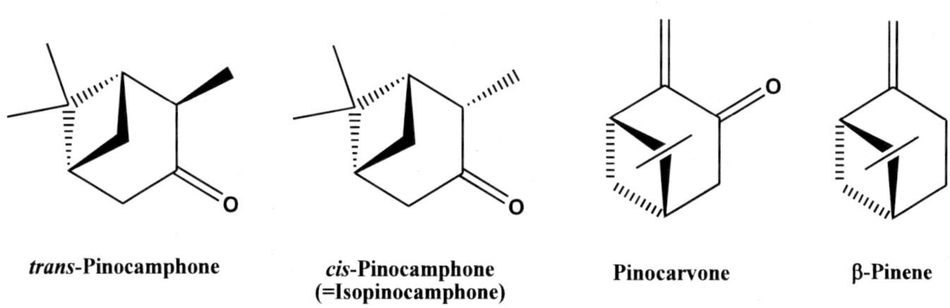

Figure 6. Chemical markers of Himalayan *Hyssopus officinalis*.

2.5. *Mentha* spp.

Mentha species (commonly known as mint) are most important materials of Industrial importance. *Mentha* spp are widely cultivated in the temperate and sub-temperate regions of the world (Chauhan et al. 2011). The genus *Mentha* is represented by 62 taxa belonging to 18 species and 11 hybrids (Baser et al., 2012). The *Mentha* essential oils and aroma chemicals are used in food, flavour, cosmetic, and pharmaceutical industries (Lawrence, 2007; Padalia et al. 2013). Mint essential oil production is one of the important and wide-ranging segments of essential oil industry mainly in Brazil, France, China, India, Argentina, Bulgaria, Angola, Australia, Hungary, Italy, Switzerland, Thailand and USA (Bahl et al. 2000). Among various *Mentha* spp., *Mentha arvensis* L. f. *piperascens* Malinvaud ex. Holmes (Cornmint or Menthol mint), *Mentha*× *piperita* L. (Peppermint), *Mentha citrata* Ehrh. (Bergamot mint), *Mentha* × *spicata* L. (Spearmint), *Mentha cardiaca* Bak. (Scotch spearmint) and *Mentha pulegium* L. (Pennyroyal) are cultivated worldwide for their essential oils and aroma chemicals. CSIR-Central Institute of Medicinal and Aromatic Plants, India has developed a large number of cultivars for large scale cultivation. The cultivars developed are CIM-Kranti, Kosi, CIMAP-Saryu, Kushal, Himalaya, Kalka, Saksham, Sambhav, Damroo, Gomti, and MAS-1 for menthol-mint; Pranjal, Tushar, CIM-Madhurus, Kukrail, CIM-Indus, and CIMAP-Patra for peppermint; MSS-5, Arka, Neerkalka and Neera of spearmint; MCAS-2 of scotch-spearmint; Kiran of bergamot-mint; and Supriya of garden mint/spearmint (Padalia et al. 2013, 2015). Due to the broad acceptability of these cultivars the mint is being cultivated on thousands of hectare of agricultural land in tropical and subtropical climates of Indian plains (Khanuja, 2007; Verma et al. 2010). India is global leader in production and export of the menthol mint essential oil due to extensive cultivation of *M. arvensis* in Indo-Gangetic plains and north-west India. The menthol mint oil, its isolates and by-product and dementholized oil (DMO) of *M. arvensis* have the highest share in the global mint trade followed by peppermint, spearmint and bergamot mint (Padalia et al. 2015). The essential oil composition of mints show a complex mixtures of variety of valuable aroma chemicals (mainly monoterpenoids) viz., menthol, menthone, isomenthone, menthyl acetate, menthofuran, carvone, linalool, linalyl acetate, pulegone, piperitone oxide, piperitenone oxide (Chauhan et al. 2011; Padalia et al. 2013, 2015). Previous reports on the chemical composition of oils from *Mentha* species showed menthol (61.92-89.30%) and carvone (59.62-76.65%) as major constituents of the essential oils of *M. arvensis* and *M. spicata,* respectively (Padalia et al. 2013; Verma et al. 2010; Chauhan et al., 2009). The essential oil composition of common cultivars and wild *M. piperita* analyzed by capillary GC and GC/MS showed menthol (30.3-47.8%), menthone (4.5-48.6%), menthyl acetate (1.0-6.5%), menthofuran (0.1-14.6%), and 1,8-cineole (4.1-8.9%) as major constituents of all the cultivars and wild accessions of *M. piperita* grown in foot and mid hills of Uttarakhand (Padalia et al. 2011). Carvone (15.3%-68.5%), piperitenone oxide (24.0%-79.2%) and α-humulene (0.1%-29.9%) rich essential oil composition of *M. spicata* were reported from ten different natural local habitats (Chauhan et al. 2011). *M. spicata* populations collected from the subtropical and temperate zones of north-west Himalayan region of India showed carvone (49.6-76.6%) and limonene (9.5-22.3%) as characteristic constituents in their essential oils (Chauhan et al., 2009). The essential oil of *M. citrata* cultivar 'Kiran' grown in mid and foothills of Uttarakhand, India showed linalool (32.8-59.7%), linalyl acetate (32.8-59.7%), and *α*-terpineol (1.5-4.6%) as major constituents at different crop stages (Padalia et al. 2013; Verma et al., 2010). *M. longifolia* is an erect or diffused, scented aromatic herb widely distributed is an ingredient of the Indian household spices (Polunin and Stainton, 1984). The carvone content of essential oil wild *M. longifolia* collected from different regions of Kumaon was screened. The results showed carvone (61.1-78.0%) along with dihydrocarveol (0.4- 9.4%), cis-carvyl acetate (0.1- 6.4%), and germacrene D (1.2-5.7%) (Mathela et al. 2005). The essential oil from aerial parts of wild *M. longifolia* growing in high altitudes of Auli (Garhwal) was characterised by *trans*-piperitone epoxide (48.7%), piperitenone oxide (21.2%), and germacrene D (9.8%) as major constituents (Shah et al. 1992; Verma et al. 2015). Similar composition was noticed in another report on oil of *M. longifolia* collected from wild from Badrinath, Chamoli and Joshimath which were also dominated *cis*-piperitone oxide and piperitenone oxide with significant quantitative variations. The major constituents of *M. longifolia* from Chamoli were piperitenone oxide (60.4%), 1,8-Cineole (6.1%), piperitenone (3.4%), and piperitone (1.6%); while *M. longifolia* collected from Joshimath showed the presence of piperitenone oxide (36.8%), cis-piperitone oxide (17.8%), and camphor (3.1%), as major constituents (Singh et al. 2011).

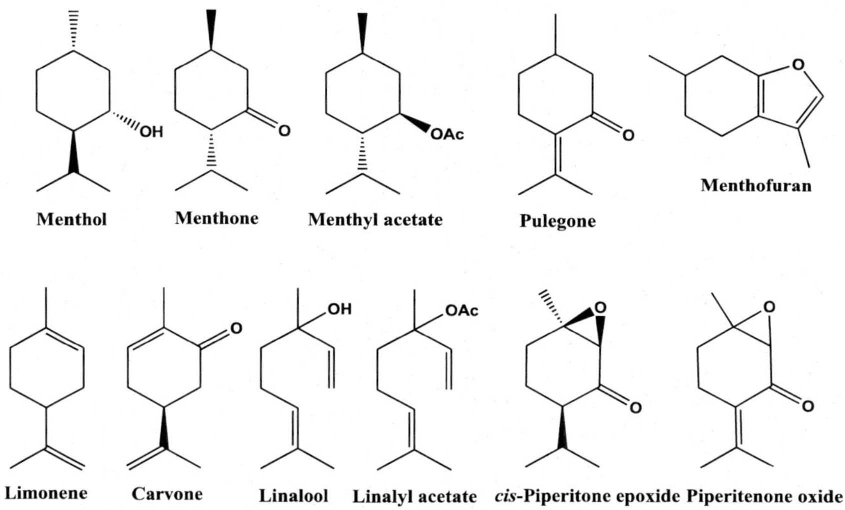

Figure 7. Chemical markers of essential oils from *Mentha* spp.

2.6. *Nepeta* Species

Nepeta is one of the largest genera of the Lamiaceae family, in the tribe *Mentheae*, subtribe *Nepetinae*. It comprises ca. 300 species distributed worldwide in central and southern Europe, central, eastern and southern Asia, North America, and some areas of Africa (Jamzad et al. 2000; Razavi et al. 2018). In India, it is represented by 35 species in western Himalayan region, of which over a dozen *Nepeta* species are distributed in Uttarakhand (Polunin and Stainton, 1984). Some of the

Nepeta species are part of folk medicines as antispasmodic, diuretic, anti-asthmatic, antioxidant, anti-inflammatory and antiseptic etc with various other pharmacological activities (Formisano et al. 2011). Additionally, *Nepeta* species are well known for their feline and canine attractant, insect repellent and arthropod defense activities possibly due to the presence of biologically active iridoids/nepetalactones/terpene alkaloids/rich essential oils and related compounds (Cigremis et al. 2010, Mathela, 2017). Our research findings on the Himalayan *Nepeta* species viz. *N. leucophylla* Benth., *N. clarkei* Hook.f., *N. govaniana* Benth., *N. discolor* Benth., *N. elliptica* Royle ex Benth., *N. erecta* (Royle ex Benth.) Benth., *N. ciliaris* Benth., *N. hindostana* (B. Heyne ex Roth) Haines and *N. laevigata* D. Don collected from different altitudes of Himalayan region of Uttarakhand showed immense diversity in their phytoconstituents. The notable finding was the occurrence of iridodial β-monoenol acetate (9.8%) and its structurally related new compounds from the essential oil of *N. leucophylla* (Bottini et al. 1987; Kumar et al. 2014; Mathela and Kumar, 2018). The essential oil of *N. clarkei* was found to be characterised by β-sesquiphellandrene (22.0%), a rare terpene alkaloid actinidine (10.0%) besides germacrene D (8.0%) (Kumar et al. 2014; Saxena and Mathela, 1996). β-Caryophyllene (18.0), β-sesquiphellandrene (15.0%), and caryophyllene oxide (14.8%) were identified as the characteristics constituents of *N. ciliaris* (Kumar et al. 2014). However, (7R)-*trans,trans*-nepetalactone (85.1%), germacrene D (4.9%), β-elemene (2.1%), α-humulene (1.8%) and spathulenol (1.2%) were identified as the major constituents of *N. elliptica* (Kumar et al. 2014). The essential oil of *N. govaniana* in the flowering stage was characterised on the basis of GC, GC-MS and X-ray crystallographic data on its major constituent. The essential oil of *N. govaniana* was marked by the presence of pregeijerene (38.4%), isoiridomynnecin (16.5%), 4aα,7α,7aα-nepetalacetone (3.9%), elemol (4.8%) and germacrene D (4.3%) (Kashyap et al. 2003; Mathela et al. 1994). The chemical composition of the essential oil of the aerial parts of *N. discolor* was found to be dominated by 1,8-cineole (20.0%),

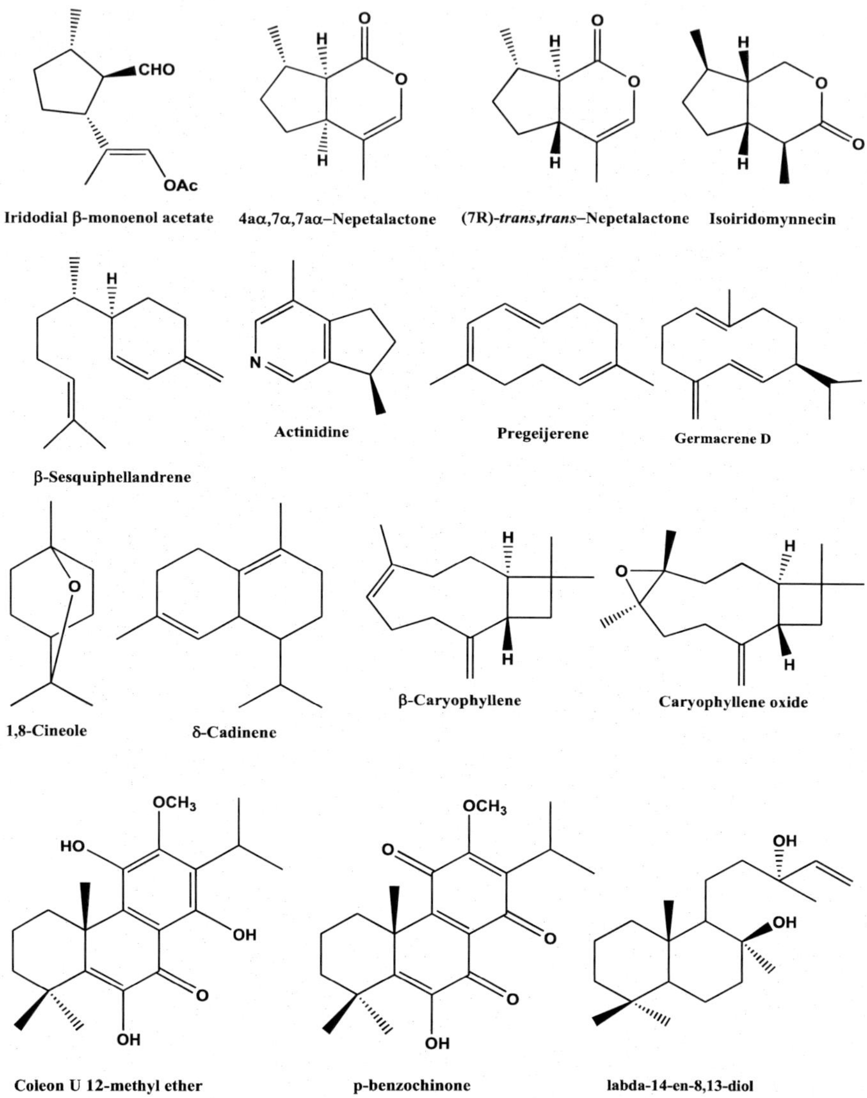

Figure 8. Chemical markers of *Nepeta* species.

β-caryphyllene (10.6%), caryophyllene oxide (9.0%), sabinene (8.3%), germacra-1(10),5-dien-4-ol (6.8%), α-cadinol (3.9%) as major constituents (Mathela et al. 1994). The essential oil of the aerial parts of *N. erecta* was found to be dominated by oxygenated monoterpenoids (74.0%), isoiridomyrmecin (70.6%) being the single major constituent. Other significant constituents identified in the essential oil of *N. erecta* were caryophyllene oxide (9.6%), β-bourbonene (2.0%), humulene epoxide II (1.7%) and linalool (1.0%). (Bisht et al. 2012). The hydrodistilled essential oil of the aerial parts of *N. laevigata* from Kumaun Himalaya was found to be dominated by caryophyllene oxide (15.2%), 1,8-cineole (11.1%), manool (7.9%), β-caryophyllene (5.7%) and pimaradiene (4.6%) as major constituents (Joshi and Mathela, 2013). β-Caryophyllene (15.6%), δ-cadinene (39.8%), caryophyllene oxide (6.9%), 2,3-dihydrofarnesol (6.4%) were identified as the major constituents of the essential oil of *N. hindostana* (Kumar et al. 2016, 2019). Several diterpenoids, triterpenoids and phenolic constituents were also isolated from the roots of *Nepeta leucophylla, N. clarkei* and *N. elliptica.* Coleon U 12-methyl ether was reported as the main coloured constituent form the roots of *N. leucophylla.* (Bottini et al., 1987; Mathela et al. 1991). Coleon U 12-methyl ether, dehydro coleon U 12-methyl ether and its 7-(1-methylethenyl) homologue were isolated and identified in chloroform extract of the roots of *N. elliptica* (Bhandari et al. 1993). Three diterpenoids p-benzochinone, labda-14-en-8,13-diol, coleon U 12-methyl ether, a triterpenoid ursolic acid and two sterols (β-sitosterol and stigmasterol) have been isolated and identified from the roots of *N. clarkei* (Bisht et al. 2010). The essential oils of *Nepeta* spp. with iridoids/nepetalactones/terpene alkaloids/ monoterpenoids and sesquiterpenoids and related compounds are found to possess various pharmacological activities. The results of our works on Himalayan *Nepeta* spp revealed that nepetalactones in *N. elliptica*, *N. govaniana*; and iridoids viz. β-monoenol acetate in *N. leucophylla* and isoiridomynnecin in *N. govaniana* and *N. erecta* are among one of the characteristic marker constituents of their essential oil. Terpene alkaloid, actinidine is the marker constituent of *N. clarkei*, while the essential oil of *N. discolor*, *N. ciliaris*, *N. hindostana* and *N. laevigata* are mono and sesquiterpenoids (Table 1). Thus, the results on *Nepeta* spp. are indicative of chemosystematic markers of these plants.

Table 1. Chemical markers of Himalayan *Nepeta* species

S.No.	Plant Name	Chemical markers
1	*N. leucophylla*	Caryophyllene oxide, iridodial β-monoenol acetate, δ-cadinene, caryophyllene oxide, 2,3-dihydrofarnesol and β-sesquiphellandrene.
2	*N. clarkei*	β-Sesquiphellandrene, actinidine and germacrene D
3	*N. elliptica*	(7R)-*trans,trans*-Nepetalactone
4	*N. erecta*	Isoiridomyrmecin
5	*N. govaniana*	Pregeijerene, isoiridomynnecin, 4aα,7α,7aα-nepetalacetone
6	*N. ciliaris*	β-Caryophyllene, β-sesquiphellandrene, and caryophyllene oxide
7	*N discolor*	1,8-Cineole, β-caryphyllene, caryophyllene oxide, sabinene and germacra-1(10),5-dien-4-ol
8	*N. laevigata*	Caryophyllene oxide, 1,8-cineole, manool, β-caryophyllene and pimaradiene
9	*N. hindostana*	β-Caryophyllene, δ-cadinene, caryophyllene oxide, 2,3-dihydrofarnesol

2.7. *Ocimum* Species

The genus *Ocimum* L., collectively called as 'Basil' is comprised of ca. 30-150 species widely distributed in tropical and subtropical regions of Asia, Africa, and Central and South America (Paton et al. 1999). The members of *Ocimum* are annual or perennial, branched herb or shrubs which are enormously reputed for their medicinal values. The word 'Basil' is replicates the Greek word '*Basilica*' which means 'Royal Plant'; while the Indian name 'Tulsi' is derived from *'Sanskrit'*, which means 'matchless one' and it is also often referred as the "King of the plants', being widely utilized due to its economic, nutritional, industrial and medicinal properties. In ancient Indian literature 'Tulsi' is considered highly sacred plant and cultivated in temples, court-yards, and gardens for various religious rituals to symbolise its spiritual and medicinal uses (Labra et al. 2004; Mathew and Sankar, 2013; Prakash and Gupta, 2005; Pushpangadan and Bradu, 1995). In India, seven species viz. *O. tenuiflorum* L., *O. basilicum* L., *O. kilimandscharicum* Gurke, *O.*

gratissimum L., *O. americanum* L., *O. africanum* Lour. and *O. filamentosum* Forssk. have been reported (Padalia et al. 2017; Balyan & Pushpangadan 1988). Among them, *O. tenuiflorum*, *O. basilicum*, *O. kilimandscharicum*, *O. gratissimum*, and *O. americanum* are cultivated for religious and medicinal purpose as well as a source of essential oils and aroma/flavour constituents for food, perfumery and cosmetic products. The essential oil composition of *Ocimum* spp. is very complex and exhibits immense compositional diversity due to existence of various genotype, chemotype, morphotypes and variable climatic and agronomical conditions. However, the chemical variability among the *Ocimum* spp. and their chemotypes are generally represented by dominance of one or two major monoterpenoids (linalool, camphor, citral, thymol, 1,8-cineole, limonene, (*Z*)-β-ocimene) and phenyl propanoids (methyl chavicol, eugenol, methyl cinnamate, (*E*)-anethole, methyl eugenol). Mixed compositions are characterised by these monoterpenoids/ phenyl propanoids, sesquiterpenoids (β-elemene, β-caryophyllene, *trans*-α-bergamotene, (*E*)-α-bisabolene, β-bisabolene, germacrene D) were also reported from different regions (Padalia and Verma, 2011; Padalia et al. 2013; Verma et al. 2013). Lawrence (1988) On the basis of essential oil composition of *O. basilicum*, established four chemotypes (methyl chavicol, linalool, methyl eugenol, and methyl cinnamate) along with subtypes. Considering the differences in chemical compositions of different geographic origins, *O. basilicum* have been classified into four types: European basil oil (from Italy, France, Bulgaria, Egypt and South Africa) with methyl chavicol (25.0%-75.0%) and linalool (25.4%-50%) as major constituents; Tropical basil oil (from India, Guatemala and Pakistan) dominated by methyl cinnamate (upto 65.0%), reunion basil oil (from Comoro Island, Thailand, Madagascar and Vietnam) with high content of methyl chavicol (≥80%); and basil oil from North Africa and USSR is rich in eugenol (Koutsos et al. 2009; Marotti et al. 1996). In addition to these, eugenol, *trans*-β-bergamotene/linalool, menthone/methyl chavicol, 1,8-cineole, camphor, citral, camphor/linalool, geraniol/linalool, (*E*)-anethole chemotypes were also reported in *O. basilicum* from different geographic regions (Alves et al. 2007; Jirovetz et al. 2003; Koba et al.

2009; Mondello et al. 2002; Purkayastha and Nath, 2006; Runyoro et al. 2010; Tchoumbougnang et al. 2006; Telci et al. 2006; Verma et al. 2013). In *O. basilicum*, eight different chemotypes rich in methyl chavicol (83.9%-88.2%); linalool (74.3%); methyl chavicol (64.5%-73.8%)/linalool (16.8%-27.1%); citral (64.5%)/nerol (9.7%)/geraniol (6.8%); methyl chavicol (46.5%-62.1%)/methyl eugenol (0.8%-14.9%)/camphor (8.2%-9.1%); β-bisabolene (25.6%)/1,8-cineole (17.5%)/α-cadinene (11.9%); linalool (35.5%)/citral (24.9%); and linalool (36.3%-45.0%)/(*E*)-methyl cinnamate (33.3-38.7%) were reported from Uttarakhand (Padalia et al. 2017; Verma et al. 2013). The constituents specific varieties of *O. basilicum* (methyl chavicol (>80%) rich 'CIM-Sharada', methyl chavicol (~60%)/linalool (~20%) rich 'CIM-Saumya', and methyl cinnamate (~72.0%) rich CIM-Snigdha varieties of *O. basilicum* are being cultivated and promoted by CSIR-CIMAP at mid and foothills agroclimatic conditions of Uttarakhand (Padalia et al., 2017). In *O. tenuiflorum*; literature reports revealed eugenol (upto 65%), and methyl eugenol chemotypes to be very common and widely grown hill region (Kothari et al. 2001; Kothari et al. 2005; Padalia and Verma, 2011). Another report shows that the essential oil from *O. tenuiflorum* cultivated in different regions of Uttarakhand, showed the existence of three chemotypes, dominated by eugenol (54.9-77.3%); methyl eugenol (79.7%); and (*E*)-caryophyllene (36.2%)/β-elemene (23.4%). Beside these, varying amount of these constituents with sesquiterpene hydrocarbons viz. (*E*)-caryophyllene, β-elemene, α-bisabolene, germacrene D, β-bisabolene etc. have also been reported in *O. tenuiflorum* (Verma et al. 2013). The essential oil from *O. kilimandscharicum* distributed in Uttarakhand was mainly dominated by camphor (63.4%-64.9%) and eugenol (47.5%)/methyl chavicol (12.7%) rich two chemotypes (Padalia and Verma, 2011, Verma et al., 2011). Eugenol (60.0%-80.0%) is the dominant constituent in *O. gratissimum* of Indian origin as industrial source of eugenol (Kothari et al. 2005; Sobti et al. 1976). Besides common eugenol rich chemotype, other chemotypes rich in thymol, geraniol, linalool, methyl eugenol, methyl cinnamate, citral are also reported in *O. gratissimum* from various origins (Agnaniet et al. 2005; Benitez et al.

2009; Charles and Simon, 1992; Dubey et al. 2000; Tchoumbougnang et al. 2006). It has been observed that eugenol (63.7%-77.2%) rich chemotype of *O. gratissimum* are mostly distributed in Uttarakhand. A eugenol-rich hybrid (RRL-og-10) of *O. gratissimum* containing 90-95% eugenol was developed for commercial cultivation in Himalayan region (Janmoni and Khan, 2013). The essential oil composition of *O. americanum* from Uttarakhand was dominated by camphor (45.6%), longipinanol (7.4%) and limonene (6.3%) (Verma et al. 2013).

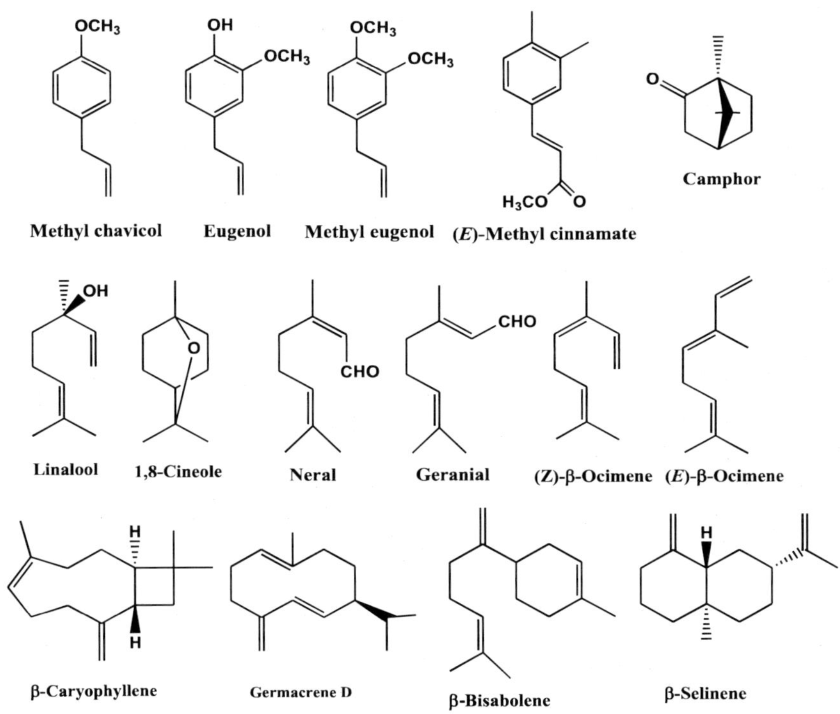

Figure 9. Chemical markers of *Ocimum* species.

In another study from Uttarakhand, six chemotypes dominated by eugenol (27.6-38.2%)/β-bisabolene (9.9-17.8%)/1,8-cineole (4.5-16.8%); methyl chavicol (28.9%)/eugenol (14.7%)/β-bisabolene (14.7%); eugenol (45.2%)/(*E*)-caryophyllene (30.2%); methyl chavicol (78.3%)/linalool (14.2%); camphor (42.5%-49.4%)/1,8-cineole (11.5-28.4%); and aliphatic hydrocarbons (44.1%)/camphor (23.0%) were identified in *O. americanum* (Singh et al. 2013). The essential oil composition of *O. africanum* was less explored. Citral chemotype (71.9%) was reported from Indian origin; while citral (36.7-70.2%) and camphor (52.9%) chemotypes were reported from Egyptian origin (Farag et al. 2016; Pisutthanan and Pisutthanan, 2009). (*Z*)-Isoeugenol (88.7-94.2%) and methyl chavicol (66.5%) rich chemotype of *O. africanum* were reported from Brazilian and Croatian origin (Stanko et al. 2011). Citral (55.0–75.5%), (*E*)-γ-bisabolene (2.6–9.5%), nerol (1.7–8.9%), geraniol (1.5–6.5%), linalool (1.1–6.0%), β-caryophyllene (0.7–3.2%), α-humulene (0.4–2.7%) and 6-methyl-5-hepten-2-one (≤0.03–2.1%) were identified as the characteristic constituents of the essential oil of *O. africanum* var. CIM-Jyoti grown at foothills agroclimatic conditions of Uttarakhand (Padalia et al. 2018). The chemical diversity in essential oil composition and existence of various chemotypes/cultivars/ subspecies with *Ocimum* spp. offers great scope for valuable flavour/pharmaceuticals in future.

2.8. *Origanum* Species

The genus *Origanum* is represents by 49 taxa (species, subspecies and varities) distributed and cultivated mainly in Mediterranean region and subtemperate to temperate regions of Asia, Europe, North Africa, and America. The plants of this genus are erect or ascending, perennial and highly aromatic (Goliaris et al. 2002; Ietswarrt, 1980; Verma et al. 2012). In Indian subcontinent, *O. vulgare* commonly known as 'Van Tulsi' or 'Oregano' or 'Himalayan Marjoram' which is one of the main species widely distributed in sum-temperate to temperate (1800-3600 m) Himalayan region (Sharma and Lal, 2005; Chauhan et al. 2011). It is reputed with trade name 'Oregano' in spice industry (Olivier, 1997). The genus, *Origanum* is known for its huge chemical diversity with occurrence of various morphotypes and chemotypes. The essential oil composition of

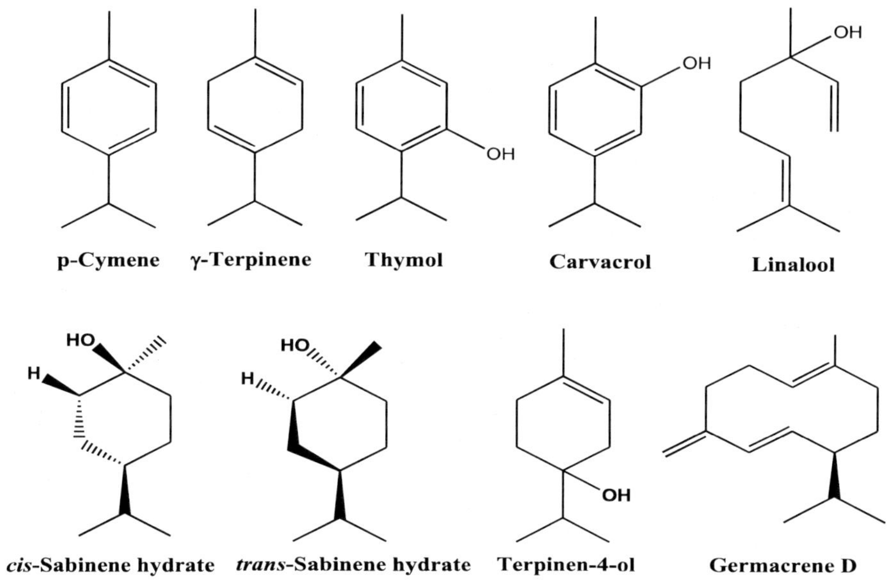

Figure 10. Chemical markers of *Origanum* spp.

Indian oregano (*O. vulgare*) has been explored from time to time. The leaf and flower essential oils of *O. vulgare*, grown in Kumaon region showed the presence of thymol (27.9-49.0%), p-cymene (13.6-25.1%), γ-terpinene (7.9- 8.7%) and 1-octen-3-ol (1.8-4.7%) as major constituents (Verma et al. 2012). In another report γ-muurolene (62.0%) rich chemotype was identified in *O. vulgare* (Pande and Mathela, 2000). Verma et al. (2010) studied seventeen different populations of *O. vulgare* and reported 6 chemotypes. The chemotype-I is characterised by high content of thymol (39.6-56.4%, followed by γ-terpinene (20.0-40.1%). The chemotype-II was carvacrol was characterised by high carvacrol content (58.2-65.3%) and γ-terpinene (15.1-29.3%) The chemotype-III was found to be dominated by high amount of γ-terpinene (43.4-44.1%) with thymol (17.8-19.0%). Chemotype-IV contained γ-terpinene (45.9%) with carvacrol (20.09%). The chemotype-V was characterised by thymol (23.30%), (*E*)-β-ocimene (16.03%), *p*-cymene (11.33%) and (*Z*)-β-ocimene (8.86%) as major constituents while the chemotype-VI was found to be rich in linalool (0.5% -34.0%), germacrene D (3.0% - 26.3%), α-terpineol (1.0% - 20.1%), borneol (<0.10 - 15.4%) and β-caryophyllene (1.0% - 14.5%). Another important species of this genus is *O. majorana* L. syn (*Majorana hortensis* Moench), commonly known as 'Marjoram' and 'sweet Marjorum'. The plant is used in cosmetics, food-flavour and in local traditional medicine. The essential oil composition *O. majorana* grown in mid and foot hills of Uttarakhand showed *cis*-sabinene hydrate (20.2-46.2%), terpinen-4-ol (9.3-23.4%), γ-terpinene (5.6-13.7%), α-terpinene (2.9-8.3%), sabinene (4.9-8.1%), *trans*-sabinene hydrate (5.0-7.3%) as major constituents (Verma et al. 2013, 2016)

2.9. *Perilla* spp.

The *Perilla* species are distributed in China, India, Japan, Korea, Thailand and other Asian countries. The taxonomic classification of *Perilla* species is controversial and confused with some *Elsholtzia* (Ahmed, 2019; Zeevaart, 1969). Based on morphological characters and uses, the genus is recognized monotypic with only one species, *P. frutescens* (L.) Britton and its cultivars distributed and cultivated in throughout East Asian countries (Asif, 2011). *P. frutescens* is an erect annual herb, native to the mountainous areas of China and India, commonly grown in waste places, crop fields, road sides upto 3000 m altitude (Ahmed, 2019; Dimri et al. 2008). It is commonly known as 'Bhanjira' in India and used for culinary and traditional medicinal uses (Gwari et al. 2016). It is cultivated in China, India, Japan, Korea, Thailand and other East Asian countries, and now introduced into other countries like Europe, Russia and USA (Youfang et al. 2012; Gwari et al. 2014). Phytochemical investigation on *P. frutescens* worldwide reported numerous bioactive secondary metabolites including phenolics, glycosides, flavonoids, anthocyanins, essential oils, triterpenes, caretenoids, phytosterols, fatty acids with various pharmacological properties (Ahmed, 2019; Bachheti et al. 2014). *P. frutescens* is highly aromatic and used as flavoring agents in local cuisines. The essential oil composition of *P. frutescens* has been the subject of numerous investigations. Based on the dominance of main components different chemotypes were reported in *P. frutescens* such as perillaketone, perillaldehyde, elsholtzia ketone isoegomaketone, citral, perillene, piperitenone, linalool and phenylpropanoids (Ahmad, 2019; Gwari et al. 2016; Verma et al. 2013; Zhang et al. 2009). The essential oil composition of *P. frutescens* grown in Himalayan region of Uttarakhand was also studied for their chemical diversity. Mathela et al. 1989 analysed the oil on the basis of extensive NMR and MS and reported compounds structurally closely related to 3-substituted-furanyl skeleton (Mathela et al. 1989). The essential oils of *P. frutescens* collected at different phenophases of the crop grown in foot hills agroclimatic conditions of Uttarakhand was found to be characterised by perilla ketone (39.5-59.3%), isoegomaketone (9.4-45.7%), and (*E*)-caryophyllene (0.1–13.4%), and caryophyllene oxide (0.4–3.8%) as major constituents. Moreover, the essential oil composition showed significant quantitative variability in terms of the content of major constituents with higher perilla ketone (59.3%) at seed maturity stage and higher isoegomaketone (39.8–43.6%) during vegetative stages of the crop (Verma et al. 2013). The essential oil composition of twelve wild populations of *P. frutescens* collected from different regions of Garhwal showed perillaketone (44.4–69.2%), isoegomaketone (7.3–27.6%), β-caryophyllene (0.1–17.8%) and linalool (0.3–5.0%) were identified as major constituents of *P. frutescens* (Gwari et al. 2016). Linalool (52.2%) and β-bourbonene (10.1%) rich chemotype of *P. frutescens* was reported from Pauri Garhwal region (Dimri et al. 2008). Besides, the essential oil of the aerial parts of *P. frutescens*, the seeds are also considered economic parts of this plant used for its nutritional and medicinal value. The seeds of *P. frutescens* are renowned as good source of fatty oil with high content of polyunsaturated fatty acids for dietary supplements. The seed oil is used for cooking purposes and oil cake consumed as raw or fed to cattle (Ahmad, 2019; Gauri et al. 2014; Longvah and Deosthale, 1991; Pandey and Bhatt, 2008). The seed oil composition of *P. frutescens* from twelve location of Garhwal Himalayan region of Uttarakhand was found to be characterised by poly unsaturated fatty acids (72.7%) followed by mono unsaturated fatty acids (17.9%), and unsaturated fatty acids (12.3%). Linolenic acid (51.2-63.3%), linoleic acid (12.2-18.6%), and oleic acid

(11.9-23.8%) were major identified fatty acids (Gauri et al. 2014). The seed oil of *P. frutescens* grown in northeast Himalayan region was found to be dominated by polyunsaturated fatty acids, including linolenic (56.8%) and linoleic (17.6%) acids as major constituents for nutritional and medicinal application (Longvah and Deosthale, 1991).

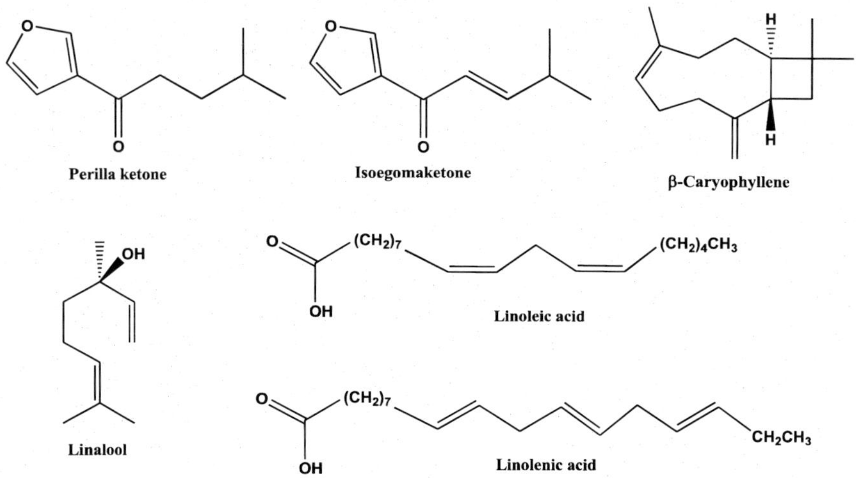

Figure 11. Chemical markers of *P. frutescens*.

2.10. *Plectranthus* and *Coleus* Species

The genus *Plectranthus* sub-family Nepetoideae and the tribe Ocimeae consist of ca. 300 species distributed over Tropical Africa, America, Asia and Australia (Lukhoba et al. 2006). The name *Plectranthus* originated from the Greek words 'Plektron' (spur) and 'Anthos' (flower), as the flowers are having basal spurs. Taxonomic characterisation of the members of genus *Plectranthus* is considered difficult because of its resemblance to closely related genera such as *Coleus*, *Rabdosia*, *Salvia*, and *Isodon* (Cantino et al. 1992; Paton et al. 2004). *Plectranthus* spp. are renowned for their ornamental and medicinal uses worldwide. In India, the genus is represented by ca. 30 spp distributed in all geographic regions, mainly in Himalaya, Southern Ghats and the Nilgiri region. *P. amboinicus*, *P.*

barbatus, *P. incanus*, *P. mollis*, *P. coetsa*, and *P. vettiveroides* are the most common species used in folk medicine for treatment of various diseases (Padalia and Verma, 2011; Waldia et al. 2011). Phytochemical studies on *Plectranthus* spp. revealed the presence of diverse secondary metabolites viz. mainly diterpenoids including abietane, labdane, kaurenes, phyllocladanes, quinones; phenolics, coleons, long-chain alkyl phenols, steroids, flavonoids, triterpenoids etc. majority of members of this genus are highly aromatic, and produce essential oils dominated by monoterpenoids and sesquiterpenoids (Abdel-Mogib et al. 2002; Gaspar-Marques et al. 2007; Lukhoba et al. 2006; Mathela et al. 1986; Waldia et al. 2011). The *Coleus* species are widely used as ornamental plants which are annual in habitat with variety of leaf shapes and colours. Phytochemical studies on *Plectranthus* spp., such as *P. rugosus, P. barbatus* (syn. *C. barbatus*), *P. incanus* and *P. amboinicus* (syn. *C. aromaticus*) from Uttarakhand region revealed significant chemical diversity in their chemical constituents. The leaf essential oil of *P. rugosus* collected from Kumaun region was found to be dominated β-caryophyllene (38.4%), germacrene D (23.8%), spathulenol (3.2%), α-cadinol (2.2%), p-cymene (3.6%), γ-terpinene (2.8%) and limonene (2.7%) as major constituents (Tiwari et al. 2009). The essential oils of *P. incanus* grown in north-western Himalayan region of Uttarakhand was characterised by *cis*-piperitone oxide (35.7%) and piperitenone oxide (45.0%) as main constituents (Shah et al. 1992). The leaf and flower essential oils of *P. rugosus* and *P. incanus* collected from Nainital was studied and compared for their chemical composition. The leaf and flower essential oil of *P. rugosus* was found to be characterised by sesquiterpene hydrocarbons, represented by β-caryophyllene (29.8-36.2%), germacrene D (25.2-28.2%), α-humulene (6.6-8.6%) and (*E*)-β-farnesene (2.3-3.8%) as the major constituents. On the contrary, the leaf and flower oil of *P. incanus* were found to be dominated by monoterpenoids with piperitenone oxide (38.5-44.2%), piperitone (8.6-12.2%) and terpinolene (10.2-14.5%) as major constituents (Padalia and Verma, 2011). Mathela et al. (1986) identified a biologically active diterpenoids, dehydroabietane, in roots of *P. barbatus* (syn. *C. barbatus, C. forskohlii*) (Mathela et al. 1986). *P. rugosus* collected from Garhwal region was found to be characterised by α-pinene (0.8-19.3%), *trans*-ferruginol (3.0-17.1%), α-phellandrene (2.6- 10.5%), (*E*)-caryophyllene (4.0-9.3%), germacrene D (1.9-8.8%), and abietatriene (1.9-5.3%) (Verma et al. 2015). Moreover, *P. amboinicus* (syn. *Coleus aromaticus*) collected from mid hill and foot hill regions of Uttarakhand was characterised by thymol (57.7-66.4%), γ-terpinene (5.8- 11.7%), and p-cymene (4.1-14.2%) (Verma et al. 2012).

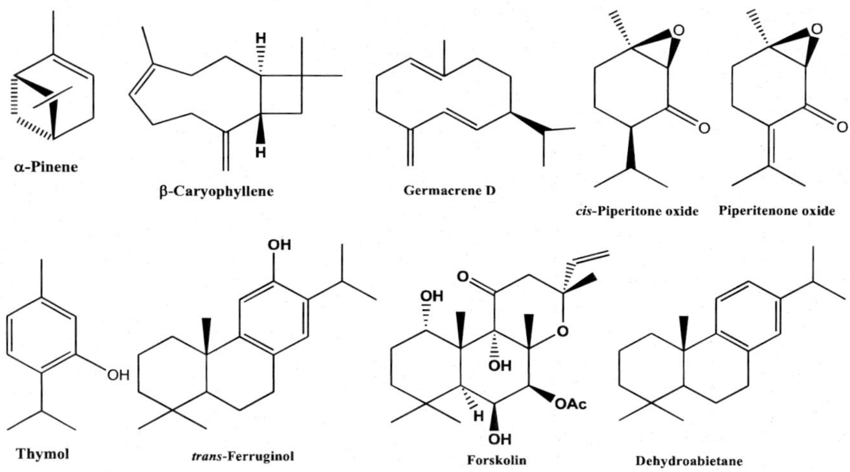

Figure 12. Chemical markers of *Plectranthus* and *Coleus* spp.

2.11. *Pogostemon* Species

The genus *Pogostemon* (tribe *Pogostemoneae*, subfamily *Lamioideae)* is represented by over 80 species distributed in tropical and subtropical Asia (Harley et al. 2004; Yao et al. 2015). The highest plants diversity is found in the Indian subcontinent represented by 56 taxa (53 species and 3 varieties) mainly in Western Ghats and plains (Bhatti & Ingrouille, 1997). Most of the *Pogostemon* spp. are widely used for fragrance, perfumery and cosmetics purpose. *P. cablin* (Blanco) Benth. commonly known as 'Patchouli' reputed as a source of patchouli essential oil of the trade. It ranks high among the essential oil bearing crops for perfumery and cosmetic products (Akhila and Tewari, 1984). *P. benghalensis* Burm. F., is another important member of this genus widely distributed in Himalayan region growing in waste places, especially in moist places. It is cultivated for medicinal use in India, Bangladesh, Sri Lanka, Nepal, Myanmar, Thailand and China. It is strongly scented pubescent gregarious herb, with a hoary pubescent stem, having stalked ovate opposite leaves, and white to tinged white pink strongly aromatic flowers (Bhuiyan et al. 2011; Gupta, 1968). Phytochemical investigations on essential oil composition of *Pogostemon* spp. viz. *P. cablin, P. benghalensis, P. amaranthoides* from Uttarakhand regions have been investigated. The essential oils of the aerial parts of patchouli (*P. cablin*) grown in foot hills agroclimatic conditions was characterised by patchouli alcohol (42.2–57.7%), α-bulnesene (9.0–15.2%), α-guaiene (6.4–17.9%), seychellene (3.4–6.9%), and pogostol (0.3–5.0%) as major constituents. However, the root essential oil of *P. cablin* was characterized by pogostone (70.2%), norpatchoulenol (5.3%) and β-pinene (4.5%) as major constituents (Verma et al. 2019). In an another report, patchouli alcohol (32.3%), α-guaiene (23.3%), α-bulnesene (21.4%), y-patchoulene (6.3%), and β-caryophyllene (4.8%) were identified as the major constituents of the essential oil of dry leaves of *P. cablin* organically grown in Tarai region of Uttarakhand (Prakash et al. 2007). The leaf and inflorescence essential oil of *P. benghalensis* from two different altitudes of Kumaon were studied and compared. β-Bisabolene (8.0-18.0%), β-caryophyllene (8.1-15.2%), elemol (1.1-11.6%), and β-copaen-4α-ol (4.2-9.6%) were the major constituents of inflorescence oils of *P. benghalensis*; while the leaf oils were dominated by high content of elemol (4.0-20.5%), β-caryophyllene (12.5-13.6%), β-copaen-4α-ol (6.0-7.7%), and germacrene B (4.8-7.0%) (Chanotiya et al. 2007). The essential oil of dried mature inflorescence of *P. benghalensis* collected from Garhwal was characterised by azulen-2-ol (32.0%), octatriene (6.5%), β-patchoulene (6.4%), germacrene D (5.1%), and germacrene B (5.0%), as major constituents (Chauhan et al. 2008). The essential oil of *P. amaranthoides* collected from high altitudes of Uttarakhand was found to be characterised by β-caryophyllene (15.4%), guaia-3,9-diene (8.2%), and δ-guaiene (=α-bulnesene) (7.1%), as major constituents (Prasad, 2018).

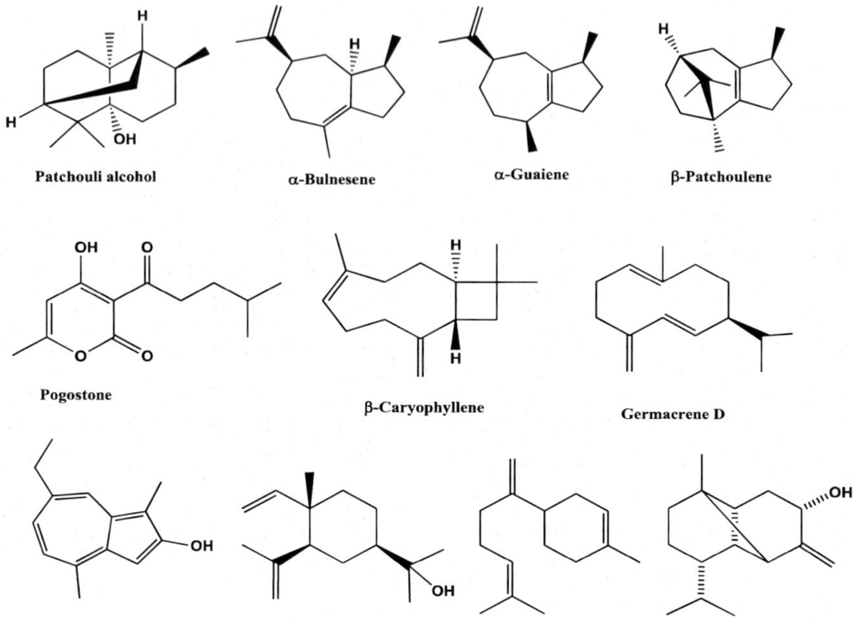

Figure 13. Chemical markers of Himalayan *Pogostemon spp.*

2.12. *Salvia* spp.

The genus *Salvia*, commonly known as 'Sage' is one of the largest genera and distributed in the temperate and subtropical regions of the world (Hedge, 1992). Most members of this genus are cultivated for ornamental, aromatic, and culinary usage, spices, flavouring agents, and for perfumery and cosmetics purpose (Senatore et al. 2004, 2006; Fu et al. 2013). This genus is known to exhibits a very high variability in morphological and genetic characters as well as in chemical constitution based upon their geographical origin (Hedge, 1984; Raina et al. 2013). In India, ca. 24 species have been reported with great variation in their habitat, growth, form and colour (Gupta 1968; Upadhyay et al. 2009). *Salvia* spp. are renowned for their pleasant essential oils and numerous biologically active secondary metabolites extensively used in folk medicine (Guenther, 1949). *S. leucantha* Cav., *S. officinalis* L., and *S.*

sclarea L. are the most common *Salvia* spp. grown in Himalayan region of Uttarakhad for ornamental purpose and as source of essential oils for fragrance and medicinal purpose. *S. officinalis,* known as garden sage, is a perennial herb or sub-shrub native to Southern Europe and Turkey. It is grown for ornamental purpose, its dry herb and essential oils, concrete and absolute in various parts of the world (Anonymous, 1972, Raina et al. 2013). *S. leucantha* is widely distributed in Himalayan region (Upadhyay et al. 2009). *S. sclarea*, popularly known as 'Clary Sage' is cultivated for its high value essential oil, concrete and absolute, and aroma molecules used in perfumery industries, traditional medicine and in aromatherapy (Shawl et al. 1999; Verma et al. 2010). Phytochemical studies on these three spp. of *Salvia* showed monoterpenoids (thujones, camphor, 1,8-cineole, linalool, linalyl acetate, pinenes, bornyl acetate, borneol, α-terpineol etc) and sesquiterpenoids (α-humulene, β-caryophyllene, spathulenol, α-himachalene, γ-cadinene, viridiflorol) as major constituents (Raina et al. 2013; Verma et al. 2010; Verma et al. 2015; Upadhyay et al. 2009). The essential oil of *S. leucantha* was found be characterised by sesquiterpenoids (57.6%) represented by spathulenol (12.1%), β-caryophyllene (10.7%), α-himachalene (10.5%) and γ-cadinene (6.5%); along with 35.7% monoterpenoids (35.7%) represented by bornyl acetate (27.8%) as major constituents (Upadhyaya et al. 2009) while essential oil of *S. officinalis* was found to be dominated by α-thujone (31.68%), camphor (10.80%), 1,8-cineole (8.31%) and β-thujone (3.67%) and β-caryophyllene (4.38%) as major constituents (Raina and Negi, 2012). In an another report, the essential oil from aerial parts of *S. officinalis* from six collections were characterised by α-thujone (21.4-40.1%), camphor (11.3-37.6%), α-humulene (4.5-9.5%),1,8-cineole (4.4-9.1%), β-thujone (2.0-7.4%), camphene (1.8-7.0%), viridiflorol (2.1- 5.5%), α-pinene (1.5-6.1%), and β-caryophyllene (1.0-5.5%) as major constituents (Raina et al. 2013). The essential oil composition of *S. officinalis* L. grown in foots hills agroclimatic conditions was characterised by *cis*-thujone (19.8-42.5%), β-caryophyllene (1.2-16.1%), manool (3.6-15.1%), viridiflorol (3.1-12.8%), 1,8-cineole (2.8-13.8%), and camphor(1.4-22.1%) as major constituents (Verma et al. 2015). The essential oil composition of *S. sclarea* var. CIM-

Chandni cultivated in mid hills of Uttarakhand was studied. The major constituents of the hydrodistilled essential oil were linalool (40.2%), linalyl acetate (34.2%), α-terpineol (5.17%), and β-myrcene (5.47%), while the essential oil of hydrophilic oil were linalool (62.0%), α-terpineol (20.5%), β-myrcene (2.03%) with no linalyl acetate (Verma et al. 2010).

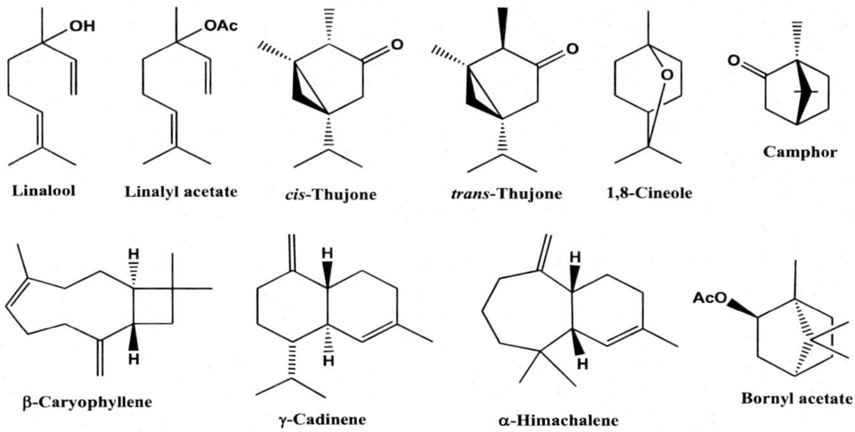

Figure 14. Chemical markers of *Salvia* spp.

2.13. *Scutellaria* spp.

The genus *Scutellaria* is widely distributed in Europe, the United States and East Asia has a long history of use in traditional and fold medicine (Sripathi and Ravi, 2017; Xiong et al. 2011). It is represented by 6 species viz. *S. scandens, S. repens, S. grossa, S. prostrata, S. discolor* and *S. linearis* in Uttarakhand. The plants of this genus are annual herbs or shrubs which grow wild in sub-temperate to temperate (upto 1300-2500-m in elevation) (Gupta, 1968; Melkani et al. 2007; Strachey, 1974). Phytochemical investigations on various *Scutellaria* spp from different geographic regions revealed numerous bioactive secondary metabolites viz. flavonoids, diterpene, phenolics, iridoids, triterpenoids, sterol and alkaloids as well as volatile constituents with mono and sesquiterpenoids (Bruno et al. 2002; Kikuchi et al. 1991; Ohno et al. 1997; Sripathi and

Ravi, 2017; Xiong et al. 2011). The essential oil composition of three widely distributed *Scutellaria* species viz. *S. scandens, S. repens*, and *S. grossa* in Kumaun Himalayan region of Uttarakhand were studied. The essential oil of the aerial parts of *S. repens* collected from Kumaun was found to be characterised by aromadendrene (30.7%), β-funebrene (15.0%), β-gurjunene (8.0%) and α-neo-clovene (5.0%) as major constituents (Melkani et al. 2013) while the essential oil of the aerial parts of *S. scandens* was characterised by β-caryophyllene (27.7%), germacrene D (19.0%), γ-amorphene (6.5%) as major constituents (Melkani et al. 2007). The chemical essential oil from the aerial part of *S. grossa* showed linalool (37.0%) and 1-octen-3-ol (32.0%) as the marker constituents (Pant et al. 2012).

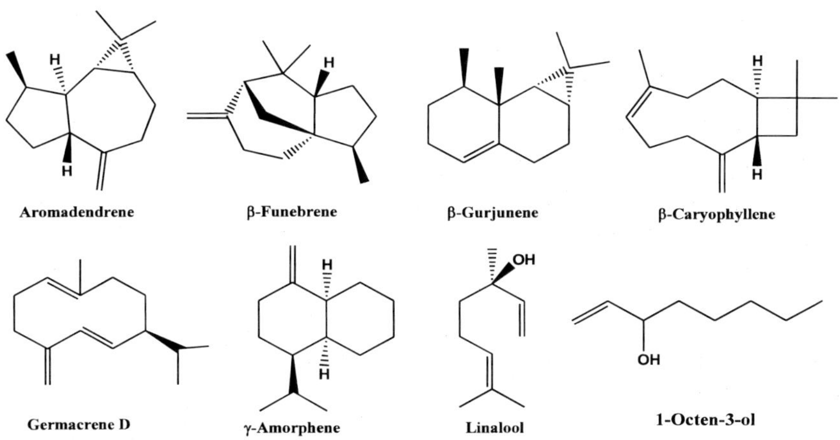

Figure 15. Chemical markers *Scutellaria* spp.

2.14. *Stachys* spp.

The genus *Stachys* L. comprises ca 300 species distributed worldwide and is considered as one of the largest genera of the Lamiaceae (Goren et al. 2011; Lindqvist and Albert, 2002). *S. sericea* Wall. is one of the widely distributed member of this genus wildly distributed in Himalayan region. It is an erect flowering herb grown on waste and rocky lands (Bisht et al. 2008; Gupta, 1968). In spite of one of the largest genera of the Lamiaceae, there are limited phytochemical investigations on this genus. The phytochemical investigation on essential oils on aerial parts of some *Stachys* species revealed sesquiterpenoids represented by β-caryophyllene, germacrene D, cadinenes, farnesene and caryophyllene oxides as marker constituents, along with lesser content of monoterpenoids (pinenes, linalool, linalyl acetate) and diterpenoids (labdane, abietane, kaurane and pimaranes) (Bisht et al. 2008; Goren et al. 2011; Piozzi and Bruno, 2009; Skaltsa et al. 2001, 2003; Kaya et al. 2001; Radulovic et al. 2007). However, the roots of several *Stachys* spp. were found to be dominated by diterpenes represented by neoclerodane, kaurane, and labdane skeletons (Piozzi and Bruno, 2011). To best of our knowledge, the phytochemical investigation from Indian origin report only study on essential oil composition of *S. sericea* grown in Kumaun region of Uttarakhand Himalaya. The essential oil of *S. sericea* collected from Nainital of Uttarakhand Himalaya showed the presence of sesquiterpene hydrocarbons viz. germacrene D (37.7%), β-caryophyllene (17.4%), δ-cadinene (6.0%) and (*E*)-β-farnesene (3.9%) as major constituents (Bisht et al. 2008).

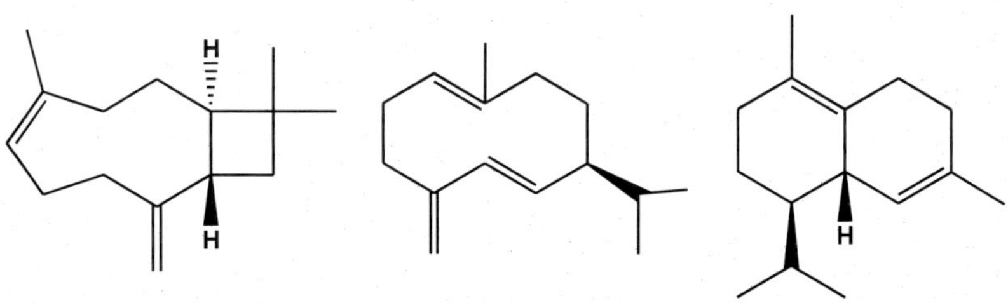

β-Caryophyllene **Germacrene D** **δ-Cadinene**

Figure 16. Chemical markers of *S. sericea.*

2.15. *Teucrium* spp.

The genus *Teucrium* includes ca. 300 species widely distributed in Europe, Asia, America, Africa and Australia with major distribution in Mediterranean area. Various members of this genus possess several biological activities and used in folk medicine (Ali et al. 2008; Khaled-

Khodja et al. 2014). *T. royleanum* Wall ex. Benth. and *T. quadrifarium* Buch.-Ham. are the most common species of this genus occurring wild in the temperate region (Mohan et al. 2010; Stracheyi, 1974). The essential oil composition of several *Teucrium* spp. reported from different parts of the globe contained sesquiterpene hydrocarbons mainly represented by β-caryophyllene and germacrene D; cadinenes, cadinols along with pinenes and limonene as the principal marker constituents of their essential oils (Aburjai et al. 2006; Elshazly and Hussein, 2004; Mohan et al. 2010; Morteza-Semnani et al. 2005; Nasser et al. 2008). β-Caryophyllene (23.6% and 38.3%), germacrene D (28.9% and 9.4%), α-humulene (5.7% and 5.9%), linalool (4.8% and 1.5%), and 1-octen-3-ol (8.5% and 0.7%) were identified as the major constituents in *T. royleanum* and *T. quadrifarium* (Mohan et al. 2010).

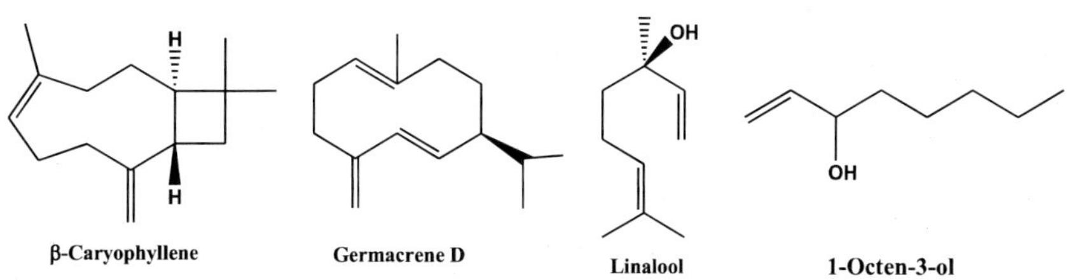

Figure 17. Chemical markers of *Teucrium* spp.

2.16. *Thymus* spp.

The genus *Thymus* L., commonly known as ‘thyme’, consists of hybrids, varieties, and ecotypes, which are small perennial herbs (Figueiredo et al., 2008; Stahl-Biskup and Saez, 2002). Thyme is cultivated mainly in European countries viz. Spain, France, Italy, and Bulgaria etc. for the production of fresh and dry herbs, and essential oil (Figueiredo et al., 2008). In India, the genus is represented by two species, viz. *T. linearis* (Benth. ex Benth) and *T. serpyllum* L. which are wild and distributed in sub-temperate to temperate Himalayan regions (Mathela et al. 1980; Verma et al. 2016). Thyme herb, thyme oil and extracts are widely used in pharmaceutical, cosmetic, perfume and flavour industries (Figueiredo et al., 2008; Stahl-Biskup and Saez, 2002). Various reported pharmacological properties such as antiseptic, antioxidative, insecticidal, preservative and anaesthetic etc are mainly due to presence of thyme oil with phenolic compounds viz. thymol, carvacrol, their acetates and ethers as major components. Earlier studies on Himalayan thyme (*T. linearis* and *T. serpyllum* L) showed great morphological and chemical variability. The essential oil of *T. serpyllum* from Kumaun region of Himalaya was found to be dominated by thymol (60%) and carvacrol (2%) as the major constituent (Mathela et al. 1980). In another study, *T. serpyllum* collected from two localities viz. Martoli and Jageshwar of Kumaun Himalayan region and *T. serpyllum* collected from western Himalayan region of Nepal were compared. The study revealed that *T. serpyllum* collected from Martoli contained thymol (68.32%), carvacrol (7.19%) as major constituents. The essential oil of *T. serpyllum* grown in the Jageshwar of Kumaun Himalayan region consists of thymol (46.2%), carvacrol (4.6%), p-cymene (7.3%) and γ-terpinene (12.2%) as major constituents. However, *T. serpyllum* collected from western Himalayan region of Nepal was characterised by higher content of thymol (74.9%) and carvacrol (5.1%), with p-cymene (2.2%) and γ-terpinene (1.5%) (Thakuri et al. 2009). The essential oil of *T. serpyllum* grown in Chakrata (Dehradun) consist thymol (37.2–55.5%) and thymyl methyl ether (3.2–12.9%) as major components (Mohan et al. 2012). The hydrodistilled essential oil of aerial parts of *T. serpyllum* cultivated in Kumaon region was characterised by thymol (58.8%), p-cymene (5.7%), thymol methyl ether (4.0%), and γ-terpinene (3.4%) as major constituents (Verma et al. 2009). *T. linearis* collected from five distinct locations of Uttarakhand were found to contain thymol (52.28–66.65%), p-cymene (1.81–21.60%) and γ-terpinene (1.94–12.48%) were the major constituents (Verma et al. 2010). In an another study, the essential oil of successfully domesticated Himalayan thyme (*T. linearis*) in foot hills of Uttarakhand Himalaya was found to dominated by thymol (34.3–67.3%), p-cymene (4.0–27.4%), γ-terpinene (4.9–24.4%), thymyl acetate (0.6–5.2%), thymol methyl ether (2.4–3.7%), and carvacrol (0.9–1.5%) as marker constituents at different crop durations and harvests (Verma et al. 2016). The essential oil quality of *T. linearis* and *T.*

serpyllum produced in Uttarakhand are comparable to the standard criterion set by European pharmacopeia with quality parameters for common thymes with thymol (36.0-55.0%), carvacrol (1.0-4.0%), p-cymene (15-28%), and γ terpinene (5-10%) (Verma et al. 2010; European Pharmacopeia, 2005).

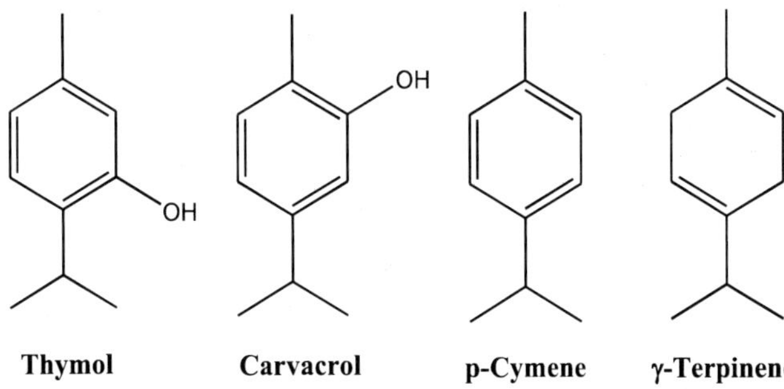

Figure 18. Chemical markers of thyme.

3. Bioactivity and Pharmacological Properties

3.1. *Ajuga* spp.

Ajuga spp. are extensively used in Asian and African folk medicine for treatment of rheumatic fevers, diabetes, malaria, dysentery, hypertension, and gastrointestinal disorders, as well as anthelmintic, astringent, febrifuge, hypoglycemic, anti-tumor, diuretic, antimicrobial and anti-inflammatory agents (Israili et al. 2009; Khaled-Khodja et al. 2014; Qing et al. 2017). *A. parviflora* was found to possess diverse medicinal uses viz. astringent and for the treatment of swollen wounds, diarrhoea, rheumatism, fever, eye trouble in indigenous systems of medicine (Perry and Metzger, 1980). *A. macrosperma* and *A. bracteosa* are used as stimulant, diuretic, expectorant, antipyretic, and as a substitute for cinchona in local traditional medical system. (Bhakuni and Kaul, 1961; Singh et al. 2014). The ecdysteroids, clerodane and neo-clerodane diterpenoids reported in *Ajuga* spp. were found to possess diverse biological activities (Coll and Tandron, 2008,

Meritt and Ley, 1992). These metabolites were found to exhibit well-established physiological activities in insects (growth, molting and other hormonal functions) and play important roles for defense against phytophagous insects with significant antifeedant activities along with insecticidal, larvicidal, antiviral, antitumor, antimalarial, antimicrobial hypoglycemic, cytotoxic, cholinesterase inhibitory, antioxidative and vasorelaxant properties (Agrawal et al. 2010; Castro et al. 2008; Coll and Tandron, 2008; Qing et al. 2017).

3.2. *Calamintha* spp.

Calamintha spp. are known to be used in traditional medicine as stimulant, digestive, tonic, antiseptic antispasmodic, emmenagogue, diaphoretic, diuretics, carminatives, expectorant and for strengthening central nervous system (Bozovic and Ragno, 2017). The aqueous decoction of is used to treat gastric and intestinal problems. The essential oil of *Calamintha* species are known to possess potent pharmacological activities viz. antioxidant, antimicrobial, anti-inflammatory, cicatrizing and insecticidal activities (Bozovic and Ragno, 2017; Mancini et al. 2014). The antibacterial activity of the essential oil of *C. umbrosa* was tested against six bacterial strains viz. *P. aeruginosa*, *E. coli*, *A. hydrophila*, *K. pneumonia, B. subtilis* and *S. scandidus.* Results showed that the essential oil of *C. umbrosa* showed potent bactericidal activity against *S. scandidus* (zone of inhibition: 10.5 mm; MIC: 25 µL/mL) and *E. coli* (zone of inhibition: 6.6 mm; MIC: 25 µL/mL). However, the essential oil of *C. umbrosa* showed varying degree of activity against *P. aeruginosa* (zone of inhibition: 8.5 mm; MIC: 30 µL/mL), *A. hydrophila* (zone of inhibition: 6.8 mm; MIC: 30 µL/mL), *K. pneumonia* (zone of inhibition: 5.1 mm; MIC: 55 µL/mL), and *B. subtilis* (zone of inhibition: 10.5 mm; MIC: 35 µL/mL) (Kumar et al. 2016). The toxicity and insecticidal activity of the essential oil of *C. umbrosa* was tested against mustard aphid, *L. erysimi.* The essential oil of C. umbrosa was found to possess significant toxicity (LC_{50} 1.07 mg/ml) as well as fumigant repellent activity (LC_{50} of 1 ml/l air) against *L. erysimi.* While, the 95% fiducial limits of the LT_{50} value for the concentration tested (0.50 ml/l air) was 17.17 and 79.42 h for the essential oil of *C. umbrosa.* Results revealed that essential oil of *C. umbrosa* from Himalayan region can be used as natural insecticidal agent against *L. erysimi* (Kumar and Mathela, 2017). The in-vitro antifungal activity of the essential oil *of C. umbrosa* was tested against *F. oxysporum, H. maydis, A. solani, R. solani, S. sclerotiorum, C. lunata* and *A. candida.* The essential oil of *C. umbrosa* showed moderate to high in vitro antifungal activity (MIC: 1500-3000 μg mL^{-1}) against tested pathogenic fungal strains, and found most effective in spore germination inhibition (92.8%) of *C. lunata* (IC_{50} : 557.4 μg mL^{-1}) (Kumar et al. 2014). The antioxidant activity of the essential oil of *C. umbrosa* was assessed by different methods viz. radical scavenging activity (RSA) with DPPH, superoxide radical scavenging, Fe+2-ion chelating power, ferric educing antioxidant power (FRAP), lipid peroxidation inhibition (LPI) and total antioxidant capacity (TAC) assays are included. The essential oil of *C. umbrosa* exhibit significant antioxidant activity (IC_{50}: 1.3 mg/mL, 2.1 mg/mL and 2.8 mg/mL) in DPPH, superoxide radical scavenging, Fe^{+2}-ion chelating assay, respectively. However, in ferric-reducing antioxidant power (FRAP) assay, the reducing power of the essential oil of C. umbrosa was found significant (2.6 mg AAE/g) antioxidant activity. The total antioxidant capacity (TAC) results showed that the *C. umbrosa* oil also exhibited good antioxidant activity that was promising for use of its essential oil as natural antioxidants (Kumar et al. 2019).

3.3. *Elsholtzia* spp.

Elsholtzia species have a long history of medicinal use in traditional herbal medicines. They are used as diuretics, carminative, astringent, antipyretics, and for treatment sores, colds, fever, dysentery, choleric diarrhea, heat stroke, rheumatic arthritis, respiratory and digestive disorder, (Chopra et al. 1956; Guo et al. 2012; Mathela et al. 2009; Thappa et al. 1999). *Elsholtzia* spp. are mostly aromatic in nature and produce pleasant essential oils and aroma chemicals used in food-flavouring, herbal tea, spices, beverages, perfumeries, cosmetics, and aromatherapies (Chauhan et al. 2019; Guo et al. 2012). *Elsholtzia* spp. are also used for phytoremediation of soil contaminated by heavy metals (Tian et al. 2008; Zhang et al. 2010). Pharmacological investigations on the extracts, essential oils and pure compounds from various *Elsholtzia* spp. revealed pharmacological properties viz. antiviral, antibacterial, antifungal, anti-oxidant, anti-inflammatory, myocardial ischemia protection and spasmolytic activities for medicinal and industrial application (Guo et al. 2012; Liu et al. 2007; Pant et al. 1992).

3.4. *Hyssopus* spp.

Hyssopus spp. has a long history of use in various traditional systems of medicine as stimulant carminative, tonic, soothing, expectorant, antiseptic, and for treatment of various respiratory, gastric and urinary disorders (Fathiazad and Hamedeyazdan, 2011). *H. officinalis* has been used traditionally for medicinal purposes and as flavouring agents in local cuisines (Charles, 2012; Kazazi et al. 2007). The aerial part of *H. officinalis* is highly aromatic with strong minty aroma and spicy pungent flavor, hence used in food products, sauces and seasonings, beverages, and liqueurs (Jahantigh et al. 2016; Tahir et al., 2018). In traditional medicine, it is used for treatment of colds, coughs, sore throats, toothaches, bronchitis, asthma, spasm, epilepsy, ulcer, gastric and nervous disorders (Tahir et al. 2016). The aerial parts of *H. officinalis* is known as 'Zufah Yabis' in the Unani system of medicine for the treatment of cough, cold, and asthma. Za'atar is its well-known herbal formulation derived from its aromatic herb and essential oil used for various ailments (Pandey et al. 2014; Tahir et al. 2016). The essential oil of *H. officinalis* is widely used as fragrance component in soaps, cosmetics and perfumes, food-flavour and pharmaceuticals (Charles, 2012; Khazaie et al. 2008). Moreover, the extract and essential oil of *H. officinalis* were found to possess various pharmacological activities viz., antibacterial, antifungal, antioxidant, antiinflammatory, anticatarrhal, antispasmodic, sedative, spasmolytic, antiviral, nematocidal, insecticidal, cytotoxic, mosquito repellent, and anti platelet activities for diverse medicinal use (Gollapudi et al. 1995; Pandey et al. 2014; Srivastava et al. 2018; Tahir et al. 2016).

3.5. *Mentha* spp.

Mentha spp. has been used as extensively in traditional medicine since ancient times for treatment of digestive tract diseases, nausea, bronchitis, flatulence, anorexia, ulcerative colitis, liver complaints due to their broad pharmacological properties. The mint oils mainly mentholmint, peppermint and spearmint essential oil and their main constituents were used extensively in food, flavor, cosmetics, pharmaceutical and allied industries (Lawrence, 2007). The essential oil of *M. arvensis* and menthol are widely utilized in the flavour, fragrance, and pharmaceutical industries. Menthol is primarily consumed by tobacco industry, followed by pharmaceutical and confectionary industries (Rao and Syamasundar, 2013). Worldwide, approximately 10,000 tonnes of natural menthol and 2,000 tonnes of synthetic menthol are used by the pharmaceutical, cosmetic and cigarette industries every year. Menthol it is also used as fragrance component of soaps, toothpastes, detergents, cosmetic, and perfumes (Kamatou et al., 2013; Lal, 2013). *M. piperita* is used in common cold, headache, nausea, skin, respiratory digestive and diseases and possess pharmacological activities such as an astringent, stimulant antiseptic, antimicrobial, rubefacient, antipyretic, antispasmodic, and anti-aging agent. Peppermint oil is used in various food products, cosmetics, dental preparations and pharmaceutical formulation due to its strong antimicrobial, antioxidant, virucidal, nematicidal insecticidal properties (Murray et al. 1988; Singh et al. 2011). *M. spicata* is used in various ailments such as nausea, anxiety, cold, flu, flatulence, diarrhea, gastric and menstrual pain (Cakilcioglu et al., 2011) The essential oil of spearmint and its major components, carvone used in food, cosmetic, confectionary, chewing gum, toothpaste, and pharmaceutical industries (Lawrence, 2007). *M. citrata* essential oil is renowned for its distinct lavender and floral scent due to the presence of linalool and linalyl acetate for food-flavor, perfumes, and cosmetic products. *M. longifolia* is used in multiple pharmacological activities (Rasooli and Rezaei, 2002). Our earlier work on the essential oil of *M. longifolia* rich in piperitenone oxide and *cis*-piperitone oxide from Himalayan region was also found to possess significant antipyretic, antibacterial, antifungal activity and moderate anti-inflammatory and analgesic activity (Singh et al. 2011). *Mentha spicata* essential oil rich carvone (15.3%-68.5%) and piperitenone oxide (24.0%-79.2%) was found to moderate to significant antioxidant activities by chelating properties of Fe^{2+}, DPPH radical-scavenging activity, and their reducing power along with good sprout suppressant activity of potato (*Solanum tuberosum* L.). in comparison to standard sprout suppressant (Chauhan et al. 2011).

3.6. *Nepeta* spp.

Nepeta spp. are known for their diverse uses in folk medicine due to the various pharmacological activities viz. antimicrobial, antiinflammatory, sedative, antioxidant, relaxant, carminative, diuretic, diaphoretic, antiasthmatic, febrifuge, vermifuge, cholesterol lowering herbicidal, insecticidal and insect repellents etc. (Baser et al. 2002; Cigremis et al. 2011). The distereomeric nepetalactones, iridoid constituents and other essential oils in various *Nepeta* spp. were found to be responsible for the diverse biological activities viz. feline attractant, canine attractant, and insect repellent and arthropod defense activity and other pharmacological activities (Tucker and Tucker, 2009; Mathela, 2017). Iridodial β-monoenol acetate and actinidine reported as marker constituents of the essential oils of *N. leucophylla* and *N. clarkei,* respectively possess significant antifungal potential against *A. flavus*, *A. ochraceus*, *P. citrinum*, and *P. viridicatum*, *S. rolfsii* and *M. phaseolina*s. Iridodial β-monoenol acetate was found most effective against *S. rolfsii*, while actinidine was found highly active against *M. phaseolina.* The essential oil of both *N. leucophylla* and *N. clarkei* from Himalayan region were promising for antifungal herbal formulations

(Saxena and Mathela, 1996). The essential oil of *N. elliptica* and its major constituent, (7R)-*trans,trans*-nepetalactone (>85%) was found to possess significant antifungal activities against soil borne plant pathogenic fungi viz. *F. oxysporum, H. maydis, R. solani, A. solani, S. sclerotiorum, C. lunata* and *A. candida*. The results our study showed that the essential oil of *N. elliptica* possessed varying degree of antifungal activity (mycelia growth inhibition: 54.1–89.0% and IC_{50}: 115.2-384.1 μg/mL) at a concentration of 500 μg/mL; with significant sensitivity against *H. maydis* (inhibition: 89.0%; IC_{50}: 115.0 μg/mL) and *F. oxysporum* (inhibition: 88.6%; IC_{50}: 122.7 μg/mL) and *R. solani* (inhibition: 80.8%; IC_{50}: 153.3 μg/mL), respectively. The major constituents of the essential oil, (7R)-*trans,trans*-nepetalactone, also showed highest activity against *H. maydis* (inhibition: 82.3%; IC_{50}: 182.7 μg/mL) and *F. oxysporum* (inhibition: 81.6%; IC_{50}: 191.7 μg/mL). The essential oil of *N. elliptica* exhibited a potent inhibitory effect on spore germination against *H. maydis, F. oxysporum* and *C. lunata* with IC_{50} value 221.07 μg/mL, 278.66 μg/mL, 242.25 μg/mL, respectively. Thus, the essential oil of *N. elliptica* and its major constituent have potential for their development as antifungal agents to suppress common plant diseases caused by these pathogenic fungal strains (Kumar et al. 2014). In our another study the essential oil of *N. leucophylla, N. ciliaris*, and *N. clarkei* were also found to possess significant in vitro antifungal activity as well as spore inhibition of the fungal strains viz. *F. oxysporum, H. maydis, A. solani, R. solani, S. sclerotiorum, C. lunata* and *A. candida*. Results indicate significant antifungal potential against pathogenic fungal strains. Furthermore, oxygenated terpenoids were found to possess higher antifungal activity than other constituents. In recent years, essential oils and their constituents are being considered as potential natural substitutes for synthetic fungicides related to plant and human pathogenic fungi (Kumar et al. 2016). The antibacterial activity of four essential oils from *N. leucophylla, N. hindostana*; *N. ciliaris* and *N. clarkei* was investigated against four gram negative bacteria (*P. aeruginosa, E. coli, A. hydrophila* subsp. *hydrophila, K. pneumonia*) and two gram positive bacterium (*B. subtilis* and *S. scandidus* subsp. *azaticus*). The essential oil of *N. leucophylla* showed significant antibacterial activity against gram-negative bacteria *P. aeruginosa* and *K. pneumonia* while *N. ciliaris* oil against *B. subtilis* and *P. Aeruginosa.* Results showed that antibacterial activity of *Nepeta* species had a strong and broad spectrum effect against (Kumar et al. 2016). The essential oil of *N. erecta*, dominated by isoiridomyrmecin, was found to possess significant in vitro antioxidant activity by β-carotene bleaching (54.6±2.52%) and by scavenging DPPH free radical (IC50 0.74±0.12 mg/mL) (Bisht et al. 2012). The antioxidant activity of the essential oils of four Himalayan *Nepeta* spp. (*N. leucophylla*, *N. clarkei*, *N. ciliaris*, and *N. hindostana*) were assessed by using in vitro methods viz. radical scavenging activity (RSA), Fe+2-ion chelating power, ferric educing antioxidant power (FRAP), lipid peroxidation inhibition (LPI) and total antioxidant capacity (TAC) methods. In DPPH scavenging assay, *N. ciliaris* oil was found to exhibit maximum antioxidant activity (IC_{50}: 0.9 mg/mL) followed by *N. leucophylla* (IC_{50}: 1.2 mg/mL), *N. hindostana* (IC_{50}: 2.0 mg/mL) and *N. clarkei* (IC_{50}: 2.1 mg/mL). While, the superoxide radical scavenging assay showed that the highest scavenging activity of the essential oil of *N. ciliaris* (IC50: 1.0 mg/mL) followed by *N. leucophylla* (IC_{50}: 1.4 mg/mL), *N. hindostana* (IC_{50}: 3.9 mg/mL) and *N. clarkei* (IC_{50}: 3.7 mg/mL). However, in Fe^{+2}-ion chelating assay, the essential oil of *N. ciliaris* and *N. leucophylla* (IC_{50}: 4.4 and 4.6 mg/mL, respectively) showed good chelating activity than *N. hindostana* (IC_{50}: 5.1 mg/mL) and *N. clarkei* (IC_{50}: 5.0 mg/mL). The total antioxidant capacity (TAC) results showed that the essential oil of *N. leucophylla* exhibited highest TAC (6.6 mg AAE/g) followed by *N. clarkei*, *N. ciliaris* and *N. hindostana* (Kumar et al. 2019). Diterpenoids viz. *p*-benzochinone, labda-14-en-8,13-diol, coleon U 12-methyl ether, and triterpenoid (ursolic acid) isolated from the roots of *N. leucophylla* and *N. clarkei* Hook.f were also found to possess antioxidant activity. Coleon U 12-methyl ether and *p*-benzochinone have shown good DPPH scavenging activity (Bisht et al. 2010). The essential oils and constituents of various *Nepeta spp.* are being of interest due to their feline attractant, canine attractant and arthropod defense properties as well as antimicrobial potential due to distereomeric nepetalactones, iridoids and related compounds for insect pest management as well as natural antimicrobial agents. Therefore, considering the various ethnobotanical uses and various reportes on pharmacological activities on various *Nepeta* spp.

3.7. *Ocimum* spp.

The genus *Ocimum* possess various biological activities such as carminative, diuretic, anti-emetic, appetite stimulant, anti-stress, anti-inflammatory, anti-arthritic, antispasmodic, antipyretic, antihelmentic, antimalarial, and antimicrobial etc. and are extensively used in traditional system of medicine (Lawrence, 1988; Makinen and Paakkonen, 1999; Saha et al. 2013; Vina and Murillo, 2003). The leaf juice of *O. tenuiflorum* along with Triphala is used in Ayurvedic preparations for treatments of glaucoma, cataract, and chronic conjunctivitis. The juice of fresh leaves of *O. tenuiflorum* is used to treat chronic fever, dysentery, haemorrhage, dyspepsia, ringworm, and skin diseases (Prakash and Gupta, 2005; Rajeshwari, 1992). Leaves of *O. tenuifolium* are considered vital as rejuvenator and to sharpen memory and to cures mouth ulcers and infections of mums. The leaf aqueous extract of *O. gratissimum* reported to inhibit tumor growth and angiogenesis on HeLa cells as well for treatments of gastrointestinal problems (Mathew and Sankar, 2013). Rosmarinic acid, chicoric acid, ursolic acid, caffeic acid esters and various flavanoids and their glycosides reported in *O. tenuiflorum*, *O. basilicum*, *O. gratissimum*, and *O. kilimandscharicum* are found to be biologically active as strong antioxidant, anti-HIV, anti-inflammatory, antirheumatic, antiviral, antitumeral agent (Khalighi et al. 2002; Kosar et al. 2005; Silva et al. 2008). The essential oils of various *Ocimum* spp. and their aroma-flavour constituents are used in various food, fragrances, cosmetics and pharmaceutical products due to their scientifically validates biological activities viz. antimicrobial, anti-inflammatory, antiviral, anti-oxidant, germicidal etc.

3.8. *Origanum* spp.

The plants of genus *Origanum* are in general powerful disinfectants, flavoring agents, in perfumes and fragrance. The seeds of *O. vulgare* are of commercial value with trade name 'Oregano' for food-flavouring purpose used especially with cooked vegetables, meat and as an ingredient in chilli powder. The worldwide consumption volume of Oregano is over 500,000 tonnes with Turkey as the leader in both production and trade, followed by Mexico, Greece and other Mediterranean countries (Bansleben et al. 2010; Ozel and Kaymaz, 2004). Dried *Origanum* leaves and its essential oils are used extensively in various liqueur formulations, tomato sauces, condiments, in baked goods such as pizzas and various other local cuisines. It is used as a traditional remedy to treat various ailments (Sharma and Lal, 2005; Oliver, 1997). The plant is also reported to possess various pharmacological properties such as diuretic, antihysterial, antiasthmatic, antiparalytic, anticancer, antioxidant, and antifungal properties (Burdock, 1995).

3.9. *Perilla* spp.

Perilla frutescens has been used as a valuable source of culinary and traditional medicine for various therapeutic applications. The leaves of *P. frutescens* are used for flavoring, food, medicine purpose in Asian countries. It has traditionally been prescribed for treatment of cough, cold, tumors, headache, allergies, intoxication, fever, anxiety, bronchitis, rheumatic arthritis, asthma, chest stuffiness, vomiting and various other gastric and depression-related diseases (Ahmed, 2019; Dimri et al. 2008; Duke 1988; Gaur, 1999). It is also used as an ornamental plant in gardens due to its wide morphological variability and attractive appearance (Pandey and Bhatt, 2008). *P. frutescens* also act as an analgesic, anti-abortive agent, and a sedative as possess various pharmacological activities such as antioxidant, anti-allergic, antidepressant, antispasmodic, expectorant, anti-allergic. antimicrobial, anti-inflammatory, anticancer, and neuroprotection etc The essential oil and seed oil of *P. frutescens* with reported pharmacological properties used in cosmetics products, such as skin creams, soaps, and dermatological medicinal preparations (Ahmed, 2019: Gwari et al. 2016; Meng et al. 2009; Negi et al. 2011). The essential oil of *P. frutescens* and its aroma constituents were used for flavouring, fragrance and cosmetic products (Misra and Husain 1977; Guenther, 1949). However, the seed oil of *P. frutescens* is of commercial significance as a source of dietary omega-3 fatty acid and protein of good quality for food supplements. The seed oil is also used as cooking oil, drying oil and as fuel. It is in used in paints, varnishes, printing ink, linoleum, and lacquers (Dimri et al. 2009; Gwari et al. 2014; Longvah and Deosthale, 1991).

3.10. *Plectranthus* and *Coleus* spp.

Plectranthus species find use for ornamental as well as for medicinal purpose due to the presence of diverse biologically active secondary metabolites. Several species are cultivated for their edible tubers or as aromatic plants for food flavoring or fragrance. In folk medicines, *Plectranthus* spp. are used for treatment of ear infection, toothache, stomach ache, vomiting, nausea, sores, allergies, respiratory and gastric disorders, as well as antiseptic, vermicidal, purgative, antimicrobial and stimulant (Ascensao et al. 1998; Catino et al. 1992). *P. rugosus*, one representative species in Himalayan region, is used extensively against worm infestations. It is also used an antiseptic, germicidal and cardiac stimulant in folk medicine (Tiwari et al. 2008; Singh, 1994). In Pakistan, *P. rugosus* is used for toothache, scabies and as an antiseptic, a hypoglycaemic, an antidiarrheal and a bronchodilator (Ajmal et al. 2012). *P. amboinicus* commonly known as ‘Pathorchur’ or ‘Ajwain patta’ is widely used in Indian traditional medicine for the treatment of burns, insect bites, also as carminative and anti-asthmatic. *P. barbatus* (syn. *C. barbatus*) is used for several medicinal purposes such as for treatment of respiratory, gastrointestinal, liver and urinary disorders (Marques et al.

2012). *P. barbatus* popularly known as ‘Makandi’ and/or ‘Mayani’ is one of important plants of Ayurvedic ‘*Materia Medica*’ used for various herbal formulations (Bhowal and Mehta, 2017). The tuberous root of this plant is used in various Ayurvedic medicines for treatment of coronary heart diseases, skin, stomach and kidney disorders. The Indian herbal industries recognize *P. barbatus* as most medicinally and economically important medicinal plants. This plant is cultivated for its tuberous roots in various parts of the country, mainly Maharashtra, Rajasthan, Himachal Pradesh, Karnataka and Tamilnadu (Bhowal et al. 2017; Paul et al. 2013). The roots are considered a valuable source of bioactive molecule, forskolin having immense value in the treatment of cardiovascular diseases, hypertension, glaucoma, asthma, skin diseases (eczema and psoriasis), menstrual problems, insomnia, convulsions, tumors, irritable bowel syndrome, urinary tract and bladder infections (Bhowal et al. 2017; Bhat, 1993; Valdes et al. 1987).

3.11. *Pogostemon* spp.

The *Pogostemon* species have been used in traditional medicine for treatment of variety of ailments as well as for fragrance and cosmetic purpose. Patchouli oil, extracted from leaves *P. cablin* is highly valued in perfumes, soaps, cosmetics and flavor industries (Verma et al. 2012). Patchouli oil is renowned for its characteristic pleasant and long lasting woody, earthy and camphoraceous odor, as well as for its fixative properties (Akhila and Tewari, 1984; Donelian et al. 2009). It is also used as a natural additive for food flavoring (Manoj et al. 2012). Indonesia and Malaysia are the major producers of patchouli oil followed by China and India (Jhunjhunwalla, 2006; Lawrence, 2014). The patchouli oil possesses various pharmacological activities, such as antidepressant, antiphlogistic, antiseptic, antirheumatic, aphrodisiac, analgesic, anti-mutagenic activity, diuretic, febrifuge, astringent, fungicidal, and insecticidal properties (Chakrapani et al. 2015; Leung, 1980). It has been used for various medicinal purposes in Asian countries to treat athlete’s foot, dandruff, constipation, worm infection, wounds and scars; and for reduction and management of stress and nervous system disorders (Chakrapani et al. 2015). *P. benghalensis* is reported to possess significant antibacterial, antifungal, antitubercularis, antirheumatic activities (Bhuyian et al. 2011). The leaves of *P. amaranthoides* are edible and used for treatment of blood diseases, anaemia, and to cure kidney problem (Prasad, 2019).

3.12. *Salvia* spp.

The *Salvia* species are cultivated for ornamental, aromatic, culinary, as herbal tea, as spices, as food flavouring, and used in fold medicine as well as in cosmetics, perfumery and the pharmaceutical industry (Demirci et al., 2003; Senatore et al., 2004). *Salvia* spp. are known for their pleasant biologically active essential oils an plethora of biologically active molecules possessing various pharmacological properties viz. carminative, antispasmodic, antiseptic, astringent, antimicrobial, antioxidant, anticholinesterase, antimutagenic, anticancer, antiinflammatory, including improvement of cognitive performance and mood, reducing work-related stress and for strengthening brain and nervous system (Lawrence, 2005; Fu et al. 2013). The essential oils of *Salvia* spp. used extensively for flavouring sausages, canned meat seasoning and liquors (Raina and Negi, 2012). Garden sage (*S. officinalis*) is cultivated in temperate regions all around the world for its medicinal, aromatic, and culinary properties. The essential oil of *S. officinalis* possesses various biological activities of pharmaceutical importance (Perry et al. 1999). *S. sclarea* is cultivated worldwide, mainly in Russia, Bulgaria, France and Morocco as a source of essential oils, concrete and absolute, and bioactive aroma molecules widely used in perfumery and cosmetics industries (Asilew, 1930; Dzamic et al. 2008; Furia and Bellanca, 1975). The essential oil of *S. sclarea* is being used for clearing the eye vision from ancient times, hence it is popularly known as 'Clary'-sage or 'Clear Eye' (Gunther, 1949). It is used in folk medicine as an antispasmodic, carminative, oestrogenic, and digestive and menstrual problems (Geschel et al. 1998; Lattoo et al. 2006). The essential oil of S. *sclarea* is also used in aromatherapy as relaxant for the treatment of stress and insomnia, and also possesses anti-inflammatory, antitumor, antituberculosis, larvicidal, and genotoxic activities (Lattoo et al. 2006). The major principal constituents of traded clary sage oil are linalool and linalyl acetate used extensively in food-flavour, fragrance, cosmetics and pharmaceutical products. However, the content of Linalool and linalyl acetate varies in commercial clary sage oil as per their origin viz. French clary sage oil (9-16% linalool and 49-73.6% linalyl acetate), American clary sage oil (20.3-28.5% linalool and 44.9-53.4% linalyl acetate) and Russian clary sage oil (10.4-19.5% linalool and 45.3-61.8% linalyl acetate), respectively (Dzumayev et al. 1995). However, the Italian Clary sage oil is rich in α-terpineol (47.4%) and *α*-terpinyl acetate (22.1%) (Peana et al. 1999). European Pharmacopoeia (EP5) standardised the commercial clary sage oil with large amounts of linalool (6.5%-24%) and linalyl acetate (56%–78%) for used in fragrance and cosmetics products (Adams and Yanke, 2007; Verma et al. 2010). Scalreol is one of the minor marker constituents of clary sage oil, however, it major component of the clary sage concrete. It is used for its peculiar aroma note as well as a starting material for Ambrox synthesis, an important synthetic for rare ambergris-like fragrance of high values ambergris secreted by sperm whales, and used extensively in exclusive perfumery products (Moulines et al. 2004; Schmiderer et al. 2008)

3.13. *Scutellaria* spp.

Scutellaria spp. also finds use as folk medicine for the treatment of inflammatory diseases, hepatitis allergy, cancer etc. The flavonoids, diterpene, and essential oils of *Scutellaria* spp. were found to possess wide pharmacological actions, such as antiarthritic, anti-inflammatory, anti-angiogenesis, antineoplastic, antioxidant, anticonvulsant, antiviral, antifeedant and antimicrobial etc (Kubo et al. 1984; Melkani et al. 2013; Munoz et al. 1997; Sripathi and Ravi, 2017). The essential oil of *S. repens* rich in aromadendrene (30.7%) and β-funebrene (15.0%) from Uttarakhand was found to possess significant antibacterial activity (Melkani et al. 2013). The essential oil of *S. grossa* was found to possess significant antibacterial activity against pathogenic strains of *B. subtilis, E. faecalis, K. pneumoniae, S. enterica* (Pant et al. 2012).

3.14. *Stachys* spp.

Stachys spp. have been used in traditional herbal remedies to treat genital tumors, sclerosis of the spleen, inflammatory diseases, cough, asthma, rheumatic and digestive disorders (Goren et al. 2011; Khanavi et al. 2009). In Iran, the decoctions or infusions of Stachys spp. named as 'mountain tea' used as tonics to treat skin or taken internally for stomach disorders (Goren et al. 2011; Ozturk et al., 2009). The whole plant of *S. sericea* is taken internally to treat epilepsy in Kashmir (Dutt et al. 2015). *Stachys* spp. are also used as sedative, antispasmodic and diuretic, and possess various pharmacological activities, such as antiinflammatory, anti-toxic, antibacterial, anti-nephritic, anti anxiety, antioxidant, anti-hepatitis and anticancer etc. (Goren et al. 2011; Khanavi et al., 2005; Ozturk et al., 2009; Skaltsa et al., 2001).

3.15. *Teucrium* spp.

Teucrium spp. are used in folk medicine for the treatment of fever, digestive disorders, hypertension, inflammations, diabetes, rheumatism, parasitic diseases, and as carminative and favouring agent (Khaled-Khodja et al. 2014; Nasser et al. 2008). *Teucrium* spp. and their reported bioactive constituents including essential oils were found to possess various pharmacological activities such as antispasmodic, diuretic, antidiabetic, diaphoretic, analgesic, antipyretic, antirheumatic, antihelmintic, antibacterial, antiinflammatory, antioxidant, antiulcer, antiallergic, and antinociceptive properties etc. (Ahmad et al. 2007; Ali et al. 2008; Khaled-

Khodja et al. 2014, Elshazly and Hussein 2004; Parsaee and Shafiee-Nick, 2007; Ricci et al. 2005).

3.16. *Thymus* spp.

Thymus spp. are well-known medicinal plants having diverse medicinal uses mainly as tonic, herbal tea, spasmolytic, antiseptic, antitussive, expectorant, and carminative. The British Herbal Pharmacopoeia describes *T. serpyllum* as one of the important medicinal plant for treatment of bronchitis, bronchial catarrh, whooping cough, and sore throats. *T. serpyllum* is used as infusions, tinctures, syrups, decoctions, tea, and as it essential oil in various traditional and modern medicine as aromatic, anthelmintic, antiseptic, antispasmodic, analgesic, diaphoretic, expectorant, diuretic, spasmolytic, carminative, sedative, stimulant, and tonic (The British Pharmacopoeia, 2015; Chevallier 1996). *T. linearis* is also extensively used in native medicines for treatment of fever, cough, cold, asthma, antiseptic, stomach and menstrual problems. The essential oil of *T. linearis* and *T. serpyllum* are being used in pharmaceutical, cosmetic, perfume and flavour industries (Jaric et al. 2015). The oil of Himalayan thyme is used as flavouring agents in chocolates, toothpastes, mouthwashes, cough medicines and as preservative for food products (Figueiredo et al., 2008; Surburg and Panten, 2006).

CONCLUSION

Plants are the valuable source of diverse natural products viz. terpenoids, alkaloids, glycosides, anthocyanins, flavonoids, lignans, tannins, phenolics, caretenoids, steroids and saponins and various other classes of constituents useful for human for diverse applications viz. biofuels, medicine, nutrition and other health-care products. Lamiaceae family is known for diverse secondary metabolites associated with pharmacological properties for their utilisation in flavours, medicine, nutrition and other health-care products. Investigations on Himalayan Lamiaceae plants revealed structurally diverse molecules with immense chemical diversity with potential for industrial use. The ethnobotanical information paved the way for phytochemical research and activity screening research. Some of the species such as basil, mints, sage, patchouli, hyssop and *Perilla* have great potential for extensive commercial exploitation. Therefore, the future bioprospection of these plants for socio-economic development still need attention on their conservation status and development of elite cultivars and cultivation strategies for further industrial applications.

Photo plates of some Himalayan Lamiaceae species

Nepeta leucophylla

Nepeta goviániana

Nepeta clarkii

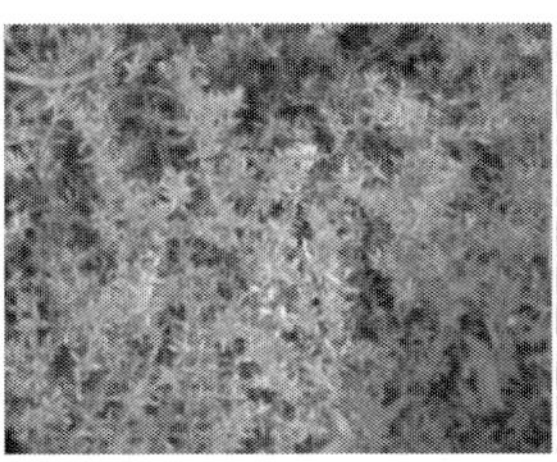

Thymus serpyllum

Thymus linearis

Ajuga parviflora

Origanum vulgare

Origanum majorana

Pogostemon cablin

Ocimum tenuiflorum 'Green'

Ocimum tenuiflorum 'Purple'

Ocimum basilicum

Ocimum grattisimum

Ocimum kilimandscharicum

Ocimum americanum

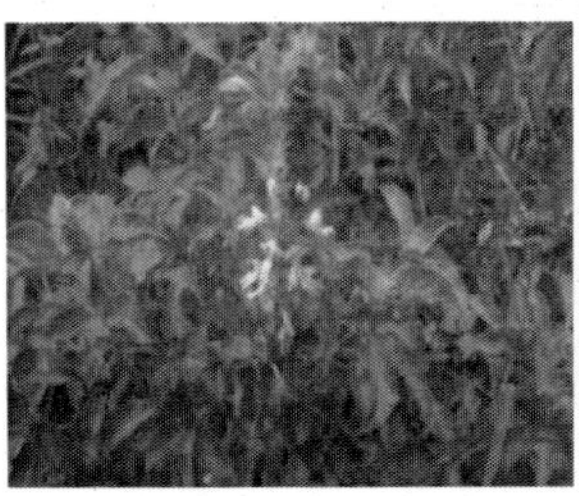

Ocimum africanum

Mentha arvensis

Mentha piperita

Mentha spicata

Mentha citrata

Mentha longifolia

Salvia leucantha

Salvia sclarea

Salvia officinalis

Plectranthus rugosus

Plectranthus incanus

Plectranthus amboinicus

Perilla frutescens cv. Green

Perilla frutescens cv. Green

Elsholtzia polystachya

Elsholtzia flava

Elsholtzia eriostachya

Elsholtzia densa

REFERENCES

Abdel-Mogib, M., Albar, H. A. and Batterjee, S. M. (2002). Chemistry of genus *Plectranthus*. *Molecules*, 7: 271-301.

Aburjai, T., Hudaib, M. and Cavrini, V. (2006). Composition of the essential oil from Jordanian germander (*Teucrium polium* L.). *Journal of Essential Oil Research*, 18: 97-99.

Adams, R. and Yanke, T. (2007). Variation in composition and enantiomeric analyses of commercial vs. new Kashmir clary sage oils. *Perfumer and Flavourist*, 32 (9): 36-39.

Agarwal, G., Pant, A. K. and Hore, S. K. (2010). In vitro evaluation of anthelmintic efficacy of *Trichilia* and *Ajuga* species on *Ascaridia galli*. *Hygeia Journal of Drug and Medicine*, 2 (2): 43-53.

Agnaniet, H., Anguillet, J. J., Bessière, J. M. and Menut, C. (2005). Aromatic plants of tropical central Africa. Part XLVII. Chemical and biological investigation of essential oils of *Ocimum* species from Gabon. *Journal of Essential Oil Research*, 17 (4): 466-470.

Ahmad, S. M., Shah, M., Khan, H. and Hassan Shah, S. M. (2007). Enzyme inhibition activities of *Teucrium royleanum*. *Journal of Enzyme Inhibition and Medicinal Chemistry,* 22: 730-732.

Ahmad, T. (2011). Geology of the Himalayan Mountain Range, with special reference to the western Himalaya. *Geology*, 63: 142-1472.

Ahmed, H. M. (2019). Ethnomedicinal, phytochemical and pharmacological investigations of *Perilla frutescens* (L.) *Britt. Molecules*, 24 (1): 1-23.

Ajmal SM, Mohammad S, Zahid K, Bakht Z (2012) Ethnomedicinal and phytoeconomic elaboration of Lilownai valley, district Shangla Pakistan. *International Research Journal of Pharmacy*, 3: 164-169.

Akhila, A. and Tewari, R. (1984). Chemistry of patchouli oil: a review. *Current Research on Medicinal and Aromatic Plants*, 6: 38-54.

Ali, N. A. W., Wurster, M., Arnold, Lindequist, U. and Wessjohan, L. (2008). Chemical composition of the essential oil of *Teucrium yemense* Deflers. *Records of Natural Products*, 2 (2): 25-32.

Alves, P. B., Filho, P. S. F., Moraes, V. R. S., Blank, A. F., Filho, J. L. S. C., Arrigoni-Blank, M. F., Oliva, G. and Thiemann, O. H. (2007). Chemical composition of essential oil from seven *Ocimum basilicum* L. accessions, Brine Shrimp Lethality bioassay and inhibitory activities against GAPDH and APRT. *Journal of Essential Oil Research*, 19 (1): 89-92.

Anonymous (1972). *The Wealth of India. A Dictionary of Indian Raw Materials Industrial Products-Vol. IX.* New Delhi, India: CSIR Publication.

Ascensao, L., Figueiredo, A. C., Barroso, J. G., Pedra, L. G., Schripsema, J., Deans, S. G. and Scheffer, J. J. C. (1998). *Plectranthus madagascariensis*: morphology of the glandular trichomes, essential oil compositions, and its biological activity. *International Journal of Plant Sciences*, 159: 31-38.

Asif, M. (2011). Health effects of omega-3,6,9 fatty acids: *Perilla frutescens* is a good example of plant oils. *Oriental Pharmacy and Experimental Medicine*, 11 (1): 51-59.

Asilew, V. W. (1930). *Salvia sclarea* L. *Bulletin of Applied Botany and Plant Breeding,* 23: 403-410.

Babu, C. R. (1977). *Herbaceous Flora of Dehradun.* New Delhi, India: CSIR Publication and Information Directorate.

Bachheti, R. K., Joshi, A. and Ahmed, T. (2014). A phytopharmaco-logical overview on *Perilla frutescens. International Journal of Pharmaceutical Sciences Review and Research*, 26 (2): 55-61.

Bahl, J. R., Bansal, R. P., Garg, S. N., Naqvi, A. A., Luthra, R., Kukreja, A. K. and Kumar, S. (2000). Qualitative evaluation of the essential oils of the prevalent cultivars of commercial mint species *Mentha arvensis, M. spicata, M. piperita, M. cardiaca, M. citrata* and *M. viridis* cultivated in indo-gangetic plains. *Journal of Medicinal and Aromatic Plant Sciences*, 22: 787-797.

Baldovini, N., Ristorcelli, D., Tomi, F. and Casanova, J. (2000). Intraspecific variability of the essential oil of *Calamintha nepeta* from Corsica (France). *Flavour and Fragrance Journal*, 15: 50-54.

Balyan, S. S. and Pushpangadan, P. (1988). A study on the taxonomical status and geographic distribution of genus *Ocimum*. *Perfumes and Flavours Association of India Journal*, 10 (2): 13-19.

Bansleben, A., Schellenberg, I., Ulrich, D. and Bansleben, D. (2010). A new and efficient sensory method for a comprehensive assessment of the sensory quality of dried aroma intensive herbs using oregano as a reference plant. *Flavour and Fragrance Journal*, 25: 214-218.

Baser, K. H. C., Kirimer, N., Kurkcuoglu, M. and Demirci, B. (2000). Essential oils of *Nepeta* species growing in Turkey. *Chemistry of Natural Compounds*, 36: 356-359.

Baser, K. H. C., Kurkcuoglu, M., Demirci, B., Ozek, T. and Tarimcilar, G. (2012). Essential oils of *Mentha* species from Marmara region of Turkey. *Journal of Essential Oil Research*, 24: 265-272.

Beauchamp, P. S., Bottini, A. T., Caselles, M. C., Dev, V., Hope, H., Larter, M., Lee, G., Mathela, C. S., Melkani, A. B., Millar, P. D., Miyatake, M., Pant, A.K, Raffle, R. J., Sharma, V. K. and Wyatt, D. (1996). Neo-clerodane diterpenoids from *Ajuga parviflora*. *Phytochemistry*, 434: 827-828.

Benitez, N. P., Melendez, E. M., Leon and Stashenko, E. F. (2009). Eugenol and methyl Eugenol chemotypes of essential oil of species *Ocimum gratissimum* L. and *Ocimum campechianum* Mill. from Colombia. *Journal of Chromatographic Sciences*, 47: 800-803.

Bestmann, H. J., Kobold, U., Claben, B., Vostrowsky, O., Sah, G. C., Pant, A. K., Melkani, A. B. and Mathela, C. S. (1988). New constituents of essential oil from *Elsholtzia pilosa*. *Zeitschrift für Naturforschung*, 43: 370-372.

Bestmann, H. J., Rauscher, J., Vostrovosky, O., Pant, A. K., Dev, V., Parihar, R. and Mathela, C. S. (1992.) Constituents of the essential oil of *Elsholtzia blanda* Benth (Labiatae). *Journal of Essential Oil Research*, 4: 121-124.

Bestmann, H. J., Rauscher, J., Vostrowsky, O., Pant, A. K. and Mathela, C. S. (1997). The volatile constituents of *Elsholtzia flava*. *Planta Medica*, 63: 88-89.

Bestmann, H. J., Rauscher, J., Vostrowsky, O., Pant, A. K., Mathela, C. S. and Singh, A. K. (1995). Terpenoids from *Elsholtzia* Species, Part VIII. The essential oil from *Elsholtzia incisa* (Labiatae). *Journal of Essential Oil Research*, 7: 85-87.

Bhakuni, D. S., Kaul, K. N. (1961). Chemical examination of *Ajuga bracteosa* Wall. *Journal of Scientific and Industrial Research,* 20 (B): 185.

Bhandari, R., Mathela, C. S., Beauchamp, P. S., Bottini, A. T. and Dev, V. (1993). Coleons from *Nepeta elliptica. Phytochemistry*, 34 (5): 1438-1439.

Bhat, S. V. (1993). Forskolin and congeners. *Fortschr Chem Org Naturst*, 62: 1-74.

Bhatti, G. R. and Ingrouille, M. (1997). Systematics of Pogostemon (Labiatae). *Bulletin of the Natural History Museum, London (Botany)*, 27: 77-147.

Bhowal, M. and Mehta, D. M. (2017). *Coleus forskohlii*: Phytochemical and pharmacological profile. *International Journal of Pharmaceutical Sciences and Research,* 8 (9): 3599-3618.

Bhuiyan, M. N., Varshney, V. K., Shiam, C. V. and Farhana, A. T. (2011). Composition of essential oil of the leaf and inflorescence of *Pogostemon benghalensis* (Burm. f.) Kuntze. *International Research Journal of Plant Sciences*, 2 (9): 271-275.

Bisht, D. S., Joshi, S. C., Padalia, R. C. and Mathela, C. S. (2012). Isoiridomyrmecin rich essential oil from *Nepeta erecta* Benth. and its antioxidant activity. *Natural Product Research*, 26 (1): 29-35.

Bisht, D. S., Joshi, S. C., Padalia, R. C., Tiwari, A. and Mathela, C. S. (2010). Constituents of *Nepeta clarkei* Hook. f. and their antioxidant activity. *Indian Journal of Chemistry,* 49B: 807-811.

Bisht, D. S., Padalia, R. C., Joshi, S. C., Singh, K. K. and Mathela, C. S. (2008) Sesquiterpene hydrocarbons rich essential oil of *Stachys sericea* Wall. *Journal of Essential Oil Bearing Plants*, 11 (6): 586-590.

Bisht, J. C., Pant, A. K., Mathela, C. S., Kobold, U. and Vostrowsky, O. (1985). Constituents of essential oil of *Elsholtzia strobilifera. Planta Medica*, 51: 412-414.

Bottini, A. T., Dev, V., Garfagnoli, D. J., Lohani, H., Mathela, C. S. and Pant, A. K. (1987). (7R)-*trans,trans*-Nepetalactone from *Nepeta elliptica. Phytochemistry*, 26 (4): 1200-1202.

Bozovic, M. and Ragno, R. (2017). *Calamintha nepeta* (L.) Savi and its main essential oil constituent pulegone: Biological activities and chemistry. *Molecules*, 22 (2): 1-50.

Bruno, M., Piozzi, F., Maggio, A. M. and Simmonds, M. S. J. (2002). Antifeedant activity of neo-clerodane diterpenoids from two Sicilian species of *Scutellaria*, *Biochemical Systematics and Ecology*, 30 (8):793-799.

Burdock, G. A. (1995). *Fenaroli's Handbook of Flavor Ingredients, Vol. I-Natural flavors*. Boca Raton: CRC Press.

Bure, C. M. and Sellier, N. M. (2004). Analysis of the essential oil of Indonesian patchouli (*Pogostemon cabin* Benth.) using GC/MS (EI/CI). *Journal of Essential Oil Research*, 16: 17-19.

Cakilcioglu, U., Khatun, S., Turkoglu, I. and Hayat, S. (2011). Ethanopharmacological survey of medicinal plants in Maden (Elazig-Turkey). Journal of Ethnopharmacology, 137 (1): 469-486.

Camps, F. and Coll, J. (1993). Insect allelochemicals from *Ajuga* plants. *Phytochemistry*, 32: 1361-1370.

Cantino, P. D., Harley, R. M. and & Wagstaff, S. J. (1992). Genera of Labiatae: Status and classification. In *Advances in Labiate Science*. Kew: Royal Botanic Gardens.

Castro, A., Coll, J., Tandron, Y. A., Pant, A. K. and Mathela, C. S. (2008). Phytoecdysteroids from *Ajuga macrosperma* var. *breviflora* Roots. *Journal of Natural Products*, 71: 1294-1296.

Chakrapani, P., Venkatesh, K, Singh, C. S., Arun, B., Jyothi, B., Kumar, P., Amareshwari, P. and Rani, A. R. (2013). Phytochemical, pharmacological importance of patchouli (*Pogostemon cablin* (Blanco) Benth) an aromatic medicinal plant. *International Journal of Pharmaceutical Science Review and Research*, 21 (2): 7-15.

Chanotiya, C. S., Yadav, A., Singh, A. K. and Mathela, C. S. (2007). Composition of the leaf and inflorescence essential oil of *Pogostemon benghalensis* Burm. F. from Kumaon. Natural Products Communications, 2(9): 941-944.

Charles, D. J. (2012). Hyssop. In: *Antioxidant Properties of Spices, Herbs and Other Sources*. New York: Springer.

Charles, D. J. and Simon, J. E. (1992). A new geraniol chemotype of *Ocimum gratissimum* L. *Journal of Essential Oil Research*, 4 (3): 231-234.

Chauhan, A., Venkatesha, K. T., Padalia, R. C., Singh, V. R., Verma, R. S. and Chanotiya, C. S. (2019). Essential oil composition of leaves and inflorescences of *Elsholtzia densa* Benth. from western Himalaya. *Journal of Essential Oil Research*, 31 (3): 217-222.

Chauhan, A., Verma, R. S., and Padalia, R. C. (2011). Aroma in plants: can it serve as a criterion in delimitation of taxa at infraspecific level. *Current Science,* 101 (4): 489.

Chauhan, N. K., Haider, S. Z., Sah, S. and Lohani, H. (2008). Constituents of the mature inflorescence essential oil of *Pogostemon benghalensis*. *Journal of Medicinal and Aromatic Plant Sciences*, 30 (3): 293-294.

Chauhan, R. S., Kaul, M. K., Shahi, A. K., Kumar, A., Ram, G. and Tawa, A. (2009). Chemical composition of essential oils in *Mentha spicata* L. accession [IIIM(J)26] from north-west Himalayan region, India. *Industrial Crops and Products*, 29: 654-656.

Chauhan, S. S., Prakash, O., Padalia, R. C., Vivekanand, Pant, A. K. and Mathela, C. S. (2011). Chemical diversity in *Mentha spicata* L.: Antioxidant and fumigation activity of essential oils. *Natural Product Communications*, 6 (9): 1373-1378.

Chen, T., Diao, Q. Y., Yu, H. Z., Jiao, C. L. and Ruan, J. (2018). Phytochemical, cytotoxic and chemotaxonomic study on *Ajuga forrestii* Diels (Labiatae). *Natural Product Research*, 32 (8): 977-981.

Chevallier, A. (1996). *The Encyclopedia of Medicinal Plants*. London, UK: University of Michigan: Dorling Kindersley Publishing.

Chevallier, A. (2001). *Encyclopedia of Medicinal Plants*. London, UK: Dorling Kindersley Publishing.

Chopra, R. N., Nayar S. L. and Chopra, I. C. (2009). *Glossary of Indian medicinal Plants*. New Delhi, India: CSIR-NISCAIR publication.

Chopra, R. N., Nayar, S. L. and Chopra, I. C. (1956). *Glossary of Indian Medicinal Plants*. New Delhi, India: CSIR Publications.

Cigremis, Y., Ulukansi, Z., Ilcim, A. and Akgoz, M. (2010). In-vitro antioxidant and antimicrobial assay of acetone extracts from *Nepeta meyeri* Benth. *European Review of Medical and Pharmacological Sciences*, 14 (8): 661-668.

Coll, J. and Tandron, Y. A. (2008). Neo-Clerodane diterpenoids from *Ajuga*: structural elucidation and biological activity. *Phytochemistry Reviews,* 7: 25-29.

Demirci, B., Baser, K. H. C., Yaldiz, B. and Bahcecioglu, Z. (2003). Composition of the essential oils of six endemic *Salvia* spp. from Turkey. *Flavour and Fragrance Journal*, 18: 116-121.

Demirci, B., Temel, H. E., Portakal, T., Kirmizibekmez, H., Demirci, F. and Baser, K. H. C. (2011). Inhibitory effect of *Calamintha nepeta* subsp. *glandulosa* essential oil on lipoxygenase. *Turkish Journal of Biochemistry*, 36: 290-295.

Dimri, M., Rawat, M. S. M. and Pant, G. (2008). Essential and fixed oils composition of *Perilla frutescens* and insecticidal activity against stored grain pest *Sitophilus oryzae*. *Indian Perfumer*, 52: 57-59.

Donelian, A., Carlson, L. H. C., Lopes, T. J. and Machado, R. A. F. (2009). Comparison of extraction of patchouli (*Pogostemon cablin*) essential oil with supercritical CO_2 and by steam distillation. *The Journal of Supercritical Fluids*, 48 (1): 15-20.

Dragland, S., Senoo, H., Wake, K., Holte, K. and Blomhoff, R. (2003). Several culinary and medicinal herbs are important sources of dietary antioxidants. Journal of Nutrition, 133: 1286-1290.

Dubey, N. K., Tiwari, T. N., Mandin, D., Adriamboavonjy, H. and Chaumoumt, J. P. (2000). Antifungal properties of *Ocimum gratissimum* essential oil (ethyl cinnamate chemotype). *Fitoterapia*, 71: 567-569.

Duke, J. A. (1988). *Handbook of Medicinal Herbs*. Boca Raton, FL, USA: CRC Press.

Dutt, H. C., Bhagat, N. and Pandita, S. (2015). Oral traditional knowledge on medicinal plants in jeopardy among Gaddi shepherds in hills of northwestern Himalaya. *Journal of Ethnopharmacology*, 168: 337-348.

Dzamic, A., Sokovic, M., Ristic, M., Grujic-Jovanovic, S., Vukojevic, J. and Marin, P. D. (2008). Chemical composition and antifungal activity of *Salvia sclarea* (Lamiaceae) essential oil. *Archives of Biological Sciences Belgrade*, 60 (2): 233-237.

Dzumayev, K. K., Tsibulskaya, I. A., Zenkevich, I. G., Tkachenko, K. G. and Satzyperova, I. F. (1995). Essential oils of *Salvia sclarea* L. produced from plants grown in southern Uzbekistan. *Journal of Essential Oil Research,* 7: 597-604.

Elshazly, A. M. and Hussein, K. T. (2004). Chemical analysis and biological activities of essential oil of *Teucrium leucocladum* Boiss. *Biochemical Systematics and Ecology*, 32: 665-674.

European Pharmacopeia (2005). 5.0. Thyme, Thymi herba, 01/2005, 0865, 2567-2570.

Farag, M. A., Ezzat, S. M., Salama, M. M., Tadros M. G. and Serya, R. A. T. (2016). Anti-acetylcholinesterase activity of essential oils and their major constituents from four *Ocimum* species. *Zeitschrift für Naturforschung*, 71: 393-402.

Fathiazad, F. and Hamedeyazdan, S. (2011). A review on *Hyssopus officinalis* L.: Composition and biological activities. *African Journal of Pharmacy and Pharmacology,* 5 (17): 1959-1966.

Figueiredo, A. C., Barroso, J. G., Pedro, L. G., Salgueiro, L., Miguel, M. G. and Faleiro, M. L. (2008). Portuguese *Thymbra* and *Thymus* species volatiles: chemical composition and biological activities. *Current Pharmaceutical Design*, 14: 3120-3140.

Formisano, C., Rigano, D. and Senatore, F. (2011). Chemical constituents and biological activities of *Nepeta* species. *Chemistry and Biodiversity*, 8: 1783-1818.

Fu, Z., Wang, H., Hu, X., Sun, Z. and Han, C. (2013). The pharmacological Properties of *Salvia* essential oils. *Journal of Applied Pharmaceutical Science*, 3 (7): 122-127.

Furia, T. E. and Bellanca, N. (1975). *Fenaroli's Handbook of Flavour Ingredients*, 2nd Ed., Vol. 2, Boca Raton FL: CRC Press.

Garg, S. N., Naqvi, A. A., Singh, A., Ram, G. and Kumar, S. (1999). Composition of essential oil from an annual crop of *Hyssopus officinalis* grown in Indian plains. *Flavour and Fragrance Journal*, 14 (3): 170-172.

Gaspar-Marques, P. R. C., Simoes, M. F., Jimeno, M. L. and Rodriguez, B. (2007). Further diterpenoids from *Plectranthus ornatus* and *P. grandidentatus*. *Biochemical Systematic and Ecology*, 35: 215-221.

Gaur, R. D. (1999). *Flora of the District Garhwal North West Himalaya*. Srinagar, Garhwal Uttarakhand, India: Trans Media.

Geschel, G. C., Maffei, P., Moretti, M. D. L., Peana, A. T. and Dementis, S. (1998). In vitro permeation through porcine buccal mucosa of *Salvia sclarea* L. essential oil from topical formulations. *STP Pharma Science*, 8: 103-106.

Goliaris, A. H., Chatzopoulou, P. S. and Katsiotis, S. T. (2002). Production of new Greek Oregano clones and analysis of their essential oils. *Journal of Herbs Spices and Medicinal Plants*, 10(1): 29-35.

Gollapudi, S., Sharma, H. A., Aggarwal, S., Byers, L. D., Ensley, H. E. and Gupta, S. (1995). Isolation of a previously unidentified polysaccharide (MAR-10) from *Hyssopus officinalis* that exhibits strong activity against human immunodeficiency virus type 1. *Biochemical and Biophysical Research Communications*, 210 (1): 145-151.

Goren, A. C., Piozzi, F., Akcicek, E., Kilic, T., Clarikc¸ S., Mozioglu, E. and Setzer, W. N. (2011). Essential oil composition of twenty-two *Stachys* species (mountain tea) and their biological activities. *Phytochemistry Letters*, 4: 448-453.

Guenther, E. (1949). *The Essential Oils*. D. Van Nostrand, New York.

Gunther, E. (1949). *Oil of Clary sage, In: The Essential Oil-Vol. III D.* New York: Van Nostrand Co.

Guo, Z., Liu, Z., Wang, X., Liu, W., Jiang, R., Cheng, R. and She, G. (2012). *Elsholtzia:* Phytochemistry and biological activities. *Chemistry Central Journal*, 6: 2-8.

Gupta, R. K. (1968). *Flora Nainitalensis: Handbook of the Flowering Plants of Nainital.* New Dehli, India: Navyug Traders.

Gwari, G., Lohani, H., Bhandari, U., Haider, S. Z., Singh, S., Andola, H. and Chauhan, N. (2016). Chemical diversity in the volatiles of *Perilla frutescens* (L.) Britt. populations from Uttarakhand Himalaya (India). *Journal of Essential Oil Research*, 28 (1): 49-54.

Gwari, G., Lohani, H., Haider, S. Z., Bhandari, U., Chauhan, N. and Rawat, D. S. (2014). Fatty acid and nutrient composition of perilla (*Perilla frutescens* L.) accessions collected from Uttarakhand. *International Journal of Phytopharmacology*, 5 (5): 379-382.

Harley, R. M., Atkins, S., Budantsev, A. L., Cantino, P. D., Conn, B. J., Grayer, R., Harley, M. M., de Kok, R., Krestovskaja, T., Morales, R., Paton, A. J., Ryding, O. and Upson, T. (2004). Labiatae. In: *The Families and Genera of Vascular Plants*, *Lamiales*, edited by Kadereit, J. W., 167–282. Berlin: Springer.

Harley, R. M., Atkins, S., Budantsev, A. L., Cantino, P. D., Conn, B. J., Grayer, R., Harley, M. M., Kok, R., Krestovskaya, T., Morales, R., Paton, A. J., Ryding, O. and Upson, T. (2004). Labiatae. In: *The Families and Genera of Vascular Plants*, edited by Kubitzki, K. and Kadereit, J. W., 167-275. Berlin & Heidelberg: Springer.

Hedge, C. (1984). *A Global Survey of the Biogeography of Labiatae*. Kew: Royal Botanical Gardens.

Hedge, I. C. (1992). A global survey of the biogeography of the Labiatae. In: *Advances in Labiate Science*, edited by Harley, R. M and Reynolds T., 85-96. Kew: Royal Botanical Garden.

Heywoo, V. H. (1978). *Flowering Plants of the World*. Oxford: Oxford University Press.

Hussein, A. A. (2018). Chemistry of South African Lamiaceae: Structures and biological activity of terpenoids. In: *Terpenes and* Terpenoids, edited by Perveen, S. and Al-Taweel, A., 13-38, https://www.intechopen.com/books/terpenes-and-terpenoids/chemistry-of-south-african-lamiaceae-structures-and-biological-activity-of-terpenoids.

Ietswarrt, J. H. (1980). *A Taxonomic Revision of the Genus Origanum (Labiatae).* Netherlands: Leiden Univ. Press, Hague.

Israili, Z. H. and Lyoussi, B. (2009). Ethnopharmacology of the plants of genus *Ajuga. Pakistanis Journal of Pharmaceutical Sciences,* 22: 425-462.

Jahantigh, O., Najafi, F., Badi, H. N., Khavari-Nejad, R. A. and Sanjarian, F. (2016). Essential oil composition of Hyssop (*Hyssopus officinalis* L.) under salt stress at flowering stage, *Journal of Essential Oil Research*, 28 (5): 458-464.

Jamzad, Z., Harley, M. M., Simmonds, M. S. J. and Jalili, A. (2000). Pollen exine and nutlet surface morphology of the annual species of *Nepeta* L. (Lamiaceae) in Iran. In: *Pollen* and *Spores*: *Morphology* and *Biology*, edited by Harley, M. M., Morton, G. M. and Blackmore, S., 385-397. London: Royal Botanic Gardens, Kew.

Janmoni, K. and Khan, M. L. (2013). Commercial potentialities of essential oil of *Ocimum* members growing in north east India. *International Journal of Pharmacy and Life Sciences*, 4 (4): 2559-2567.

Jaric, S., Mitrovic, M. and Pavlovic, P. (2015). Review of ethnobotanical, phytochemical, and pharmacological study of *Thymus serpyllum* L. *Evidence-Based Complementary and Alternative Medicine*, 101978: 1-10.

Jhunjhunwalla, A. (2006). Materials imported to India. *Indian Perfumer*, 50: 21-27.

Jirovetz, L., Buchbauer, G., Shafi, M. P. and Kaniampady, M. M. (2003). Chemotaxonomical analysis of the essential oil aroma compounds of four different *Ocimum* species from southern India. *European Food Research and Technology*, 217: 120-124.

Joshi, R. K. and Mathela, C. S. (2013). Chemical constituents of leaf essential oil of *Nepeta laevigata* (D. Don) Hand.-Mazz from Kumaun Himalaya. *American Journal of Essential Oils and Natural Products*, 1 (1): 7-10.

Kamatou, G. P. P., Vermaak, I., Viljoen, A. M. and Lawrence, B. M. (2013). Menthol: A simple monoterpene with remarkable biological properties. *Phytochemistry*, 96, 15-25.

Kashyap, T. K., Melkani, A. B., Mathela, C. S., Dev, V., Olmstead, M. M., Hope, H. and Bottini, A. T. (2003). Essential oil from *Nepeta govaniana* Benth. Study of its major terpenes. *Journal of Essential Oil Research*, 15: 28-30.

Kaya, A., Demirci, B. and Baser, K. H. C. (2001). The composition of the essential oil of *Stachys siberica* subsp. *stenostachya* growing in Turkey. *Chemistry of Natural Compounds*, 37: 326-328.

Kazazi, H., Rezaei, K., Ghotb-Sharif, S. J., Emam-Djomeh, Z. and Yamini, Y. (2007). Supercritical fluid extraction of flavors and fragrances from *Hyssopus officinalis* L. cultivated in Iran. *Food Chemistry*, 105: 805-811.

Khaled-Khodja, N., Boulekbache-Makhlouf, L. and Madan, K. (2014). Phytochemical screening of antioxidant and antibacterial activities of methanolic extracts of some Lamiaceae. *Industrial Crops and Products*, 61: 41-48.

Khalighi, J. J., Kashi, A., Bais, H. P. and Vivanco, J. M. (2002). Chemical characterization of Basil (*Ocimum basilicum* L.) found in local accessions and used in traditional medicines in Iran. *Journal of Agriculture and Food Chemistry*, 50 (21): 5878-5883.

Khan, R., Shawl, A. S. and Tantry, M. A. (2012). Determination and seasonal variation of chemical constituents of essential oil of *Hyssopus officinalis* growing in Kashmir valley as incorporated species of western Himalaya. *Chemistry of Natural Compounds*, 48 (3): 502-505.

Khanavi, M., Hadjimahmood, M., Cheraghi-Niroomand, M., Kargar, Z., Ajani, Y., Hadjiakhoondia, A. and Oevesi, M. R. (2009). Comparison of the antioxidant activity and total phenolic contents in some *Stachys* species. *African Journal of Biotechnology*, 8: 1143-1147.

Khanuja, S. P. S. (2007). Aroma vision 2020: Technology path of cultivating today for a fragrant tomorrow. *Paper presented at National Convention and Seminar on Business Enabling of Aromatic Plants & Products*, Dehradun, India, November, 21-23.

Khazaie, H. R., Nadjafi, F. and Bannayan, M. (2008). Effect of irrigation frequency and planting density on herbage biomass and oil production of thyme (*Thymus vulgaris*) and hyssop (*Hyssopus officinalis*). *Industrial Crops and Products*, 27 (3): 315-321.

Kikuchi, Y., Miyaichi, Y. and Tomimori, T. (1991). Studies on the Nepalese crude drugs XIII: on the flavonoids and iridoid constituents of the roots of Scutellaria grossa Wall. *Chemical and Pharmaceutical Bulletin*, 39: 1051-1054.

Koba, K., Poutouli, P. W., Raynaud, C., Chaumont, J. P. and Sanda, K. (2009). Chemical composition and antimicrobial properties of different basil essential oils chemotypes from Togo. *Bangladesh Journal of Pharmacology*, 4: 1-8.

Kosar, M., Dorman, H. J. D. and Hiltunen, R. (2005). Effect of an acid treatment on the phytochemical and antioxidant characteristics of extracts from selected Lamiaceae species. *Food Chemistry*, 91: 525-533.

Kothari, S. K., Bhattacharya, A. K., Ramesh, S., Garg, S. N. and Khanuja, S. P. S. (2005). Volatile constituents in oil from different plant parts of methyl eugenol-rich *Ocimum tenuiflorum* L.f. (syn. *O. sanctum* L.) grown in south India. *Journal of Essential Oil Research*, 17 (6): 656-658.

Kothari, S. K., Bhattacharya, A. K., Singh, C. P., Singh, K., Syamasundar, K. V., Ramesh, S. and Kumar, S. (2001). A new methyl eugenol-rich and short duration strain of *Ocimum tenuiflorum*. *Journal of Medicinal and Aromatic Plants Sciences*, 23 (1), 385-388.

Kothari, S. K., Bhattacharya, A. K., Singh, K., Ramesh, S. I., Rao, E. V. S. P. and Garg, S. N. (2005). Pre-flowering harvesting of *Ocimum gratissimum* for higher essential oil and eugenol yields under semi-arid tropics. *Journal of Essential Oil Research*, 17 (2): 212-216.

Koutsos, T. V., Chatzopoulou, P. S. and Katsiotis, S. T. (2009). Effects of individual selection on agronomical and morphological traits and essential oil of a "Greek basil" population. *Euphytica*, 170: 365-370.

Kubo, I., Lee, Y. W., Balogh-Nair, V., Nakanishi, K. and Chapya, A. (1976). Structure of ajugarins. *Journal of the Chemical Society, Chemical Communications,* 22: 949–950.

Kumar, V. and Mathela, C. S. (2017). Toxicity and repellent effect of essential oils and a major component against *Lipaphis erysimi. Journal of Crop Protection*, 6 (1): 15-23

Kumar, V., Mathela, C. S. and Panwar, A. (2016). In vitro antibacterial activity of essential oils from Lamiaceae species. *The Journal of Phytopharmacology*, 5(4): 141-144.

Kumar, V., Mathela, C. S., Kumar, M. and Tewari, G. (2019). Antioxidant potential of essential oils from some Himalayan Asteraceae and Lamiaceae species. *Medicine in Drug Discover*y, 100004: 1-7.

Kumar, V., Mathela, C. S., Tewari, A. K. and Bisht, K. S. (2014). In vitro inhibition activity of essential oils from some Lamiaceae species against phytopathogenic fungi. *Pesticide Biochemistry and Physiology*, 114: 67-71.

Kumar, V., Mathela, C. S., Tewari, G. and Singh, D. (2014). Antifungal activity of *Nepeta elliptica* Royle ex Benth. oil and its major constituent (7R)-*trans,trans*-Nepetalactone: A comparative study. *Industrial Crops and Products*, 55: 70-74.

Kumari, P., Tewari, L. M. and Singh, B. K. (2009). Biodiversity in Uttrakhand Himalaya region. *Nature and Science*, **7**: 113-125.

Labra, M., Miele, M., Ledda, B., Grassi, F., Mazzei, M. and Sala, F. (2004). Morphological characterization, essential oil composition and DNA genotyping of *Ocimum basilicum* L. cultivars. *Plant Sciences*, 167: 725-731.

Lal, R. K. (2013). Adaptability patterns and stable cultivar selection in menthol mint (*Mentha arvensis* L.). *Industrial Crops and Products*, 50: 176-181.

Lattoo, S. K., Dhar, R. S., Dhar, A. K., Sharma, P. R. and Agarwal, S. G. (2006). Dynamics of essential oil biosynthesis in relation to inflorescence and glandular ontogeny in *Salvia sclarea. Flavour and Fragrance Journal*, 21 (5): 817-821.

Lawrence, B. M. (1988). A further examination of the variation of *Ocimum basilicum* L. In: *Flavors and Fragrances: A World Perspective*, edited by Lawrence, B. M., Mookhejee, B. D. and Willis, B. J., 161-170. Amsterdam: Elsevier Science Publisher.

Lawrence, B. M. (2005). *The Antimicrobial/Biological Activity of Essential Oils.* Carol Stream, IL, USA: Allured Publishing Corporation.

Lawrence, B. M. (2007). *Mint: The Genus Mentha.* Boca Raton: CRC Press.

Lawrence, B. M. (2014). Progress in essential oils. *Perfumer and Flavourist*, 39: 46-52.

Lee, C. J., Chen, L. J., Chang, T. L., Ke, W. M., Lo, Y. F. and Wang. C. C. (2011). The correlation between skin-care effects and phytochemical contents in Lamiaceae plants. *Food Chemistry*, 124: 833-841.

Leung, A. Y. (1980). *Encyclopedia of Common Natural Ingredients Used In Food, Drugs and Cosmetics*. New York: John Wiley and Sons Inc.

Lindqvist, C. and Albert, V. A. (2002). Origin of the Hawaiian endemic mints within North American *Stachys* (Lamiaceae). *American Journal of Botany*, 89: 1709-1724.

Liu, A. L., Lee, S. M. Y., Wang, Y. T. and Du, G. H. (2007). *Elsholtzia:* Review traditional uses, chemistry and pharmacology. *Journal of Chinese Pharmaceutical Sciences*, 16: 73-78.

Longvah, T. and Deosthale, Y. G. (1991). Chemical and nutritional studies on Hanshi (*Perilla frutescens*) traditional oil seed from North India. *Journal of American Oil Chemists Society*, 68: 781-784.

Lukhoba, C. W., Simmonds, M. S. J. and Paton, A. J. (2006). *Plectranthus*: A review of etthnobotanical uses. *Journal of Ethnopharmacology,* 103: 1-24.

Mancini, E., De Laura, M., Malova, H. and De Vincenzo, F. (2013). Chemical composition and biological activities of the essential oil from *Calamintha nepeta* plants from the wild in southern Italy. *Natural Product Communications*, 8: 139-142.

Manoj, G., Manohar, S. H. and Murthy, H. N. (2012). Chemical constituents, antioxidant and antimicrobial activity of essential oil of

Pogostemon paniculatus (Willd.). *Natural Product Research*, 26: 2152–2154.

Marotti, M., Piccaglia, R. and Giovanelli, E. (1996). Differences in essential oil composition of basil (*Ocimum basilicum* L.) Italian cultivars related to morphological characteristics. *Journal of Agriculture and Food Chemistry*, 14: 3926-3929.

Marques, A. M., Lima, M. C. H. P, Costa Fo, H. A., Passinho, H. S. and Kaplan, M. A. C. (2012). Chemical profile analysis of three different species of *Plectranthus*. *Emirates Journal of Food and Agriculture*, 24 (2): 137-141.

Mathela, C. S. (2017). Herbal Chemo-Prospecting for New Phytomedicines. *Annals Pharmacology and Pharmaceutics*, 2 (7): 1-3.

Mathela, C. S. and Kumar, V. (2018). Antifungal activities of essential oils from Himalayan Plants. In: *Natural Antimicrobial Agents, Sustainable Development and Biodiversity*, edited by Merillon, J. M. and Riviere, C., 75-94. Switzerland: Springer, Cham.

Mathela, C. S., Agarwal, I. and Taskinen, J. (1980). Composition of essential oil of *Thymus serpyllum* L. *Journal of Indian Chemical Society,* 57 (12): 1249-1250.

Mathela, C. S., Gupta, A., Upreti, P., Pant, A. K., Olmsted, M., Hope, H. and Bottini, A. T. (1991). Coleon-U 12-methyl ether from *Nepeta leucophylla*. *Journal of Natural Products*, 54 (3): 910-912.

Mathela, C. S., Kharkwal, H. and Laurent, R. (1994), Investigations on Himalayan *Nepeta* species V: Essential oil of *Nepeta govaniana* Benth. *Journal of Essential Oil Research*, 6: 425-428.

Mathela, C. S., Kharkwal, H. and Laurent, R. (1994). Investigations on Himalayan *Nepeta* Species. VI. Essential oil of *Nepeta discolor* Benth. *Journal of Essential Oil Research*, 6 (5): 519-521.

Mathela, C. S., Melkani, A. B., Bisht, J. C., Pant, A. K., Bestmann, H. J., Erler, J., Kobold, U., Rauser, R. and Vostrowsky, O. (1992). Chemical varieties of essential oils from *Elsholtzia polystachya* from two different locations in India. *Planta Medica*, 58: 376-379.

Mathela, C. S., Padalia, R. C., Chanotiya, C. S. and Tiwari, A. (2005). Carvone rich *Mentha longifolia (Linn.):* Chemical variation and commercial potential. *Journal of Essential Oil Bearing Plants*, 8: 132-134.

Mathela, C. S., Padalia, R. C., Joshi, S. C., Singh, K. K., Bisht, D. S., Pant, A. K. and Kharkwal, H. (2009). Chemical diversity in Himalayan *Elsholtzia* species. *Chemistry and Biodiversity*, 6: 2212-2226.

Mathela, C. S., Shah, G. C., Melkani, A. B. and Joshi, P. (1989). Sustituted furanyl compounds of *Perilla ocimoides*. *Journal of Indian Chemical Society,* 66: 183-184.

Mathela, D. K., Kharkwal, H. B. and Mathela, C. S. (1986). Terpenoids of roots of *Coleus forskohlii*. *Fitoterapia*, 57: 299-301.

Mathew, R. and Sankar D. P. (2013). Plant cell culture technology and its entrée into the world of *Ocimum*. *International Journal of Pharmacy and Pharmaceutical Sciences*, 5 (2): 6-13.

Melkani, A. B., Beauchamp, P. S., Dev, V., Whalen, C. and Mathela, C. S. (1994). Rosefuran epoxide from *Elsholtzia densa* Benth. *Journal of Essential Oil Research*, 6: 475-479.

Melkani, A. B., Nailwal, M., Mohan, L., Pant, C. C. and Dev, V. (2013). Steam volatile oil from *Scutellaria repens* Buch-Ham. ex D. Don; its composition and antibacterial activity. *Journal of Essential Oil Research*, 25 (5): 368-371.

Melkani, A. B., Negi, A., Bisht, C. M. S. and Dev, V. (2007). Constituents of the essential oil from *Scutellaria scandens* D. Don. *Indian Perfumer*, 51: 37-40.

Meng, L., Lozano, Y., Bombarda, I., Gaydou, E. M. and Li, B. (2009). Polyphenols extraction from eight *Perilla frutescens* cultivars. *Comptes Rendus Chimie*, 12: 602-611.

Merritt, A. T. and Ley, S. V. (1992). Clerodane diterpenoids. *Natural Product Reports*, 9: 243-245.

Misra, L. N. and Husain, A. (1987). The essential oil of *Perilla ocimoides*: A rich source of rosefuran. *Planta Medica*, 53: 379-390.

Mittal, R. and Singh, S. P. (2007). Shifting from agriculture to agribusiness: The case of aromatic plants. *Agricultural Economics Research Review*, 20: 541-550.

Mohan, L., Pant, C.C., Melkani, A. B. and Dev. V. (2010). Terpenoid composition of the essential oils of *Teucrium royleanum* and *T. quadrifarium*. *Natural Product Communications*, 5 (6): 939-942.

Mohan, M., Seth, R., Singh, P., Lohani, H. and Gupta, S. (2012). Composition of the volatiles of *Hyssopus officinalis* (L.) and *Thymus serpyllum* (L.) from Uttarakhand Himalaya. *National Academy of Science Letters*, 35(5), 445-448.

Mondello, L., Zappia, G., Cotroneo, A., Bonaccorsi, I., Chowdhury, J. U., Yusuf, M., Dugo, G. (2002). Studies on the essential oil bearing plants of Bangladesh, Part VIII. Composition of some *Ocimum* oils *O. basilicum* L. var. *purpurascens*, *O. sanctum* L. green, *O. sanctum* L. purple, *O. americanum* L. citral type, *O. americanum* L. camphor type. *Flavour and Fragrance Journal*, 17: 335-340.

Moro, A., Zalacain, A., Hurtado de Mendoza, J., Carmona, M. (2011). Effects of agronomic practices on volatile composition of *Hyssopus officinalis* L. essential oils, *Molecules*, 16 (5): 4131-4139.

Morteza-Semnani, K., Akbarzadeh, M. and Rostami, B. (2005). The essential oil composition of *Teucrium chamaedrys* L. from Iran. *Flavour and Fragrance Journal*, 20: 544-546.

Moulines, J., Bats, J. P., Lamidey, A. M. and Da Silva N. (2004). About a practical synthesis of Ambrox from scalreol: a new preparation of a ketone key intermediate and a close look at its Baeyer-Villiger oxidation. *Helvetica Chimica Acta*, 87 (10): 2695-2705.

Munoz, D. M., De La Torre, M. C., Rodriguez, B., Simmonds, M. S. J. and Blaney, W. M. (1997). Neo-clerodane insect antifeedant from *Scutellaria alpine* ssp. *javalambarensis*. *Phytochemistry*, 44: 593-597.

Murray, M. J., Marble, P., Lincoln, D. and Hefendehl, F. W. (1988). Peppermint oil quality differences and the reasons for them. In: *Flavors and Fragrances: Proceeding of the 10th International Congress of Essential oils, Fragrances and Flavors*, 189-208, Washington, DC, USA, November, 16-20.

Naithani, B. D. (1985). *Flora of Chamoli*. Calcutta: Botanical Survey of India Publication.

Nandy, S. N., Dhyani, P. P. and Samal, P. K. (2006). *Resource Information Database of the Indian Himalaya. ENVIS Monograph 3*. Kosi-Katarmal, Almora, Uttarakhand India: Environmental Information System on Himalayan Ecology, G.B. Pant Institute of Himalayan Environment and Development, Publication.

Nasser, A., Ali, A., Wurster, M., Arnold, N., Lindequist, U. and Wessjohan, L. (2008). Chemical composition of the essential oil of *Teucrium yemense* Deflers. *Records of Natural Products*, 2 (2): 25-32.

Negi, V. S., Rawat, L. S., Phondani, P. C. and Chandra, A. (2011). *Perilla frutescens* in transition: A medicinal and oil yielding plant need instant conservation, a case study from Central Himalaya, India. *Environment & We. An International Journal of Science and Technology*, 6: 193-200.

Ohno, A., Kizu, H. and Tomimori, T. (1997). Studies of Nepalese drug crude drugs XXIII: on the diterpenoid constituents of the aerial parts of *Scutellaria grossa* Wall. *Chemical and Pharmaceutical Bulletin*, 45: 1097-1100.

Oliver, G. W. (1997). The world market of oregano. Proceedings of the IPGRI international workshop on Oregano, Bari, Itlay, 142-146, May 8-12.

Ono, M., Furusawa, C., Ozono, T., Oda, K., Yasuda, S., Okawa, M., Kinjo, J., Ikeda, T., Miyashita, H., Yoshimitsu, H. and Nohara, T. (2011). Four New Iridoid Glucosides from *Ajuga reptans*. *Chemical and Pharmaceutical Bulletin*, 59 (8): 1065-1068.

Ozel, M. Z. and Kaymaz, H. (2004). Superheated water extraction, steam distillation and soxhlet extraction of essential oils of *Origanum onites*. *Analytical and Bio-analytical Chemistry*, 379: 1127-1133.

Ozturk, M., Duru, M. E., Aydogmus-Ozturk, F., Harmandar, M., Mahlicli, M., Kolak, U. and Ulubelen, A. (2009). GC–MS analysis and antimicrobial activity of essential oil of *Stachys cretica* subsp. *smyrnaea*. *Natural Product Communication*, 4 (1): 109-114.

Padalia, R. C. (2012). Medicinal and aromatic plants: chemical goldmines. *Journal of Medicinal and Aromatic Plants*, 1 (3): 1.

Padalia, R. C. and Verma, R. S. (2011). Comparative study of volatile oil compositions of two *Plectranthus* species from northern India. *Natural Product Research*, 25 (18): 1727-1732.

Padalia, R. C. and Verma, R. S. (2011). Comparative volatile oil composition of four *Ocimum* species from northern India. *Natural Product Research*, 25: 569-575.

Padalia, R. C., Singh, V. R., Bhatt, G., Chauhan, A., Upadhyay, R. K., Verma, R. S. and Chanotiya, C. S. (2018). Optimization of harvesting and postharvest drying methods of *Ocimum* x *africanum* Lour. for production of quality essential oil. *Journal of Essential Oil Research*, 30 (8): 437-443.

Padalia, R. C., Verma, R. S. and Chanotiya, C. S. (2011). Variability in volatile terpenoid compositions of peppermint cultivars and some of the wild accession from northern India. *Journal of Essential Oil Reserach*, 23: 29-33.

Padalia, R. C., Verma, R. S., Chauhan, A. and Chanotiya, C. S. (2013). Changes in aroma profiles of 11 Indian *Ocimum* taxa during plant ontogeny. *Acta Physiologiae Plantarum*, 35: 2567-2587.

Padalia, R. C., Verma, R. S., Chauhan, A. and Goswami, P. (2015). Quality evaluation of elite mint cultivars in two cropping seasons in subtropical conditions of Uttarakhand, India. *Journal of Essential Oil Research*, 27 (1): 45-51.

Padalia, R. C., Verma, R. S., Chauhan, A., Singh, V. R., Goswami, P., Singh, S., Verma, S. K., Luqman, S., Chanotiya, C. S. and Darokar, M. P. (2018). *Zingiber zerumbet* (L.) Roscoe ex Sm. from northern India: Potential source of zerumbone rich essential oil for antiproliferative and antibacterial applications. *Industrial Crop and Products,* 112: 749-754.

Padalia, R. C., Verma, R. S., Chauhan, A., Sundaresan, V. and Chanotiya, C. S. (2013). Essential oil composition of sixteen elite cultivars of *Mentha* from western Himalayan region, India. *Maejo International Journal of Science and Technology*, 7: 83-93.

Padalia, R. C., Verma, R. S., Chauhan, A., Sundaresan, V. and Chanotiya, C. S. (2013). Essential oil composition of sixteen elite cultivars of

Mentha from western Himalayan region, India. *Maejo International Journal of Science and Technology*, 7 (1): 83-93.

Padalia, R. C., Verma, R. S., Chauhan, A., Venkatesha, K. T. and Singh, V. R. (2017). Pharmaceutical values and essential oils of Indian *Ocimum* taxa: A review. In: *Indian Ethnomedicinal Plants: Traditional Practices to Cure Diseases*, edited by Baruah, A., 116-138. Jaipur, India: Avishkar Publishers, Distributors.

Pande, C. and Mathela, C. S. (2000). Essential oil composition of *Origanum vulgare* L. ssp. *vulgare* from the Kumaon Himalaya. *Journal of Essential Oil Research*, 12: 441-442.

Pandey, A. and Bhatt, K. C. (2008). Diversity distribution and collection of genetic resources of cultivated and weedy type in *Perilla frutescens* (L.) Britton var. *frutescens* and their uses in Indian Himalayas. *Genetic Resources Crop Evolution,* 55: 883-892.

Pandey, V., Verma, R. S., Chauhan, A. and Tiwari, R. (2014). Compositional variation in the leaf, flower and stem essential oils of Hyssop (*Hyssopus officinalis* L.) from western-Himalaya. *Journal of Herbal Medicine*, 4 (2):89-95.

Pant, A. K., Dev, V., Parihar, R., Mathela, C. S., Rauscher, J., Vostrowsky, O. and Bestmann, H. J. (1992). The essential oil from *Elsholtzia eriostachya var. pusilla. Journal of Essential Oil Research*, 4: 547-549.

Pant, C. C., Melkani, A. B., Mohan, L. and Dev, V. (2012). Composition and antibacterial activity of essential oil from *Scutellaria grossa* Wall ex Benth. *Natural Product Research*, 26 (2): 190-192.

Parsaee, H. and Shafiee-Nick, R. (2006). Anti-Spasmodic and anti-nociceptive effects of *Teucrium polium* aqueous extract. *Iranian Biomedical Journal*, 10: 145-149.

Paton, A. J., Harley, R. M. and Harley, M. M. (1999). *Ocimum*: An overview of classification and relationships. In: *Ocimum, Medicinal and Aromatic Plants-Industrial Profiles*, edited by Hiltunen, R. and Holm, Y., 1–38. Amsterdam: Harwood Academic.

Paton, A. J., Springate, D., Suddee, S., Otieno, D., Grayer, R. J., Harley, M. M., Willis, F., Simmonds, M. S. J., Powell, M. P. and Savolainen,

V. (2004). Phylogeny and evolution of basils and allies (Ocimeae, Labiatae) based on three plastid DNA regions. *Molecular Phylogenetics and Evolution*, 31: 277-299.

Paul, M., Radha, A. and Kumar, D. S. (2013). On the high value medicinal plant, *Coleus forskohlii* Briq. *Hygeia Journal of Drug and Medicine*, 5 (1): 64-73.

Peana, A. T., Moretti, M. D. L. and Juliano, C. (1999). Chemical composition and antimicrobial action of the essential oils from *Salvia desoleana* and *S. sclarea*. *Planta Medica,* 65: 752-754.

Perry L. M. and Metzger, J. (1980). *Medicinal Plants of East and Southeast Asia*. Cambridge, London: The MIT Press.

Perry, N. B., Anderson, R. E., Brennan, N. J., Douglas, M. H., Heaney, A. J. and McGimpsey, J. A. (1999). Essential oil from Dalmatian sage (*Salvia officinalis* L.): variations among individuals plant parts, seasons, and sites. *Journal of Agriculture and Food Chemistry*, 47 (5): 2048-2054.

Piozzi, F. and Bruno, M. (2009). Diterpenoids in the essential oils from the Genus *Stachys*. *Records of Natural Products*, 3 (3): 120-125.

Piozzi, F. and Bruno, M. (2011). Diterpenoids from roots and aerial parts of the Genus *Stachys*. *Records of Natural Products*, 5 (1): 1-11.

Pisutthanan N. and Pisutthanan, S. (2009). Variability of essential oil constituents of Ocimum africanum. *Naresuan University Journal*, 17: 269-274.

Polunin, O. and Stainton, A. (1984). *Flowers of the Himalaya*. Delhi, India: Oxford University Press.

Prakash, O., Joshi, S., Shukla, A. K. and Pant, A. K. (2007). Sesquiterpenoid rich essential oil from the leaves of *Pogostemon patchouli* pellet grown organically under tarai conditions. *Journal of Essential Oil Bearing Plants*, 10 (2): 157-161.

Prakash, P. and Gupta, N. (2005). Therapeutic uses of *Ocimum sanctum* Linn (Tulsi) with a note on eugenol and its pharmacological actions: a short review. *Indian Journal of Physiology and Pharmacology*, 49 (2): 125-131.

Prasad, K. (2018). Chemical composition, antioxidant and antifungal activity of essential oils of *Pogostemon amaranthoides* from (Raya-Bajeta valleys) Pithoragarh, Uttarakhand Himalayas, India. *Journal of Pharmacology and Clinical Research*, 6 (4): 1-12.

Purkayastha, J. and Nath, N. C. (2006). Composition of the camphor-rich essential oil of *Ocimum basilicum* L. native to northeast India. *Journal of Essential Oil Research*, 18 (3): 332-334.

Pushpangadan, P. and Bradu, L. (1995). Basil. In: *Medicinal and Aromatic plants of India*, edited by Chadha K. L. and Gupta, R., 627-658. New Delhi, India: Malhotra publishing House.

Qing, X., Yan, H. M., Ni, Z. Y., Vavrickaa, C. J., Zhang, M. L., Shi, Q. W., Gu, Y. C. and Kiyota, H. (2017). Chemical and pharmacological research on the plants from genus *Ajuga. Heterocyclic Communications*, 23 (4): 245-268.

Radulovic, N., Lazarevic, J., Ristic, N. and Palic, R. (2007). Chemotaxonomic significance of the volatiles in the genus *Stachys* (Lamiaceae): essential oil composition of four Balkan Stachys species. *Biochemical Systematic and Ecology*, 35: 196-208.

Raina, A. P. and Negi, K. S. (2012). Chemical characterization of the essential oils of some aromatic plants of industrial interest from Uttarakhand, India. *Medicinal Plants- International Journal of Phytomedicines and Related Industries*, 4 (2): 1-6

Raina, A. P., Negi, K. S. and Dutta, M. (2013). Variability in essential oil composition of sage (*Salvia officinalis* L.) grown under north western Himalayan Region of India. *Journal of Medicinal Plants Research*, 7 (11): 683-688.

Raja, R. R. (2012). Medicinally potential plants of Labiatae (Lamiaceae) family: an overview. *Research journal of Medicinal Plants*, 6 (3): 203-213.

Rajeshwari, S. (1992). *Ocimum sanctum.* The Indian home remedy. In: Current Medical Scene, edited by Rajeshwari, S, Cipla Ltd., Bombay Central, Mumbai.

Ramakrishna, A. and Ravishankar, G. A. (2011). Influence of abiotic stress signals on secondary metabolites in plants. *Plant Signalling and Behavior*, 6 (11): 1720-1731.

Rao, B. R. R. and Syamasundar, K. V. (2013). Menthol mint (*Mentha canadensis* L. synonym *Mentha arvensis* L. forma *piperascens* Malinvaud ex Holmes): cultivation and chemical profile of the essential oil. In: Natural Essential Oils-Fragrances and Flavours, edited by Baruah, A. and Nath, S. C., 200-227. Jaipur, India: Aavishkar Publishers.

Rasooli, I. and Rezaei, M. B. (2002). Bioactivity and chemical properties of essential oils from *Zataria multiflora* Boiss and *Mentha longifolia* (L.) Huds. *Journal of Essential Oil Research*, 14: 141-146.

Razavi, S. M., Ghaffari, A. A. and Bidarlord, M. (2018): Pollen morphology of some *Nepeta* L. species in NW Iran. *Botany Letters*, 165 (2): 286-291.

Ricci, D., Fraternale, D., Giamperi, L., Bucchini, A., Epifano, F., Burini, G. and Curini, M. (2005). Chemical composition, antimicrobial and antioxidant activity of the essential oil of *Teucrium marum. Journal of Ethnopharmacology*, 98: 195-200.

Runyoro, D., Ngassapa, O., Vagionas, K., Aligiannis, N., Graikou, K. and Chinou, I., (2010). Chemical composition and antimicrobial activity of the essential oils of four *Ocimum* species growing in Tanzania. *Food Chemistry*, 119 (1): 311-316.

Sah, S., Lohani, H., Haider, S. Z. and Chauhan, N. K. (2008). Essential oil composition of *Hyssopus officinalis* L. grown in Malari region of Garhwal Himalaya. *Journal of Medicinal and Aromatic Plant Sciences*, 30 (3): 291-292.

Saha, S., Dhar, T. N., Sengupta, C. and Ghosh, P. (2013). Biological activities of essential oils and methanol extracts of Five *Ocimum* species against pathogenic bacteria. *Czech Journal of Food Sciences*, 31 (2): 194-202.

Samal, P. K., Shah, A., Tiwari, S. C. and Agrawal, D. K. (2004). Indigenous medicinal practices and their linkages in resource conservation and physical well being of the locals in central Himalayan region of India. *Indian Journal of Traditional Knowledge,* **3**: 12-26.

Saxena, J. and Mathela, C. S. (1996). Antifungal activity of new compounds from *Nepeta leucophylla* and *Nepeta clarkei*. *Applied and Environmental Microbiology*, 62 (2): 702–704.

Schmiderer, C., Grassi, P., Novak, J., Weber, M. and Franz, C. (2008). Diversity of essential oil glands of clary sage (*Salvia sclarea* L., Lamiaceae). *Plant Biology*, 10: 433-440.

Senatore, F., Arnold, N. A. and Piozzi, F. (2004). Chemical composition of the essential oil of *Salvia multicaulis* Vahl. var. *simplicifolia* Boiss. *Journal of Chromatography A*, 1052: 237-240.

Senatore, F., Arnold, N. A., Piozzi, F. and Formisano, C. (2006). Chemical composition of the essential oil of *Salvia microstegia* Boiss. et Balansa growing wild in Lebanon. *Journal of Chromatography A*, 1108: 276-278.

Shah, G. C., Bhandari, R. and Mathela, C. S. (1992). 1,2-Epoxy-p-menthane derivatives from some Labiatae species. *Journal of Essential Oil Research*, 4: 57-59.

Shah, N. C. (1991). Chemical constituents of *Hyssop officinalis* L.: 'Zufah Yabis' A Unani drug from U.P. Himalaya, India. *Indian Perfumer*, 35: 49-52.

Sharma, A. and Cannoo, D. S. (2013). Phytochemical composition of essential oils isolated from different species of genus *Nepeta* of Labiatae family: A review. *Pharmacophores*, 4 (6): 181-211.

Sharma, P. K. and Lal, B. (2005). Ethanobotanical notes on some medicinal and aromatic plants among of Himachal Pradesh. Indian *Journal of Traditional Knowledge*, 4 (4): 424-428.

Shawl, A. S., Raina, V. K. and Srivastava, S. K. (1999). GC/MS analysis of Clary Sage oil of commerce from Kashmir, India. *Journal of Medicinal and Aromatic Plants Sciences*, 21 (4): 999-1001.

Silva, M. G. V., Vieira, I. G. P., Mendes, F. N. P., Albuquerque, I. L., Santos, R. N. D., Silva, F. O. and Morais, S. M. (2008). Variation of ursolic acid content in eight *Ocimum* species from Northeastern Brazil. *Molecules*, 13: 2482-2487;

Singh, O., Kumar, R., Prakash, O., Kumar, M., Pant, A. K., Isidorov, V. A. and Szczepaniak, L. (2011). Reinvestigation of chemical composition, pharmacological, antibacterial and fungicidal activity of essential oil from *Mentha longifolia* (L.) Huds. *Research Journal of Phytochemistry*, 11 (3): 129-141.

Singh, P., Prakash, O. and Pant, A. K. (2014). Sterol composition of *Ajuga bracteosa* Wall Benth., *Ajuga macrosperma* Wall ex Benth. and *Ajuga parviflora* Benth. *Journal of Indian Chemical Society*, 91: 951-952.

Singh, R., Shushni, M. A. M. and Belkheir, A. (2011). Antibacterial and antioxidant activities of *Mentha piperita* L. *Arabian Journal of Chemistry*, 8 (3): 322-328.

Singh, S., Tewari, G., Pande, C. and Singh, C. (2013). Variation in essential oil composition of *Ocimum americanum* L. from north-western Himalayan region. *Journal of Essential Oil Research*, 25 (4): 278-290.

Singh, V. (1994). Herbal remedies for worm infestation in Kashmir Himalaya. *Fitoterapia*, 65: 354-357.

Skaltsa, H. D., Demetzos, C., Lazari, D. and Sokovic, M. (2003). Essential oil analysis and antimicrobial activity of eight *Stachys* species from Greece. *Phytochemistry*, 64: 743-752.

Skaltsa, H. D., Mavrommati, A. and Constantinidis, T. (2001). A chemotaxonomic investigation of volatile constituents in *Stachys* subsect. Swainsonianeae (Labiatae). *Phytochemistry*, 57, 235-244.

Sobti, S. N., Pushpangadan, P. and Atal, C. K. (1976). Genus *Ocimum*-A potential source of new essential oils. *Indian Perfumer*, 20: 59-68.

Sripathi, R. and Ravi, S. (2017). Ethnopharmacology, Phytoconstituents, essential oil composition and biological activities of the genus *Scutellaria*. *Journal of Pharmaceutical Sciences and Research*, 9 (3): 275-287.

Srivastava, A., Awasthi, K., Kumar, B., Misra, A. and Srivastava, S. (2018). Pharmacognostic and pharmacological evaluation of *Hyssopus officinalis* L. (Lamiaceae) collected from Kashmir Himalayas, India. *Pharmacognosy Journal*, 10 (4): 690-693.

Stahl-Biskup, E. and Saez, F. (2002). *Thyme: The genus Thymus*. London: Taylor & Francis,

Stanko, K. C., Liber, Z., Politeo, O., Strikic, F., Kolak, I., Milos, M. and Satovic, Z. (2011). Molecular and chemical characterization of the most widespread Ocimum species. *Plant Systematics and Evolution*, 294: 253-262.

Stanko, K. C., Petek, M., Grdisa, M., Pintar, J., Bedekovic, D., Custic, M. H. and Satovic, Z. (2016). Medicinal plants of the family Lamiaceae as functional foods-a Review. *Czech Journal of Food Sciences,* 34 (5): 377–390.

Stracheyi, R. (1975). *Catalogue of the Plants of Kumaon and the Portions of Garhwal and Tibet*. New Delhi, India: Periodical Experts Book Agency.

Surburg, H. and Panten, J. (2006). *Common Fragrance and Flavor Materials: Preparation Properties and Uses*, 5th Ed. Weinheim: Wiley-VCH Verlag GmbH & Co. KGaA.

Tahir, M., Khushtar, M., Fahad, M. and Rahman, M. A. (2018). Phytochemistry and pharmacological profile of traditionally used medicinal plant Hyssop (*Hyssopus officinalis* L.). *Journal of Applied Pharmaceutical Science*, 8(7): 132-140.

Tchoumbougnang, F., Zollo, P. H. A., Avlessi, F., Alitonou, G. A., Sohounhloue, D. K., Ouamba, J. M., Tsomambet, A., Okemy-Andissa, N., Dagne, E., Agnaniet, H., Bessiere, J. M. and Menut, C. (2006). Variability in the chemical compositions of the essential oils of five *Ocimum* species from tropical African area. *Journal of Essential Oil Research*, 18 (2): 194-199.

Telci, I., Bayram, E., Yılmaz, G. and Avci, B. (2006). Variability in essential oil composition of Turkish basils (*Ocimum basilicum* L.). *Biochemical Systematic and Ecology*, 34: 489-497.

Thakuri, B. C., Padalia, R. C. and Mathela, C. S. (2009). Comparison of the volatile constituents of *Thymus serpyllum* from the altitude of far Western Nepal & north India. *Scientific World*, 7 (7): 92-93.

Thappa, R. K., Agarwal, S. G., Kapahl, B. K. and Srivastava, T. N. (1999). Chemosystematics of the Himalayan *Elsholtzia*. *Journal of Essential Oil Research*, 11: 97-103.

The British Pharmacopoeia (2015). *Commission Secretariat of the Medicines and Healthcare Products Regulatory Agency*.

Tian, S. K., Peng, H. Y., Yang, X. E., Lu, L. L. and Zhang, L. (2008). Phytofiltration of copper from contaminated water: growth response, copper uptake and lignin content in *Elsholtzia splendens* and *Elsholtzia argyi*. *Bulletin of Environmental Contamination and Toxicology*, 81: 85-89.

Tiwari, A., Padalia, R. C. and Mathela, C. S. (2008) Sesquiterpene rich essential oil from *Plectranthus rugosus* Wall. *Journal of Essential Oil Bearing Plants*, 11: 58-61.

Tripathi, Y. C. and Pandey, A. K. (2017) Bioprospecting of Phytodiversity for New Therapeutic Products: Trends, Potential and Challenges. *Organic and Medicinal Chemistry Journal*, 2 (1): 1-5.

Tucker, O. and Tucker, S. S. (2009). Catnip and catnip response. *Economic Botany*, 42: 214-230.

Upadhaya, K., Dixit, V. K., Padalia, R. C. and Mathela, C. S. (2009). Terpenoids composition and anti-oxidant activity of the essential oil from the aerial parts of *Salvia leucantha* Cav. *Journal of Essential Oil Bearing Plants*, 12 (5): 551-556.

Uritu, C. M., Mihai, C. T., Stanciu, G., Dodi, G., Alexa-Stratulat, T., Luca, A. Leon-Constantin, M. M., Stefanescu, R., Bild, V., Melnic, S. and Tamba, B. T. et (2018). Medicinal plants of the family Lamiaceae in pain therapy: A review. *Pain Research and Management*, 1: 1-44.

Valdes, L. J., Mislankar, S. G. and Paul, A. G. (1987). *Coleus barbatus* (*C. forskohlii*) (Lamiaceae) and the potential new drug forskolin (Coleonol). *Economic Botany*, 44: 474-583.

Valdiya, K. S. (2001). *Himalaya: Emergence and Evolution*. Hyderabad, India: University Press Publication.

Verma, R. S. (2010). Chemical investigation of decanted and hydrophilic fractions of *Salvia sclarea* essential oil. *Asian Journal of Traditional Medicines*, 5 (3): 102-108.

Verma, R. S., Bisht, P. S., Padalia, R. C., Saikia, D. and Chauhan, A. (2011). Chemical compositions and antibacterial activity of essential oils from two *Ocimum* spp. Grown in sub-tropical India during spring-summer cropping season. *Asian Journal of Traditional Medicine*, 6 (5): 201-217.

Verma, R. S., Padalia, R. C. and Chauhan, A. (2010). Chemical Profiling of *Mentha spicata* L. var. 'viridis' and *Mentha citrata* L. cultivars at different stages from the Kumaon region of western Himalaya. *Medicinal and Aromatic Plant Sciences and. Biotechnology*, 4: 73-76.

Verma, R. S., Padalia, R. C. and Chauhan, A. (2012). Assessment of similarities and dissimilarities in the essential oils of patchouli and Indian Valerian. *Journal of Essential Oil Research*, 24 (5): 487-491.

Verma, R. S., Padalia, R. C. and Chauhan, A. (2012). Compositional analysis of the leaf and flower essential oils of Indian Oregano (*Origanum vulgare* L). *Journal of Essential Oil Bearing Plants,* 15 (4): 651-656.

Verma, R. S., Padalia, R. C. and Chauhan, A. (2012). Essential Oil Composition of *Coleus aromaticus* Benth. from Uttarakhand. *Journal of Essential Oil Bearing Plants*, 15 (2): 174-179.

Verma, R. S., Padalia, R. C. and Chauhan, A. (2013). Essential oils of Indian Oregano species. In: *Natural Essential Oils: Fragrance and Flavours*, edited by Baruah, A. and Nath, S. C., 78-95. Jaipur, India: Avishkar Publisher & Distributor.

Verma, R. S., Padalia, R. C. and Chauhan, A. (2013). Volatile oil composition of Indian Perils [*Perilla frutescens* (L.) Britton collected at different phenophases. *Journal of Essential Oil Research*, 25: 92-96.

Verma, R. S., Padalia, R. C., Chanotiya, C. S. and Chauhan, A. (2010). Chemical investigation of the essential oil of *Thymus linearis* (Benth. ex Benth) from western Himalaya, India. *Natural Product Research*, 24 (2): 1890-1896.

Verma, R. S., Padalia, R. C., Chauhan, A. and Chanotiya, C. S. (2015). Harvesting season and plant part dependent variations in the essential oil composition of *Salvia officinalis* L. grown in northern India. *Journal of Herbal Medicine*, 5: 165-172.

Verma, R. S., Padalia, R. C., Chauhan, A. and Singh, V. R. (2019). Chemical composition of leaves, inflorescence, whole aerial-parts and root essential oils of patchouli {*Pogostemon cablin* (Blanco) Benth.}. *Journal of Essential Oil Research*, 31 (4): 319-325.

Verma, R. S., Padalia, R. C., Chauhan, A. and Thul, S. (2013). Exploring compositional diversity in the essential oils of 34 *Ocimum* taxa from Indian flora. *Industrial Crops and Products*, 45: 7-19.

Verma, R. S., Padalia, R. C., Chauhan, A., Verma, R. K., Rahman, L. and Singh, A. (2016). Changes in the essential oil composition of *Origanum majorana* L. during post harvest drying. *Journal of Essential Oil Bearing Plants*, 19 (6): 1547-155.

Verma, R. S., Padalia, R. C., Chauhan, A., Verma, R. K., Yadav, A. K. and Singh, H. P. (2010). Chemical diversity in Indian Oregano (*Origanum vulgare* L.). *Chemistry and Biodiversity,* 7: 2054-2064.

Verma, R. S., Padalia, R. C., Goswami, P., Upadhyay, R. K., Singh, V. R., Chauhan, A. and Tiwari, A. K. (2016). Assessing productivity and essential oil quality of Himalayan thyme (*Thymus linearis* Benth.) in the subtropical region of north India. *Industrial Crops and Products*, 94: 557-561.

Verma, R. S., Pandey, V., Chauhan, A. and Tiwari, R. (2015). Characterization of the essential oil composition of *Isodon rugosus* (Wall. ex Benth.) Codd. from Garhwal region of western Himalaya. *Journal of Medicinal and Aromatic Plants*, S3: 1-4.

Verma, R. S., Pandey, V., Chauhan, A. Tiwari, R. (2015) Essential oil composition of *Mentha longifolia* (L.) L. collected from Garhwal region of western-Himalaya, *Journal of Essential Oil Bearing Plants*, 18 (4): 957-966.

Verma, R. S., Rahman, L., Chanotiya, C. S., Verma, R. K., Singh, A., Yadav, A., Chauhan, A., Yadav, A. K. and Singh, A. K. (2009). Essential oil composition of *Thymus serpyllum* cultivated in the Kumaon region of western Himalaya, India. *Natural Product Communications*, 4 (7): 987-988.

Vina, A. and Murillo, E. (2003). Essential oil composition from twelve varieties of basil (*Ocimum* spp) grown in Colombia. *Journal of Brazilian Chemical Society*, 14 (5): 744-749.

Waldia, S., Joshi, B. C., Pathak, U. and Joshi, M. C. (2011). The genus *Plectranthus* in India and its chemistry. *Chemistry and Biodiversity*, 8: 244-252.

Xiong, Z., Jiang, B. and Wu, P. F. (2011). Antidepressant effects of a plant-derived flavonoid baicalein involving extracellular signal-regulated Kinases cascade. *Biological and Pharmaceutical Bulletin*, 34: 253-259.

Yao, G, Deng, Y. F. and Ge, X. Z. (2015). A taxonomic revision of *Pogostemon* (Lamiaceae) from China. *Phytotaxa*, 200 (1): 1–67.

Youfang, D., Mokgoldi, N. C., Yan, H. U., Lingling, S. H. I., Chao, M. A., Yu-Jun, L. I. U. (2012). Characterization of fatty acid composition of five Perilla seed oils in China and its relationship to annual growth temperature. *Journal of Medicinal Plants Research*, 6: 1645-1651.

Zeevaart, J. A. D. (1969). Perilla. In: *The Induction of Flowering: Some Case Histories*, edited by Evans, L. T., 116-155. New York: Cornell University Press.

Zhang, H. X., Zhang, F. Q., Xia, Q., Wang, G. P. and Shen, Z. G. (2010). Excess copper induces production of hydrogen peroxide in the leaf of *Elsholtzia haichowensis* through apoplastic and symplastic Cu Zn-superoxide dismutase. *Journal of Hazard Matters*, 178: 834-843.

Zhang, X., Wu, W., Zheng, Y., Chen, L. and Qianrong, C. (2009). Essential oil variations in different *Perilla* L. accessions: Chemotaxonomic implications. *Plant Systematics and Evolution*, 281: 1-10.

BIOGRAPHICAL SKETCHES

Chandra S. Mathela, MSc, PhD, F.I.C.S.

Affiliation: Hony Emeritus Professor, Chemistry Department, Kumaun University, NAINITAL, India

Business Address: Consulting Advisor, DS Group, NOIDA

Research and Professional Experience:

Former-Dean Science Faculty, Head Chemistry & Director, Research & Extn, Emeritus Scientist (CSIR) Kumaun University.

More than 45 years' research experience on chemistry of natural Products, essential oils & aroma Chemicals, hyphenated chromatographic-spectroscopic techniques, chemosystematics of Himalayan plants.

Professional Appointments:

1. Professor/Reader/Lecturer (Kumaun University): 41 Years
2. Emeritus Scientist (CSIR): 5 years
3. Visiting Professor in Kathmandu University, Kathmandu 2000.
4. Visited Royal College of Pharmacy, London University, London July, 2003.
5. Consulting Advisor (R&D): 3 Years DS Group, NOIDA (2016-2019)

Honors:

1. Visiting Associateship (CSIR), 1990-91.
2. National Associateship (U.G.C.) 1983-85.
3. Fulbright Indo-US Fellowship, California University, Davis 1986-87.
4. Member of Chemistry Faculty (Visiting Professor), California State Polytechnic University, Pomona, California (1985, 1991 and 1997).

5. Dr. DR Dhingra Medal for Essential Oil Research, 1985 & 1989 (EOAI).
6. Acharya Narendra Dev Medal for Meritorious Services, 2003.
7. President, Chemical Sciences Section, Indian Science Congress, 2005-2006.
8. Indian Chemical Society Award: Prof. Suresh C. Ameta Award 2005.

Publications from the Last 3 Years:

1. Kumar, V., C. S. Mathela and A. Panwar (2016), *In-vitro* antibacterial activity of essential oils from Lamiaceae species. *The Journal of Phytopharmacology,* 5(4), 141-144.
2. Kumar, V. and C. S. Mathela (2017). Toxicity and repellent effect of essential oils and a major component against *Lipaphis erysimi. Journal of Crop Protection*, 6 (1): 15-23.
3. Kumar, V., C. S. Mathela, G. Tewari, A. Panwar and V. Pandey (2017). *In vitro* antimicrobial activity of essential oils and their acetylenic constituents. *Indian J. Nat. Prod. & Res.*, 8(1), 63-68.
4. Kumar, V., M. Tripathi, C. S. Mathela and Y. Joshi (2017), *In vitro* antimicrobial activity of Himalayan lichenised fungi. *Journal of Pharmacognosy & Natural Products,* 3(2), 2-4.
5. Kumar, V. and C. S. Mathela (2017). Herbal chemo-prospecting for new phytomedicines. *Annals of Pharmacol & Pharmaceutics:* 2 (13) on line 12 August, 2017.
6. Kumar, V. and C. S. Mathela (2018), Chemical constituents of essential oils of Himalayan *Nepeta ciliaris* Benth. and *Senecio nudicaulis* Buch-Ham. Ex D. Don. *Jour. Essential Oil Res.* 30(3), 207-213.
7. Mathela, C. S. and V. Kumar (2018). Antibacterial activities of essential oils from Himalayan Plants, A book chapter, "Natural antimicrobial agents" (Springer: Editor: V. Rinaudo), 75-94. Chapter First Online: 09 February 2018. *Part of the Sustainable Development and Biodiversity* book series (SDEB, volume 19).

8. Kumar, V., C. S. Mathela, M. Kumar and G. Tewari (2019). Antioxidant potential of essential oils from some Himalayan Asteraceae and Lamiaceae species. *Medicine in Drug Discovery*, 1, online (July 2019).

Rajendra Chandra Padalia, PhD

Affiliation: Principal Scientist, CSIR-Central Institute of Medicinal and Aromatic Plants, Pantnagar, Uttarakhand, India

Business Address: CSIR-Central Institute of Medicinal and Aromatic Plants, Research Center, Pantnagar, Uttarakhand, India

Research and Professional Experience: Phytochemistry and R&D on Quality evaluation, Agrotechnology and Post Harvest technologies of aromatic plants

Professional Appointments:

S.No.	*Position*	*Period*	*Institution*
1	Scientist	09/02/2009-09/02/2013	CSIR-CIMAP, Research Center, Pantnagar
2	Senior Scientist	09/02/2013-09/02/2017	CSIR-CIMAP, Research Center, Pantnagar
3	Principal Scientist	09/02/2017-till date	CSIR-CIMAP, Research Center, Pantnagar

Honors:

1. Editorial Board Member of Journal of Medicinal and Aromatic Plants (OMICS, USA); European Journal of Medicinal Plants (UK, London); Journal of Medicinal and Aromatic Plants Sciences (India); American Journal of Essential Oils and Natural Products
2. Members in selected delegation of Scientist/Academician of Uttarakhand by Uttarakhand Council of Science and Technology

(UCOST), Dehradun for participation/training in DAE-BRNS symposium on "Emerging Trends in Separation Science and Technology" (SESTEC-2008).

3. Experience in working and maintenance of Analytical instruments GC, GC/MS, FTIR, HPLC etc. Experience of working with non-polar and polar capillary columns for flavor components. Operational experience in EI, CI, DIP, Full scan, MS/MS, single ion monitoring, segment ion monitoring mode under experimental condition. Life member of Medicinal Plant Association of India

Publications from the Last 3 Years:

1. Goswami, P. A., Chauhan, R. S. Verma, R. C. Padalia, S. K. Verma, M. P. Darokar and C. S. Chanotiya (2016). Composition and antibacterial potential of the essential oil of *Artemisia nilagirica var. septentrionalis* from India. *Journal of Essential Oil Research* 28(1), 71-76.
2. Padalia, R. C., R. S. Verma, A. Chauhan, Prakash Goswami, C. S. Chanotiya (2016). Compositional and enantiomeric analysis of the essential oil of *Taxodium distichum* from India. *Natural Product Communication,* 11(3), 419-422.
3. Padalia, R. C., R. S. Verma, A. Chauhan, C. S. Chanotiya and S. T. Thul (2016). Phytochemical diversity in essential oil of *Vitex negundo* L. populations from India. *Records of Natural Products,* 10(4), 452-464.
4. Padalia, R. C., R. S. Verma, A. Chauhan, Prakash Goswami, C. S. Chanotiya (2016). Essential oil composition of *Artemisia stelleriana* Besser from India. *Journal of Essential Oil Research*, 28(5), 400-405.
5. Goswami, Prakash, A. Chauhan, R. S. Verma, R. C. Padalia (2016). Chemical constituents of floral volatiles of *Plumeria rubra* L. from India. *Med Aromat Plants* S3: 005. doi:10.4172/2167-0412.S3-005.
6. Padalia, R. C., R. S. Verma and A. Chauhan (2016). Influence of harvest cut heights on aroma profile of *Ocimum basilicum* L., *O. kilimandscharicum* Guerke and *O. tenuiflorum* L. *Journal of Essential Oil Bearing Plants,* 19(6), 1442-1453.

7. Tiwari, P. Goswami, B. S. Bisth, A. Chauhan, R. S. Verma, R. C. Padalia (2016). Essential oil composition of African marigold (*Tagetes minuta* L.) harvested at different growth stages in foothills agroclimatic conditions of North India. *American Journal of Essential Oils and Natural Products*, 4(3), 4-7.
8. Padalia, R. C., R. S. Verma, R. K. Upadhyay, A. Chauhan, V. R. Singh (2017). Productivity and essential oil quality assessment of promising accessions of *Ocimum basilicum* L. from north India. *Industrial Crop and Products,* 97, 79-86.
9. Padalia, R. C., R. S. Verma, A. Chauhan, P. Goswami, V. R. Singh, S. K. Verma, M. P. Darokar, A. Kurmi, N. Singh, D. Saikia, C. S. Chanotiya (2017). Essential oil composition and antimicrobial activity of methyl cinnamate-linalool chemovariant of *Ocimum basilicum* L. from India. *Records of Natural Products*, 11(2), 193-204.
10. Padalia, R. C., R. S. Verma, A. Chauhan (2017). Diurnal variations in aroma profile of *Ocimum basilicum* L., *O. grattisimum* L., *O. americanum* L., and *O. kilimandscharicum* Guerke. *Journal of Essential Oil Reserach* 3, 248-261.
11. Goswami, P., S. K. Verma, A. Chauhan, K. T. Venkatesha, R. S. Verma, V. R. Singh, M. P. Darokar, C. S. Chanotiya and R. C. Padalia, (2017). Chemical composition and antibacterial activity of *Melaleuca bracteata* essential oil from India: A natural source of methyl eugenol. *Natural Product Communication*, 12(6), 965-968.
12. Padalia, R. C., R. S. Verma, A. Chauhan, V. R. Singh, P. Goswami, S. Singh, S. K. Verma, S. Luqman, C. S. Chanotiya and M. P. Darokar (2018). *Zingiber zerumbet* (L.) Roscoe ex Sm. from northern India: Potential source of zerumbone rich essential oil for antiproliferative and antibacterial applications. *Industrial Crop and Products,* 112, 749-754.
13. Padalia, R. C., R. S. Verma, A. Chauhan, P. Goswami, V. R. Singh, S. K. Verma, N. Singh, A. Kurmi, M. P. Darokar and D. Saikia (2018), p-Menthenols chemotype of *Cymbopogon distans* from India: Composition, antibacterial and antifungal activity of the essential oil against pathogens, *Journal of Essential Oil Research*, 30(1), 40-46.

14. Padalia, R. C., R. S. Verma, A. Chauhan, A. Tiwari and N. Joshi (2018). Variability in essential oil composition of different plant parts of *Heracleum candicans* Wall. Ex DC from North India. *Journal of Essential Oil Research,* 30(4), 293-301.
15. Padalia, R. C., V. R. Singh, G. Bhatt, A. Chauhan, R. K. Upadhyay, R. S. Verma and C. S. Chanotiya (2018). Optimization of harvesting and postharvest drying methods of *Ocimum* x *africanum* Lour. for production of quality essential oil. *Journal of Essential Oil Research*, 30(8), 437-443.

In: The Lamiaceae Family: An Overview
Editor: Alexander Adler
ISBN: 978-1-53617-078-8

Chapter 2

LAMIACEAE AS A SOURCE OF NATURAL ANTIOXIDANTS: DIFFERENT PROCEDURES FOR THE EXTRACTION OF POLYPHENOLIC COMPOUNDS FROM *ORIGANUM VULGARE* AND *O. HERACLEOTICUM*

Vanja M. Tadić[1,*], Snežana Brašanac Vukanović[2], Ana Žugić[1], Vesna Pešić Vukašinović[2] and Nada Blagojević[2]

[1]Institute for Medicinal Plant Research "Dr. Josif Pančić," Belgrade, Serbia

[2]Faculty of Metallurgy and Technology, University of Montenegro, Podgorica, Montenegro

* Corresponding Author's Email: vtadic@mocbilja.rs.

Abstract

Different species belonging to *Oregano* genera have been traditionally used in folk medicine to alleviate conditions such as asthma, bronchitis, cough, diarrhea, indigestion, stomachache, menstrual disorders, general infections, inflammation-related illnesses and diabetes. Besides, they have been widely used in agriculture, cosmetic and food industries. For centuries, they have been recognized as culinary herbs, flavoring substances in food products, alcoholic beverages and perfumery for their spicy fragrance. The benefits of *Oregano* species on human health have been attributed to their phytochemical composition. The most important pharmacologically active phytochemicals found in species of this genus have been categorized into two main groups, namely essential oils and phenolic compounds.

Most of the studies dealing with *Oregano* species have been focused on essential oils, frequently ignoring more polar fractions represented by polyphenolic compounds. Therefore, we used different extraction methods and solvents varying in polarity to establish suitable conditions for obtaining extracts rich in polyphenolic compounds presumingly possessing traditionally established effects. Hence, our research focused on the correlation between the investigated antioxidant properties and chemical profile of different types of extracts obtained from *Origanum vulgare* L. (*O. vulgare* ssp. *viride* (Boiss.) Hayek) and *O. heracleoticum* L. (*O. vulgare* L. ssp. *hirtum*) belonging to Lamiaceae family.

It is known that the extraction protocol might affect the qualitative and quantitative composition of plant extracts. Therefore, it is important to define the extraction conditions, including among the others the type of solvent and temperature, in order to achieve the satisfactory efficacy of the process, as well as to ensure the extraction of the compounds of interest avoiding their chemical modification. In this context, the aim of the current study was to establish correlation of chemical composition and antioxidant activity of the extracts obtained from the abovementioned species applying infusion, maceration and Soxhlet extraction (OVEI, OVEM and OVES, for *O. vulgare*, and OHEI, OHEM and OHES for *O. heracleoticum*, respectively). Taking into account that water herbal infusions are most often used traditional remedies, this type of conventional extraction was our first choice in the performed investigation. However, bearing in mind that industrial scale production usually uses ethanol as extraction vehicle, Soxhlet extraction with this solvent was employed as a model method of an exhaustive extraction of the plant material. But, as the Soxhlet extraction requires high temperature that might be disadvantageous for thermolabile substances, maceration was employed, as well, as the procedure performed at ambient temperature.

This research provided systematic and thorough information on the phenolic composition of two *Origanum* species extracts obtained using different solvents and/or methods shown to possess antioxidant properties that might be correlated to their traditionally established usage.

Keywords: *Origanum vulgare, O. heracleoticum,* Maceration, Soxhlet extraction, infusion, non-volatile phenolic secondary metabolites, Rosmarinic acid

1. Introduction

Plants possess capacity to synthesize secondary metabolites with complex chemical composition, capable of effectively responding to stress situations imposed by biotic and abiotic factors. They are not essential for the functioning of the plant but are important for the plant to interact with its environment for adaptation and defense [1, 2]. Recent research indicated these substances to serve as plant mechanism against pathogens, pests, herbivores, UV-light and oxidative stress [3-5]. Their content in plants depends on factors such as cultivar, geographical localization, weather, daylight, temperature, soil conditions, water stress, harvesting time and others [3, 6]. It is commonly accepted that in oxidative stress, increase of production of free radicals and other reactive oxygen species (ROS) can cause oxidative damage to biomolecules, playing an important role in the pathogenesis of various serious diseases, such as neurodegenerative disorders, cancer, cardiovascular diseases, atherosclerosis, cataracts, and inflammation [7-9]. These radicals can be scavenged by the natural antioxidant agents. Medicinal plants are well-known as significant sources of natural antioxidants owing to their high phenolic content [10]. Therefore, they have been proven to be an effective therapy offering a variety of benefits, such as moderate reduction in manifestation of different health impairments due to the oxidative stress. Also, in modern medicine, they are recognized as a source of lead compounds for the production of medications for treating various diseases [11].

The mint family (Lamiaceae) contains about 236 genera and more than 6000 species [12]. It is a family of great diversity and variety with a cosmopolitan distribution, i.e., numerous species inhabiting different natural ecosystems, and many of them being cultivated, as well. Among the aromatic plant species from family Lamiaceae, genus *Origanum* occupies a special position. Within this genus, Ietswaart [13, 14] recognized three groups, 10 sections, 38 species, 6 subspecies and 17 hybrids based on morphological criteria [13, 15]. *Oregano* species have been traditionally used in folk medicine to alleviate conditions such as asthma, bronchitis, coughs, diarrhea, indigestion, stomachache, menstrual disorders, general infections, inflammation-related illnesses and diabetes [3, 16]. The benefits of plants of this genera on human health have been attributed to their phytochemical composition [3, 16, 17]. Their most important phytochemicals are grouped depending on their hydrophilic and hydrophobic properties into two categories: essential oils and phenolic compounds. Although research has often focused on the study of essential oils [3, 18], they represent only one of the main groups of phytochemicals found in *Oregano* species. The study of hydrophilic substances such as phenolic compounds is frequently ignored. Therefore, we used different extraction methods and solvents with different polarity to establish suitable conditions for obtaining extracts rich in polyphenolics that might be satisfactory in providing the good effects traditionally confirmed. Our research focused on the correlation between the investigated antioxidant properties and chemical profile of different types of plants extracts. It is known that quantity and quality of plant extracts are dependent on extraction protocol [19-24]. Many factors including type of solvent, the temperature, pH, the number of extraction steps, liquid-to-solid ratio and the particle size of the solute contribute to the efficacy of the extraction process [25]. The selected extraction parameters must be ensured in order to complete extraction of the compounds of interest, while at the same time avoiding their chemical modification [21, 26]. In this context, the goal of the current study was to establish correlation of chemical composition and antioxidant activity of the *Origanum vulgare* L. (*Origanum vulgare* ssp. *viride* (Boiss.) Hayek) and *Origanum heracleoticum* L. (*Origanum vulgare*

L. ssp. *hirtum*) extracts obtained by infusion, maceration and Soxhlet extraction. Taking into account that water herbal infusions are most often used traditional preparations, we used this type of conventional extraction in the performed investigation. On the other hand, bearing in mind that industrial scale production usually uses ethanol as an extraction vehicle, Soxhlet extraction with this solvent was employed as a model method of an exhaustive extraction of the plant material. Additionally taking into account that Soxhlet extraction requires high temperature, what might be disadvantageous for thermolabile substances, maceration was employed, as well, as extraction on ambient temperature. In the stated extracts total phenolic (TP), total tannins (TT), total flavonoid (TF) content were determined. In order to identify and quantify major phenolic compounds present in the investigated samples a HPLC method was employed. To evaluate the antioxidant properties, two different assays were used to avoid any possible incorrectness in the interpretation of the obtained results regarding the antioxidant properties of the investigated extracts, having in mind the differences in their working principles. Namely, the antioxidant potential of all extracts was evaluated by means of DPPH and FRAP tests. To establish the correlation between the chemical composition and the antioxidant properties of the investigated extracts, suitable statistical analysis was employed.

2. About Our Research

2.1. Materials and Methods

2.1.1. Chemicals

Analytical grade reagents acetate buffer, 2,4,6-tripyridyl-s-triazine (TPTZ), HCl, $FeCl_3$, Folin-Ciocalteu reagent, acetone, ethyl acetate, sodium bicarbonate, 1,1´-diphenyl-2-picrylhydrzyl (DPPH), absolute ethanol (96%, v/v) and methanol were purchased from Sigma-Aldrich. Acetonitrile, water, and methanol (HPLC grade) were purchased from Merck. Reference HPLC standards, protocatechuic acid, gallic acid, rosmarinic acid, p-coumaric acid, caffeic acid, chlorogenic acid, protocatechuic acid methyl ether, luteolin-7-*O*–glucoside, vitexin, luteolin, isovitexin, apigenin-7-*O*-glucoside, apigenin, vitexin-2''-*O*-rhamnoside, chrysaeriol, chrysin, kaempferol, morin, kaempferol-3-*O*-glucoside, quercetin, tiliroside, isorhamnetin, rutin, catechin, procyanidin B2, epicatechin, pyrogallol, scoparone, scopoletin (purity $\geq$ 99%) were purchased from Extrasynthese.

2.1.2. The Plant Material

The voucher specimens (No OVP_020317 (*O. vulgare*), and OHP_030317 (*O. heracleoticum*)) of the plant material used in the investigation were deposited at the Faculty of Pharmacy, Department of Botany, University of Belgrade, where the identification was performed. Shortly before the extraction, samples were ground in an electric mill (all samples powdered, dimension of the particles 355, as recommended by Ph. Eur. 9.0) [28].

2.1.3. Extraction and Analytical Methods

Applying the classical methods of maceration, infusion and continuous Soxhlet extraction of *O. vulgare* and *O. heracleoticum* (described bellow), the following samples were provided: OVEM and OHEM (macerates of *O. vulgare* and *O. heracleoticum*, respectively), OVEI and OHEI (infusion of *O. vulgare* and *O. heracleoticum,* respectively) and OVES and OHES (Soxhlet extracts of *O. vulgare* and *O. heracleoticum,* respectively).

2.1.3.1. Maceration

Ground plant material (50 g) was mixed with 300 mL of 70% (v/v) ethanol. The extraction was carried out at room temperature with occasional stirring. After 72 h, sample was filtered through filter paper and precipitate was re-extracted with 20 mL of solvent. After 48 h, the process was repeated. Solvent from the combined supernatants was evaporated using rotary evaporator Buchi R 114, USA. The yield of extraction was expressed as a percentage.

2.1.3.2. Soxhlet Extraction

10 g of the plant material was placed in a Soxhlet thimble in a Soxhlet apparatus and extracted with 325 mL of solvent (70% (v/v) ethanol) for 72 hours at at the boiling point of the solvent. After that, the sample was cooled to room temperature and the solvent was evaporated using rotary evaporator Buchi R 114, USA. The yield of extraction was expressed as a percentage.

2.1.3.3. Infusion

Boiling water (150 mL) was poured over 10 g of plant material, covered with the watch glass and left to boil further for 2 minutes. After this time, container was removed from the heater, covered and left to sit for 30 minutes. Sample was filtered through filter paper, and then water was evaporated using rotary evaporator Buchi R 114, USA. The yield of extraction was expressed as a percentage.

2.1.4. Determination of Total Phenolics, Tannins and Flavonoids

The total phenolic content was determined by the Folin-Ciocalteu method [29]. One hundred microliters of MeOH solutions of the investigated extracts OVEI, OVEM, OVES, OHEI, OHEM and OHES (0.55, 0.23, 0.45, 0.50, 0.56 and 0.59 mg/mL, respectively) was mixed with 0.75 mL of Folin-Ciocalteu reagent (previously diluted 10-fold with distilled water) and allowed to stand at 22^{0}C for 5 min; 0.75 mL of sodium bicarbonate (60 g/L) solution was added to mixture. After 90 min at 22^{0}C, absorbance was measured at 725 nm (UV-VIS Spectrophotometer HP 8453, Agilent Technologies, USA). Gallic acid (0-100 mg/L) was used for calibration of a standard curve. The calibration curve showed the linear regression at $r > 0.99$, and the results were expressed as milligrams of gallic acid equivalents per gram of plant extracts dry weight (mg GAE/g DW). The content of total phenolics presented the mean of three determinations.

The percentage content of TT was calculated using the method described in the European Pharmacopoeia 9.0 [28]. Briefly, decoctions prepared from the investigated extracts were treated with phosphomolybdotungstic reagent in alkaline medium after and without treatment with hide powder. The absorbance was measured by UV-VIS Spectrophotometer HP 8453 (Agilent Technologies, USA), at λ_{max}=760 nm. The percentage of tannins (the mean of three determinations) expressed as pyrogallol (%, w/w), was calculated from the difference in absorbance of total polyphenols (A_1) and polyphenols not adsorbed to hide powder (A_2), using following expression:

$$TT\ (\%) = 62.5(A_1 - A_2) \times m_2 / (A_3 - m_1)$$

where m_1 represented mass of the sample to be examined, in grams; and m_2 mass of pyrogallol, in grams.

The content of TF was calculated using the method described in the European Pharmacopoeia 9.0 [26]. Shortly, the sample was extracted with acetone/HCl under reflux condenser; the $AlCl_3$ complex of flavonoid fraction extracted by ethyl acetate was measured by UV-VIS Spectrophotometer HP 8453 at λ_{max}=425 nm. The content of flavonoid (mean of three determinations), expressed as hyperoside percentage, was calculated using following expression:

$$TF\ (\%) = A \times 1.25 / m$$

where A was absorbance at 425 nm and (m) was mass of the extracts to be examined in grams.

2.1.5. HPLC Analysis

"Fingerprinting" of the investigated phenolic compounds was achieved by an Agilent Technologies 1200 HPLC, equipped with Lichrospher 100RP 18e column, applying gradient elutions of two mobile phases, i.e., "A/B" ("A"-0.2 M solution of phosphoric acid, and "B"-being a pure acetonitrile) at flow–rates 1 mL/min, with photodiode-array (PDA) detection (UV at 360, 280 nm), always within 70 min. Winning combinations were 89-75% A (0-35 min); 75-60% A (35-55 min); 60-35% A (55-60 min) and 35-0% A (60-70 min). The concentrations of investigated samples were 45.35, 49.47, 43.87, 45.59, 50.26, 42.69 mg/mL for OVEI, OVEM, OVES, OHEI, OHEM and OHES, respectively. Prior to injection, samples were filtered through PTFE membrane filter. For standards used in the investigation, the concentrations were: 0.34 mg/ml for protocatechuic acid and gallic acid, 0.14 mg/ml for rosmarinic acid, 0.74 mg/ml for p-coumaric acid, 0.52 mg/ml for caffeic acid, 0.56 mg/ml for chlorogenic acid, 0.52 mg/ml for protocatechuic acid methyl ether, 0.34 mg/ml for luteolin-7-*O*–glucoside, 0.3 mg/ml for vitexin, 0.17 mg/ml for luteolin, 0.36 mg/ml for isovitexin, 0.4 mg/ml for apigenin-7-*O*-glucoside, 0.54 mg/ml for apigenin, 0.4 mg/ml for vitexin-2''-*O*-rhamnoside, 0.32 mg/ml for chrysaeriol, 0.46 mg/ml for chrysin, 0.30 mg/ml for kaempferol, 0.28 mg/ml for kaempferol-3-*O*-glucoside, 0.26 mg/ml for morin, 0.36 mg/ml for quercetin, 0.42 mg/ml for tiliroside, 0.15 mg/ml for isorhamnetin, 0.40 mg/ml for rutin, 0.44 mg/ml for catechin, 0.01 for procyanidin B2, 0.40 mg/ml for epicatechin, 0.46 mg/ml for pyrogallol, 0.56 mg/ml for scoparone, 0.38 mg/ml for scopoletin. The volume of standard solutions being injected, as well as for the tested sample extracts, was 4 μL.

Identification was based on retention times and overlay curves. Once spectra matching succeeded, results were confirmed by spiking with respective standards to achieve a complete identification by means of the so-called peak purity test. Those peaks not fulfilling these requirements were not quantified. Quantification was performed by external calibration with standards.

2.1.6. Methods for Determination of Antioxidant Activity

2.1.6.1. Radical-Scavenging Activity (DPPH Test)

Diluted extract (300 μL of the stock solution with the concentration 45.35, 49.47, 43.87, 45.59, 50.26, 42.69 mg/mL for OVEI, OVEM, OVES, OHEI, OHEM and OHES, respectively) and 2.7 mL of 0.1 mM ethanol

DPPH solution were mixed. The absorbance was recorded at 517 nm after 30 min incubation at room temperature in the dark, against ethanol as a blank [29]. Free radical scavenging activity was calculated with respect to control solution, containing ethanol instead of test solution using the formula:

$$\text{DPPH radical scavenging capacity (\%)} = 100 - [(A_S - A_B) \times 100 / A_C]$$

A_S was absorption of ethanol solution of the extract treated with DPPH radical solution; A_B was absorption of ethanol solution of the extract which was not treated with DPPH radical solution; A_C was absorption of ethanol solution of the DPPH.

The inhibition percentage was plotted against concentration of the samples, and IC_{50} values were determined by linear regression analysis.

2.1.6.2. Ferric-Reducing Antioxidant Power (FRAP)

Diluted extract (100 µL of the stock solution with the concentration 45.35, 49.47, 43.87, 45.59, 50.26, 42.69 mg/mL for OVEI, OVEM, OVES, OHEI, OHEM and OHES, respectively) and 3.0 mL of freshly prepared FRAP reagent (25 mL of 300 mM acetate buffer pH 3.6 plus 2.5 mL of 10 mM TPTZ solution in 40 mM HCl plus 2.5 mL of 20 mM $FeCl_3x6H_2O$) were mixed. The absorbance was recorded at 593 nm against a blank, containing 100 µL of resembling solvent, after 30 min incubation at 37^0C. The FRAP-value was calculated from the calibration curve of $FeSO_4x7H_2O$ standard solutions, covering the concentration range 100-1000 mmol/L and expressed as mmol Fe^{2+} /g extracts [29].

2.1.7. Statistical Analysis

Linear regression analyses and correlation coefficients to determine the relationship between two variables were calculated using MS-Windows software (Excel, 2007).

2.2. Results and Discussion

2.2.1. Yield and Content of Polyphenolic Compounds in the Extracts

It has been shown that the differentiation of phenolic compounds in the plants is associated with climatic conditions, geographical location of collection sites and other ecological and genetic factors. Besides, the variability of phenolic content may also be related to the extraction method and/or the solvent type used in the extraction procedure. Extraction is the crucial first step in the analysis of medicinal plants, because it is necessary to extract the desired chemical components from the plant materials for further separation and characterization [30]. Selection of a solvent used for the extraction depends on the specific nature of the bioactive components of interest [31]. As the target compounds may be non-polar, polar and/or thermally labile, the suitability of the methods of extraction must be considered in order to assure that potential active constituents are not lost, distorted or destroyed during the preparation of the extract from plant samples [32]. In our investigation, for the preparation of the extracts we used water in accordance to prescriptions from the folk medicine aiming to test a traditionally used phytopreparation. Besides, in the performed investigation we used 70% (v/v) ethanol as a solvent non-toxic to cells, additionally taking into account that industrial-scale extractions comprise this solvent as an optimal extragens.

Our results revealed that extraction yield was the highest when 70% (v/v) ethanol was employed in a Soxhlet extraction (OVES 27.87%, OHES 35.02%, Figure 1a). Significantly lower yields were obtained with the same solvent, but applying maceration (OVEM 10.58%, OHEM 9.48%). The advantage of the Soxhlet extraction compared to the cold maceration using the same solvent indicated that increasing the extraction temperature might increase the solubility of extractives, leading to an increase in yield as well. The extraction yield obtained with water as a solvent, was higher than the yield achieved employing maceration (with 70% (v/v) ethanol), but lower in comparison to Soxhlet extraction using 70%(v/v) ethanol (19.61% and 25.26% in OVEI and OHEI, respectively). Better yields of dry matter resulting from Soxhlet extraction with 70% (v/v) ethanol compared to the extraction using water were probably a consequence of the circulation flow of the solvent up to the complete exhaustion of plant material. Better extraction yield when using water extraction compared to maceration with ethanol, could be the result of lower selectivity of water and the consequent presence of other plant metabolites (e.g., carbohydrates and proteins) in the obtained extracts, in addition to the targeted phenolic compounds.

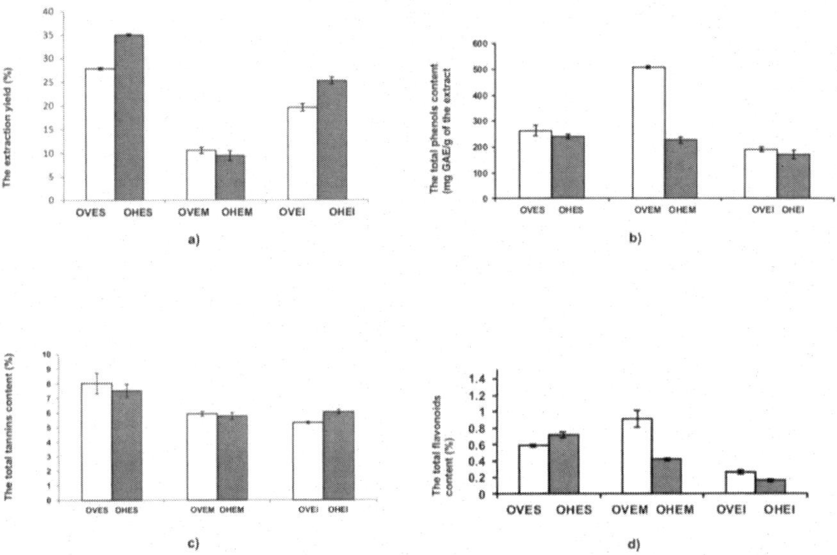

Figure 1. a) The extraction yields (%): OVES—*O. vulgare* extract obtained using Soxhlet extraction with 70% (v/v) ethanol; OHES—*O. heracleoticum* extract obtained using Soxhlet extraction with 70% (v/v) ethanol; OVEM—*O. vulgare* extract obtained by maceration with 70% (v/v) ethanol; OHEM—*O. heracleoticum* extract obtained by maceration with 70% (v/v) ethanol; OVEI—*O. vulgare* extract obtained by infusion with water; OHEI—*O. heracleoticum* extract obtained by infusion with water; b) content of total phenolic compounds (TP); c) total tannins (TT); d) total flavonoids (TF) in the investigated extracts.

It could also be observed (Figure 1b, 1c, 1d) that in the case of *O. heracleoticum* extracts, the highest content of TP, TF and TT was recorded in the extract obtained by Soxhlet extraction (240.03 mg GAE/g, 0.72%, 7.51%, respectively in OHES). Slightly lower concentrations of TP and TF content were measured when using the same solvent without increasing the temperature in the sample OHEM (226.67 mg GAE/g, 0.42%, respectively), while the lowest content was measured in the water extract (170.23 mg GAE/g, 0.16%, respectively). TT content was higher after extraction of this plant with water (6.04%), compared to maceration with 70% (v/v) ethanol (5.76%). On the other hand, in the case of *O. vulgare*, the highest content of TP and TF was measured in the extracts obtained applying maceration (507.02 mg GAE/g, 0.91%, respectively in OVEM), followed by Soxhlet (263.69 mg GAE/g and 0.59%, respectively in OVES) and water extracts (189.57 mg GAE/g and 0.26%, respectively in OVEI). TT contents represented 8.02%, 5.93% and 5.31% of OVES, OVEM and OVEI, respectively. TP content measured in investigated extracts were higher than the values in previously published studies [33-36]. Admittedly, in the stated investigations, different extraction methods and solvents were used. On the other hand, using ethanol as an extraction solvent Benchikha obtained results similar to ours [37].

The results of our study indicated the concentration of phenolics to be higher when ethanol was employed as an extraction solvent compared to water. An increase in the extraction temperature can promote higher solubility of analytes by increasing both solubility and mass transfer rate. In addition, higher temperature can reduce the viscosity and the solvent can easily penetrate the pores of the matrix, which enables more close contact of the analyte and the solvent. However, many phenolic compounds are easily hydrolyzed and oxidized. Long extraction times and high temperature can increase the chance of oxidation of phenolics which might decrease the yield of phenolics in the extracts [38, 39]. In our experiments, TP and TF contents were the highest in the extract obtained at ambient temperature (maceration) in the case of *O. vulgare* extracts (Figure 1b,d). This was especially pronounced for TP content (Figure 1b) i.e., in the sample OVEM notably higher content of TP was found compared to the extractions employing high temperatures with the same solvent (70% (v/v) ethanol in OVES) and different solvent (water in OVEI). In the case of tannin extraction (Figure 1c), the greatest TT content for both investigated *Origanum* species was slightly higher in Soxhlet extracts compared to other two extraction methods employed (maceration and infusion). The same trend was noticed for content of TP and TF in the case of *O. heracleoticum*, i.e., the highest amount of TP, TT and TF were found in the Soxhlet extracts (OHES).

2.2.2. Antioxidant Activity in the Investigation Extracts

Oxidative stress is a phenomenon caused by an imbalance between production and accumulation of ROS in cells and tissues and the ability of a biological system to detoxify these reactive products [40, 41]. Increase of ROS production may lead to cell and tissue damage. Biological systems have enzymatic and nonenzymatic antioxidant mechanisms that enable ROS inactivation. When an organism is exposed to a high concentration of ROS, the endogenous antioxidant system (superoxide dismutase (SOD), glutathione peroxidase (GPx) and catalase (CAT)) is compromised and, consequently, it fails to guarantee complete protection of the organism [42]. To compensate this deficit of antioxidants, the body can use exogenous antioxidants supplied through food, nutritional supplements, or pharmaceuticals. Among the most important exogenous antioxidants are phenolic compounds found in medicinal plants [43]. Hence, these compounds of natural origin exerting antioxidant properties as a feature often connected to their many biological effects have been intensively investigated during the past years [29, 44-49].

Recent investigations have shown that there is no universal method to quantitatively and accurately evaluate antioxidant activity [50, 51]. In this study, the analysis of antioxidant potential of the investigated extracts was performed using two methods, DPPH and FRAP. These methods are distinguished by their mechanism of action but are complementary in the evaluation of antioxidant potential of plants extracts [51]. DPPH assay is a simple inexpensive assay which utilizes a stable, nitrogen containing, purple colored radical, which is reduced by natural antioxidants or reducing compounds to a pale yellow colored hydrazine [52]. Results for DPPH test in our research were expressed as IC_{50} with the higher IC_{50} values indicating lower antioxidant capacity. The FRAP method is also a quick bioassay in which yellow colored ferric tripyridyltriazine complex

(Fe(III)-TPTZ) is reduced by phenolic compounds to a blue colored ferrous complex (Fe(II)-TPTZ) due to the electron donating abilities of its compounds [29]. Result for FRAP test were expressed as mmol Fe^{2+}/g of extracts used.

In our investigation, the ethanol extracts compared to water extracts had a higher antioxidant activity measured by both DPPH and FRAP assays (Figure 2a and 2b). Similar values for DPPH test also obtained by Benchikha [37].

In the case of *O. heracleoticum* extracts (Figure 2), the lowest IC_{50} recorded in DPPH test i.e., the highest FRAP value were measured in the sample obtained by Soxhlet extraction (OHES). Extract obtained by maceration (OHEM) revealed a slightly lower antioxidant activity, followed by water extract (OHEI) with the highest IC_{50} value and the lowest FRAP value. Overall, *O. heracleoticum* extracts had IC_{50} values in the range of 0.60 to 13.00 μg/mL, for OHES and OHEI, respectively, while FRAP values ranged from 0.0028 mmol/g for OHEI to 0.0033 mmol/g for OHES (Figure 2).

In the case of *O. vulgare* extracts (Figure 2), the highest antioxidant activity measured by DPPH and FRAP test was found in the extract obtained by maceration (OVEM). Using the DPPH method for evaluation of antioxidant activity, we found that water extract (OVEI) had a higher antioxidant potential than the extract obtained by Soxhlet extraction (OVES). In contrast, using the FRAP method, the higher antioxidant activity was found in the Soxhlet extract (OVES) compared to the one prepared with water (OVEI). IC_{50} values of the *O. vulgare* extracts ranged from 7.62 to 11.48 μg/mL for OVEM and OVES, respectively, while the FRAP values was lowest for OVEI (0.0035 mmol/g) and the highest for the OVEM (0.0049 mmol/g).

Both investigated *Origanum* species revealed similar antioxidant activity when applying the same extraction method and/or solvent (Figure 2). The exception was observed for IC_{50} values (DPPH assay) for the samples produced using Soxhlet extraction with 70% (v/v) ethanol. Although obtained employing the same procedure, the determined IC_{50}

value for the sample OVES (11.48 μg/ml) was notably higher in comparison to OHES (0.60 μg/ml).

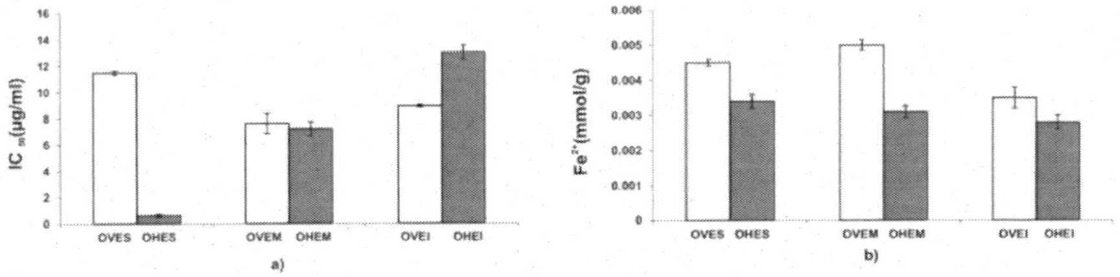

Figure 2. Antioxidant activity in the investigation extracts OVES—*O. vulgare* extract obtained using Soxhlet extraction with 70% (v/v) ethanol; OHES—*O. heracleoticum* extract obtained using Soxhlet extraction with 70% (v/v) ethanol; OVEM—*O. vulgare* extract obtained by maceration with 70% (v/v) ethanol; OHEM—*O. heracleoticum* extract obtained by maceration with 70% (v/v) ethanol; OVEI—*O. vulgare* extract obtained by infusion with water; OHEI—*O. heracleoticum* extract obtained by infusion with water measured by (a) DPPH assay; and (b) FRAP test.

It may be noticed that in our investigation antioxidant activity measured by both FRAP and DPPH method followed the same trend as the total content of phenolic compounds. This is not surprising bearing in mind that the antioxidant activity of phenolic compounds is often attributed to the capacity of scavenging free radicals, donating hydrogen atoms, electrons, or chelate metal cations [53, 54].

2.2.3. HPLC Analysis

Numerous studies have confirmed the presence of a significant number of phenolic compounds in the plant species of *Origanum* genera [55-67]. In our investigation we confirmed the presence of twenty-nine individual phenolic compounds (Table 1, Figures 3, 4).

Determined phenolic acids consisted of two subgroups being the derivatives of hydroxybenzoic and hydroxycinnamic acids. Hydroxybenzoic acids have the C6–C1 structure in common while hydroxycinnamic acids are aromatic compounds with a three-carbon side chain (C6–C3) [68, 69]. It has been shown that hydroxycinnamic acids exhibit higher antioxidant activity compared to the corresponding hydroxybenzoic acids [68, 70]. The higher activity of the hydroxycinnamic acid could be due to the CH=CH–COOH group, which ensures greater H-donating ability and radical stabilization than the –COOH group in the hydroxybenzoic acids [68, 71]. The predominant phenolic acid in *O. vulgare* and *O. heracleoticum* extracts was protocatechuic acid with content ranging from 79.88 in OVEM to 196.37 mg/g DWE in OVES and from 137.05 in OHEM to 290.31 mg/g DWE in OHES. The second most abundant acid in the tested extracts was gallic acid (from 54.57 in OVEI to 80.60 mg/g DWE in OVES and from 91.1 in OHEM to 137.5 mg/g DWE in OHES), followed by rosmarinic acid (from 19.39 in OVEI to 42.43 mg/g DWE in OVEM and from 24.81 in OHEI to 38.39 mg/g DWE in OHEM). Rosmarinic acid (RA) repsents depside, being the ester of caffeic acid and 3,4-dihydroxyphenyllactic acid. The presence of RA is of an importance in the investigated extracts, taking into account the numerous biological and pharmacological activities it possesses. RA, known as Labiatengerbstoff, a tannin-like compound in Lamiaceae, exhibits strong antioxidant, anti-inflammatory, cytotoxic and antimicrobial activities [72, 73]. RA has significant antioxidant activity, being stronger than that of vitamin E, and it might help in preventing the cell damage caused by free radicals, thereby reducing the risk for cancer and atherosclerosis. In addition, recent research has shown that the oral administration of RA might be effective in treating allergic asthma. Unlike antihistamines, RA prevents the activation of immune responder cells, which cause swelling and fluid formation [74]. It has been demonstrated that RA suppressed synovitis in mice and that it might be beneficial for the treatment of rheumatoid arthritis. In folk medicine, the applications of plants rich in RA are various - it has been used to treat peptic ulcers, arthritis, cataract, cancer, rheumatoid arthritis and bronchial asthma [72].

Determined polyphenolc acids, present in all investigated extracts, but in lower amounts, were *p*-coumaric (from 4.19 in OVEM to 7.53 mg/g DWE in OHES), caffeic (from 0.95 in OVEM to 1.75 mg/g DWE in OHEM) and chlorogenic acid (from 1.20 in OVEI to 2.74 mg/g DWE in OHES). Also, in all investigated extracts protocatechuic acid methyl ether was found (from 0.50 in OHES to 3.90 mg/g DWE in OVEM).

Table 1. Content of phenolic compounds determined by HPLC in the investigated extracts OVES—*O. vulgare* extract obtained using Soxhlet extraction with 70% (v/v) ethanol; OHES—*O. heracleoticum* extract obtained using Soxhlet extraction with 70% (v/v) ethanol; OVEM—*O. vulgare* extract obtained by maceration with 70% (v/v) ethanol; OHEM—*O. heracleoticum* extract obtained by maceration with 70% (v/v) ethanol; OVEI—*O. vulgare* extract obtained by infusion with water; OHEI—*O. heracleoticum* extract obtained by infusion with water [presented as mg/g DWE (dry weight of extract)]

		O. vulgare			*O. heracleoticum*		
	Samples	**OVES**	**OVEM**	**OVEI**	**OHES**	**OHEM**	**OHEI**
1	gallic acid	80.60	54.62	54.57	137.52	91.10	101.8
2	pyrogallol	2.63	1.59	1.60	2.97	1.35	1.99
3	protocatechuic acid	196.37	79.88	90.74	290.31	137.05	191.81
4	catechin	0.58	0.48	0.44	0.75	0.70	0.59
5	chlorogenic acid	1.91	1.53	1.20	2.74	2.49	1.78
6	procyanidin B2	0.03	0.06	0.01	0.24	0.26	0.20
7	caffeic acid	1.68	0.95	1.02	1.30	1.75	1.36
8	epicatechin				0.95	1.07	0.83
9	*p*-coumaric acid	7.46	4.19	7.05	7.53	4.70	4.70
10	scopoletin			0.52	2.45	1.65	1.54
11	vitexin	0.33	0.46	0.39	6.64	8.83	6.13
12	vitexin-2''-*O*-rhamnoside	0.39			1.32	1.14	0.83
13	isovitexin	1.51	1.43	0.07	1.79	1.10	1.12
14	rutin				0.56	1.68	0.62
15	luteolin-7-*O*-glucoside	15.56	15.19	7.46	0.43	0.49	0.30
16	protocatechuic acid, methyl ether	2.97	3.90	2.78	0.50	0.48	0.95
17	scoparone	1.47	1.83	1.30	1.20	0.92	0.99
18	kaempferol-3-*O*-glucoside	0.58	0.64	0.35	0.37	0.32	0.18
19	apigenin-7-*O*-glucoside	2.43	2.48	1.55	1.03	0.86	0.80
20	rosmarinic acid	35.54	42.39	19.39	27.32	38.39	24.81
21	morin	1.14	1.33	0.78	0.98	0.83	0.87
22	quercetin	1.39	0.47	0.49	0.46	0.62	0.37
23	luteolin	1.52	1.78	0.39	1.55	0.84	0.54
24	tiliroside	3.96	2.03	1.14	4.52	3.26	1.51
25	apigenin	0.47	0.53	0.30	0.88	1.07	0.36
26	chrysaeriol				0.28	0.42	0.21
27	kaempferol	0.60	0.47	0.38	0.74	0.59	0.51
28	isorhamnetin				0.43	0.40	0.47
29	chrysin				2.28	1.02	0.42

It has been established that the antioxidant activity of phenolic acid depends on the numbers and positions of the hydroxyl groups in relation to the carboxyl functional group [68, 71, 75]. Monohydroxy benzoic acids with the –OH moiety at the *ortho*- or *para*-position to the –COOH show no antioxidant activity, though the same is not true for the *m*-hydroxybenzoic acid. The antioxidant activity of phenolic acids increases with increasing degree of hydroxylation. However, substitution of the hydroxyl groups at the 3- and 5-position with methoxyl groups reduces the activity [68, 71].

One of the most studied groups of polyphenols are the flavonoids, which are structurally based on a 15-carbon skeleton arranged in a C6–C3–C6 configuration. The flavonoids can be sub-divided into groups, including flavan-3-ols, flavonols, flavones, isoflavones, flavanones and anthocyanins. Flavones were found to be the major class of flavonoids detected in all investigated extracts. In the *O. vulgare* extracts the most abundant was luteolin-7-*O*-glucoside (from 7.46 to 15.56 mg/g DWE in OVEI and OVES, respectively) identified in *O. heracleoticum* extracts, as well, with content ranging from 0.30 to 0.49 mg/g DWE in OHEI and OHEM, respectively. In the *O. heracleoticum* extracts, vitexin was present in the highest percentage (from 6.1 to 8.8 mg/g DWE in OHEI and OHEM, respectively). Content of vitexin in *O. vulgare* extracts ranged from 0.33 to 0.46 mg/g DWE in OVES and OVEM, respectively and in *O. heracleoticum* extracts from 6.13 to 8.83 in OHEI and OHEM, respectively. In all the investigated extracts following flavones were also detected: luteolin, isovitexin, apigenin-7-*O*-glucoside and apigenin (Table 1). Vitexin-2''-*O*-rhamnoside was found in all the investigated extracts of *O. heracleoticum,* while in investigated *O. vulgare* extracts this compound was identified only in the extract obtained by Soxhlet extraction (OVES, 0.39 mg/g DWE extract). Chrysaeriol and chrysin were found only in the *O. heracleoticum* extracts. In the case of flavonols, kaempferol, kaempferol-3-*O*-glucoside, morin, quercetin and tiliroside were detected in all the investigated extracts (Table 1). Isorhamnetin and rutin were only found in *O. heracleoticum* extracts.

Catechin and procyanidin B2, flavonoid compounds belonging to the subclass of flavan-3-ols, were also detected in all the investigated plant extracts (Table 1). The highest content of catechin was found in OHES (0.75 mg/g DWE), while the content of procyanidin B2 was the highest in OHEM (0.26 mg/g DWE). Epicatechin was found only in *O. heracleoticum* extracts (Table 1). The antioxidant activity of flavonoids was shown to increase with increasing degree of hydroxylation. However, substitution of the hydroxyl groups at the 3- and 5-position with methoxyl groups usually reduces the stated activity. Some structural features and nature of substitutions on rings B and C determine the antioxidant activity of flavonoids. The degree of hydroxylation and the positions of the -OH groups in the B ring, as an *ortho*-dihydroxyl structure of ring B (catechol group) leads to higher activity, or acts as the preferred binding site for trace metals [76, 77]. The presence of hydroxyl groups at the 3'-, 4'-, and 5'-positions of ring B (a pyrogallol group) was found to enhance the antioxidant activity. A double bond between C-2 and C-3, conjugated with the 4-oxo group in ring C improves the radical scavenging capacity of flavonoids [76, 77] and the double bond between C-2 and C-3, combined with a 3-OH (ring C), also increases the active radical scavenging capacity of flavonoids. The substitution of hydroxyl groups in ring B by methoxyl groups alters the redox potential, which affects the radical scavenging capacity of flavonoids [76-78]. It was demonstrated that in flavonoids the O-glycosylation decreases their antioxidant potential in *in vitro* assays, whereas the C-glycosylation in most cases increases their antioxidant potential [76, 79, 80]. Blocking the hydroxyl group at the C-3 position or removing the 3-OH group decreases antioxidative properties of flavonoids.

In our investigation pyrogallol was found in all the investigated extracts, the determined quantity varying from 1.35 to 2.97 mg/g DWE in OHEM and OHES, respectively. In the case of coumarins, scoparone was detected in all investigated extracts (from 0.92 in OHEM to 1.83 mg/g DWE in OVEM) while scopoletin was found in all extracts of *O. heracleoticum* (1.54 in OHEI to 2.45 mg/g DWE in OHES) and only in water extract of *O. vulgare* (0.52 mg/g DWE in OVEI). For phenolic acids and coumarins, it has been shown that vicinal diol groups are important for radical scavenging capacity and that methoxylation or glycosylation of ohydroxy group in the coumarins and esterification of phenolic acids reduce the antioxidant activity of these compounds [81, 82].

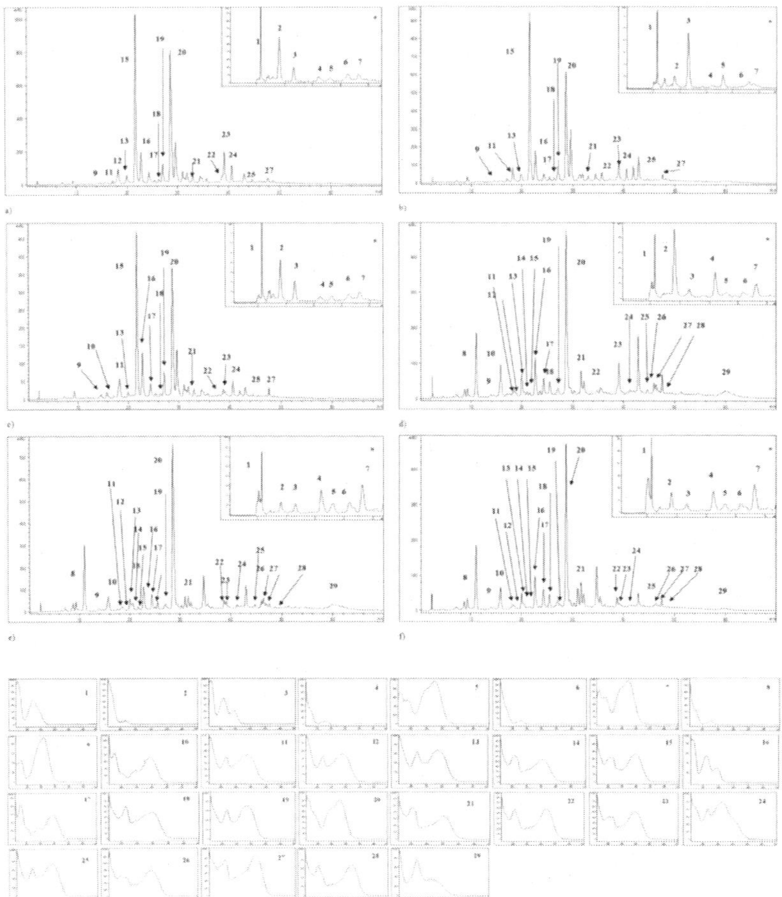

Figure 3. The HPLC "fingerprint" of analyzed samples OVES—*O. vulgare* extract obtained using Soxhlet extraction with 70% (v/v) ethanol; OVEM—*O. vulgare* extract obtained by maceration with 70% (v/v) ethanol; OVEI—*O. vulgare* extract obtained by infusion with water; OHES—*O. heracleoticum* extract obtained using Soxhlet extraction with 70% (v/v) ethanol; OHEM—*O. heracleoticum* extract obtained by maceration with 70% (v/v) ethanol; OHEI—*O. heracleoticum* extract obtained by infusion with water (a-f, respectively). The numbers at figures a-f represented the identified compounds at 360 nm and 280 nm (the small figure in the upper right corner of each chromatogram) corresponded to numbers of the specific compound given in Table 1. On the left side the corresponding UV spectra of the determined secondary metabolites were presented.

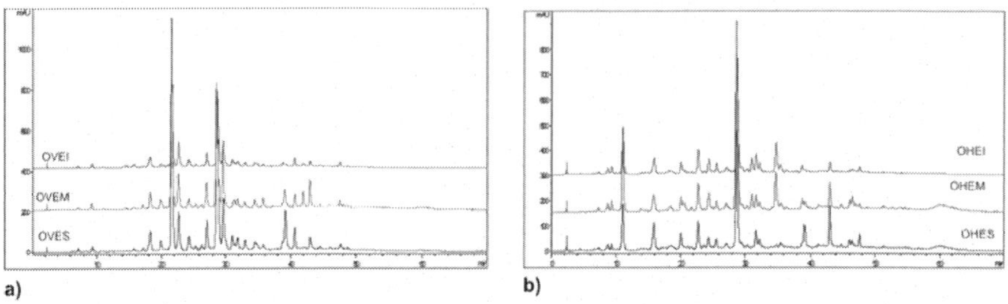

Figure 4. The comparative overly of the investigated extracts from a) *O. vulgare* and b) *O. heracleoticum*.

To the best of our knowledge, this is the first time that the presence of scopoletin, scoparone, pyrogallol, procyanidin B2, and protochatechuic acid methyl ether was confirmed in the investigated plant samples.

Comparing the fingerprints of the investigated extracts (Table 1, Figure 4a,b) it was noticed that epicatechin was determined in significant quantity only in *O. heracleoticum* extracts, while *O. vulgare* extracts revealed notable content of luteolin-7-*O*-glucoside. Rosmarinic acid was quantified in great amount in both investigated species.

2.2.4. Correlation between Content of Polyphenolic Compounds and Antioxidant Activity

In order to connect the determined antioxidant properties and chemical composition of the investigated extracts we performed the correlation analysis, applying the appropriate statistical tools. We found a very strong correlation between the DPPH and FRAP values (Table 2) for the *O. heracleoticum* extracts ($r = 0.9993$) while in *O. vulgare* extracts the same correlation was very weak ($r = 0.1503$). The explanation might be connected with the chemical profile of the investigated extracts, where the different compounds might contribute to antioxidant potential through different mechanisms [29]. TF content in *O. heracleoticum* extracts showed a very strong correlation with DPPH and FRAP assays ($r = 1$ and $r = 0.9991$, respectively), whereas in *O. vulgare* the correlation of TF and FRAP was very strong ($r = 0.9838$) in contrast to correlation of DPPH method and TF that was moderate ($r = 0.3261$). The strong correlations between TT content in *O. heraceoticum* extracts and both DPPH and

FRAP values were established (r = 0.8040, r = 0.7819, respectively). TT contents in the *O. vulgare* extracts exhibited a strong correlation with DPPH (r = 0.8466), while it was weak in the case of FRAP values (r = 0.3989). The correlation between TP content in *O. vulgare* extracts and antioxidant activity measured by the FRAP test (r = 0.8830) was strong, while the correlation with DPPH assays was moderate (r = 0.5970). In the case of *O. heracleoticum* extracts, the correlations between TP content and performed DPPH and FRAP assays were very strong (r = 0.9292 and r = 0.9420, respectively). Based on the high correlation of total TP, TF and TT content with antioxidant activity it was possible to suggest that these compounds were highly responsible for the antioxidant activity of the selected plant extracts. Their antioxidant activity seemed to be related to their molecular structure, more precisely to the presence and number of hydroxyl groups, double bond conjugation and resonance effects [83]. However, this activity is not limited only to phenolic content because the total phenolic content does not comprise all the antioxidants. In addition, the synergism between the antioxidants in the mixture makes the antioxidant activity not only dependant on the concentration, but also on the structure and the interaction between the antioxidants [84]. These different correlation coefficients support the necessity to use several methods to comprehensively evaluate antioxidant activity of plants [85].

Table 2. Correlation between TT, TF, and TP and antioxidant activity; correlation between the applied tests DPPH and FRAP in the investigation extracts extracts

	O. vulgare		*O. heracleoticum*	
	DPPH	**FRAP**	**DPPH**	**FRAP**
TP	y= -0.007x+11.602 R^2=0.3564	y= 4E-06x+0.003 R^2=0.7796	y= -0.1556x+39.964 R^2=0.8634	y= 8E-06x+0.0015 R^2=0.8873
TF	y= -1.9648x+10.503 R^2=0.1064	y= 0.0023x+0.003 R^2=0.9675	y= -22.138x+16.523 R^2=1	y= 0.0011x+0.0026 R^2=0.9983
TT	y= 1.1671x+1.8573 R^2=0.7167	y= 0.0002x+0.003 R^2=0.1591	y= -5.3064x+41.086 R^2=0.6464	y= 0.0002x+0.0015 R^2=0.6114
DPPH		y=-6E-05x +0.0049 R^2=0.0226		y=-5E-05x +0.0034 R^2=0.9987

Table 3. Correlation between identified individual phenolic compounds and antioxidant activity in the investigation extracts

	O.vulgare		*O. heracleoticum*	
Phenolic compound	**DPPH**	**FRAP**	**DPPH**	**FRAP**
Gallic acid	y=1.2277x +1.5832 R^2=0.887	y=1E-04x +0.0037 R^2=0.0363	y=-1.9361x +28.258 R^2=0.5751	y=9E-05x +0.0021 R^2=0.5391
Pyrogallol	y=30.938x +3.3254 R^2=0.8958	y=0.0023x +0.0039 R^2=0.0312	y=-47.861x +17.003 R^2=0.3956	y=0.0022x +0.0026 R^2=0.3604
Protocatecuic acid	y=0.2942x +5.7506 R^2=0.9356	y=1E-05x +0.0042 R^2=0.0111	y=-0.5287x +17.841 R^2=0.4379	y=2E-05x +0.0026 R^2=0.4021
Catechin	y=216.66x -1.5387 R^2=0.6635	y=0.0468x +0.002 R^2=0.2033	y=-703.26x +55.181 R^2=0.9303	y=-703.26x +55.181 R^2=0.9303
Chlorogenic acid	y=37.764x +3.5023 R^2=0.4643	y=0.0134x +0.0023 R^2=0.3857	y=-119.36x +34.89 R^2=0.9095	y=0.0058x +0.0017 R^2=0.9293
Procyanidin B2	y=-405.62x +10.968 R^2=0.2677	y=0.2826x +0.0032 R^2=0.8535	y=-1295.1x +38.249 R^2=0.4037	y=0.0653x +0.0015 R^2=0.4395
Caffeic acid	y=47.54x +3.5416 R^2=0.9365	y=0.002x +0.0041 R^2=0.0107	y=43.883x +0.4591 R^2=0.0299	y=-0.0017x +0.0033 R^2=0.0187
Epicatechin			y=-221.45x +28.096 R^2=0.1859	y=0.0115x +0.002 R^2=0.215
p-Coumaric acid	y=9.1417x +3.6481 R^2=0.6906	y=-0.0029x +0.0061 R^2=0.4556	y=-33.526x +25.865 R^2=0.7787	y=0.0016x +0.0022 R^2=0.7478
Scopoletin			y=-116x +28.829 R^2=0.8681	y=0.0055x +0.0021 R^2=0.8426
Vitexin	y=-288.16x +20.822 R^2=0.947	y=0.0432x +0.0026 R^2=0.1399	y=-6.1724x +11.378 R^2=0.0204	y=0.0004x +0.0028 R^2=0.032
Vitexin-2''-O-rhamnoside			y=-247.74x +34.265 R^2=0.969	y=0.012x +0.0018 R^2=0.9803
Isovitexin	y=5.4959x +8.7954 R^2=0.0521	y=0.0087x +0.0035 R^2=0.8615	y=-138.71x +25.545 R^2=0.7664	y=0.0066x +0.0022 R^2=0.735
Rutin			y=7.7945x +6.1818 R^2=0.0063	y=-0.0002x +0.0031 R^2=0.0018
Luteolin-7-0-glucoside	y=0.945x +8.1458 R^2=0.0489	y=0.0016x +0.0024 R^2=0.8666	y=-417.59x +24.181 R^2=0.4238	y=0.021x +0.0022 R^2=0.4599
Protocatechuic acid methyl ether	y=-21.133x +16.156 R^2=0.4147	y=0.0109x +0.0008 R^2=0.7281	y=194.37x -5.6943 R^2=0.677	y=-0.0096x +0.0037 R^2=0.7104
Scoparone	y=-38.238x +15.228 R^2=0.276	y=0.0261x +0.0003 R^2=0.8469	y=-314.43x +39.614 R^2=0.5399	y=0.0147x +0.0016 R^2=0.5036
Kaempferol-3-O-glucoside	y=-0.4996x +9.3764 R^2=2E-05	y=0.0485x +0.0018 R^2=0.9785	y=627.11x +25.405 R^2=0.9078	y=0.0307x +0.0022 R^2=0.9277
Apigenin-7-O-glucoside	y=4.9572x +8.2795 R^2=0.0176	y=0.014x +0.0013 R^2=0.9215	y=-496.45x +51.638 R^2=0.952	y=0.0238x +0.001 R^2=0.9353
Rosmarinic acid	y=-0.1848x +9.9495 R^2=0.0124	y=0.0006x +0.0022 R^2=0.9984	y=-1.1842x +10.504 R^2=0.019	y=7E-05x +0.0029 R^2=0.0302
Morin	y=-12.231x +10.681 R^2=0.0297	y=0.0277x +0.0013 R^2=0.9995	y=-582.12x +59.116 R^2=0.5401	y=0.0272x +0.0007 R^2=0.5038

Comparison of element contents in water and ethanol extracts of bilberry plant parts. *Arch. Biol. Sci.* 71, 145–157.

[10] Guerriero, G., Berni, R., Muñoz-Sanchez, J. A., Apone, F., Abdel-Salam, E. M., Qahtan, A. A., Alatar, A. A., Cantini, C., Cai, G., Hausman J. F., Siddiqui, K. S., Hernández-Sotomayor, S. M. T., Faisal, M. (2018). Production of plant secondary metabolites: Examples, tips and suggestions for biotechnologists. *Genes.* 9:309.

[11] Gu, R., Wang, Y., Long, B., Kennelly, E., Wu, S., Liu, B., Li, P., Long, C. (2014). Prospecting for bioactive constituents from traditional medicinal plants through ethnobotanical approaches. *Biol. Pharm. Bull.* 37, 903–915.

[12] Carović-Stanko, K., Petek, M., Grdiša, M., Pintar, J., Bedeković, D., HerakĆustić, M., Satovic, Z. (2016). Medicinal plants of the family Lamiaceae as functional foods – a review. *Czech J. Food Sci.* 34, 377–390.

[13] Čabarkapa, I., Milovanović, I. D., Plavsić, D., Palić, D. V., Kokić, B., Arsic, I. (2012).Antimicrobial activity of *Origanum heracleoticum* L. essential oil from Serbia. *Agro. Food Ind. Hi Tec.* 23, 55-58.

[14] Ietswaart, J. H. A taxonomic revision of the genus *Origanum* (Labiatae). The Hague, Boston, London: Leiden University Press; 1980.

[15] De Martino, L., De Feo, V., Formisano, C., Mignola, E., Senatore, F. (2009). Chemical composition and antimicrobial activity of the essential oils from three chemotypes of *Origanum vulgare* L. ssp. hirtum (Link) Ietswaart growing wild in Campania (Southern Italy). *Molecules.* 14, 2735–2746.

[16] Pascual, M. E., Slowing, K., Carretero, E, Sánchez Mata, D., Villar, A. (2001). Lippia: traditional uses, chemistry and pharmacology: A review. *J. Ethnopharmacol.* 76, 201–14.

[17] Leyva-López, N., Nair, V., Bang, W. Y., Cisneros-Zevallos, L., Heredia, J. B. (2016). Protective role of terpenes and polyphenols from three species of Oregano (*Lippia graveolens, Lippia palmeri* and *Hedeoma patens*) on the suppression of lipopolysaccharide-

induced inflammation in RAW 264.7 macrophage cells. *J. Ethnopharmacol.* 187, 302–312.

[18] Leyva-López, N, Gutiérrez-Grijalva, E. P., Vazquez-Olivo, G., Heredia, J. B. (2017). Essential oils of oregano: Biological activity beyond their antimicrobial properties. *Molecules*. 22, 989.

[19] Hussain, A. I., Chatha, S. A. S., Noor, S., Khan, Z. A., Arshad, M. U., Rathore, H. A., Sattar, M. Z. A. (2012). Effect of extraction techniques and solvent systems on the extraction of antioxidant components from peanut (*Arachishypogaea* L.) hulls. *Food Anal. Methods*. 5, 890–896.

[20] Sultana, B., Anwar, F., Ashraf, M. (2009) Effect of extraction solvent/technique on the antioxidant activity of selected medicinal plant extracts. *Molecules*. 14, 2167–2180.

[21] Majeed, M., Hussain, A. I., Chatha, S. A. S., Khosa, M. K. K., Kamal, G. M., Kamal, M. A., Zhang, X., Liu, M. (2016). Optimization protocol for the extraction of antioxidant components from *Origanum vulgare* leaves using response surface methodology. *Saudi J. Biol. Sci.* 23, 389–96.

[22] Arsić, I. A., Tadić, V. M., Đorđević, S. M., Žugić, A. R., Vujić, Z. B., Petrović, S. D. (2014). Optimization of extraction of antioxidant components from Yarrow herb. *Hem. Ind.* 68, 511-517.

[23] Arsic, I., Zugic, A., Runjajic Antic, D., Zdunic, G., Dekanski, D., Markovic, G., Tadic, V. (2010). *Hypericum Perforatum* L. Hypericaceae/Guttiferae sunflower, olive and palm oil extracts attenuate cold restraint stress – induced gastric lesions. *Molecules,* 15, 6688-6698.

[24] Arsic, I., Zugic, A., Tadic, V., Tasic-Kostov, M., Misic, D., Primorac, M., Runjajic Antic, D. (2012). Estimation of dermatological application of creams with St. John's Wort oil extracts. Molecules, 17, 275-294.

[25] Brasanac-Vukanovic, S., Mutic, J., Stankovic, D., Arsic, I., Blagojevic, N., Vukasinovic-Pesic, V., Tadic, V. (2018).Wild bilberry (*Vaccinium myrtillus* L., Ericaceae) from Montenegro as a

source of antioxidants for use in the production of nutraceuticals. *Molecules*. 23, 1864.

[26] Wu, X., Yu, X., Jing, H. (2011). Optimization of phenolic antioxidant extraction from Wuweizi (Schisandrachinensis) pulp using random-centroid optimazation methodology. *Int. J. Mol. Sci.* 12, 6255–66.

[27] Filipović, V., Popović, V., Aćimović, M., Marković, T., Protić, R., Ugrenović, V., Sikora, V. (2014). *Origanum heracleoticum* L. and *Origanum vulgare* L. seeds germination stimulators. *Lek.Sirovine.* 34, 81-91.

[28] Council of Europe. *European Pharmacopeia*. 9th edition; Strasbourg, France: Council of Europe: 2016.

[29] Zugic, A., Djordjevic, S., Arsic, I., Markovic, G., Zivkovic, J., Jovanovic, S., Tadic, V. (2014). Antioxidant activity and phenolic compounds in 10 selected herbs from Vrujci Spa, Serbia, *Ind. Crops. Prod.* 52, 519–527.

[30] Sasidharan, S., Chen, Y., Saravanan, D., Sundram, K. M., Yoga Latha, L. (2011) Extraction, isolation and characterization of bioactive compounds from plants' extracts. *Afr. J. Tradit. Complement. Altern. Med.* 8, 1–10.

[31] Thouri, A., Chahdoura, H., El Arem, A., Omri Hichri, A., Ben Hassin, R., Achour, L. (2017). Effect of solvents extraction on phytochemical components and biological activities of Tunisian date seeds (var. Korkobbi and Arechti). *BMC Complement. Altern. Med.* 17, 248.

[32] Peter, A., Solomon, A. (2018). Isolation and anti-bacterial activity of the active components from the stem- -back of *Enantial chlorantha. European J. Med. Plants*. 22, 1–7.

[33] Spiridon, I., Bodirlau, R., Teaca, C. A. (2011). Total phenolic content and antioxidant activity of plants used in traditional Romanian herbal medicine. *Cent. Eur. J. Biol.* 6, 388–396.

[34] Proestos, C., Lytoudi, K., Mavromelanidou, O. K., Zoumpoulakis, P., Sinanoglou, V. J. (2013). Antioxidant capacity of selected plant extracts and their essential oils. *Antioxidants*. 2, 11–22.

[35] Proestos, C., Varzakas, T. (2017). Aromatic Plants: Antioxidant capacity and polyphenol characterisation. *Foods.* 6, 28.

[36] Chrpová, D., Kouřimská, L., Harry Gordon, M., Heřmanová, V., Roubíčková, I., Pánek, J. (2010). Antioxidant activity of selected phenols and herbs used in diets for medical conditions. *Czech J. Food Sci.* 28, 317-325.

[37] Benchikha, N., Menaceur, M., Barhi, Z. (2015). Extraction and antioxidant activities of two species Origanum plant containing phenolic and flavonoid compounds. *J. Fundam. Appl. Sci.* 5, 120-128.

[38] Ivanovic, J., Tadic, V., Dimitrijevic, S., Stamenic, M., Petrovic, S., Zizovic, I. (2014). Antioxidant properties of the anthocyanin-containing ultrasonic extract from blackberry cultivar "Čačanska Bestrna." *Ind. Crops. Prod.* 53, 274–281.

[39] Dai, J., Mumper, R. J. (2010). Plant phenolics: extraction, analysis and their antioxidant and anticancer properties. *Molecules.* 15, 7313–52.

[40] Pizzino, G., Irrera, N., Cucinotta, M., Pallio, G., Mannino, F., Arcoraci, V., Squadrito, F., Altavilla, D., Bitto, A. (2017). Oxidative stress: Harms and benefits for human health. *Oxid. Med. Cell. Longev.* 8416763.

[41] Tadić, V. M., Žižović, I., Živković, J., Arsić, I., Roj, E. Chemical Characterization and antioxidant potential of diterpenoids rich fractions isolated from selected plant species. In: Jones B, editor. *Diterpenoids: Types, functions and research.* NOVA Science Publishers; 2017.

[42] Santos-Sánchez, N. F., Salas-Coronado, R., Villanueva-Cañongo, C., Hernández-Carlos, B. Antioxidant Compounds and their antioxidant mechanism. *Antioxidants* [Working Title]. IntechOpen; 2019.

[43] Anwar, H., Hussain, G., Mustafa, I. Antioxidants from natural sources. In: Shalaby E, editor. *Antioxidant in Foods and Its Applications.* IntechOpen; 2018.

[44] Tadić, V., Arsić, I., Zvezdanović, J., Žugić, A., Cvetković, D., Pavkov, S. (2017). The estimation of the traditionally used yarrow (*Achillea millefolium* L. Asteraceae) oil extracts with anti-inflamatory potential in topical application. *J. Ethnopharmacol.* 199, 138–148.

[45] Zugic, A., Jeremic, I., Isakovic, A., Arsic, I., Savic, S., Tadic, V. (2016). Evaluation of anticancer and antioxidant activity of a commercially available CO_2 supercritical extract of old man's beard (*Usnea barbata*). *PlosONE.* 11, e0146342.

[46] Nikolic, I., Lunter, D. J., Randjelovic, D., Zugic, A., Tadic, V., Markovic, B., Cekic, N., Zivkovic, L., Topalovic, D., Spremo-Potparevic, B., Daniels, R., Savic, S. (2018). Curcumin-loaded low-energy nanoemulsions as a prototype of multifunctional vehicles for different administration routes: Physicochemical and in vitro peculiarities important for dermal application. *Int. J. Pharmaceut.* 550, 333-346.

[47] Paunovic, V., Kosic, M., Djordjevic, S., Zugic, A., Djalinac, N., Gasic, U., Trajkovic, V., Harhaji-Trajkovic, J. (2016). *Marrubium vulgare* ethanolic extract induces proliferation block, apoptosis, and cytoprotective autophagy in cancer cells in vitro. *Cell. Mol. Biol.* 62, 107-113.

[48] Tasić-Kostov, M., Arsić, I., Pavlović, D., Stojanović, S., Najman, S., Naumović, S., Tadić, V. (2019). Towards a modern approach to traditional use: *in vitro* and *in vivo* evaluation of *Alchemilla vulgaris* L. gel wound healing potential. *J. Ethnopharmacol.* 238, 111777-111789.

[49] Tadic, V., Jeremic, I., Dobric, S., Isakovic, A., Markovic, I., Trajkovic, V., Bojovic, D., Arsic, I. (2012). Anti-inflammatory, Gastroprotective and Cytotoxic Effects of *Sideritis scardica* Extracts. *Planta Med.* 78, 415 – 427.

[50] Prior, R. L., Wu, X., Schaich, K. (2005). Standardized methods for the determination of antioxidant capacity and phenolics in foods and dietary supplements. *J. Agri. Food Chem.* 53, 4290-4302.

[51] Rebaya, A., Igueld Belghith, S., Baghdikian, B., Mahiou Leddet, V., Mabrouki, F., Olivier, E., Cherif, J. K., Trabelsi Ayadi, M. (2014). Total phenolic, total flavonoid, tannin content, and antioxidant aapacity of *Halimium halimifolium* (Cistaceae). *J. Appl. Pharm. Sci.* 5, 52-057.

[52] Zeghad, N., Ahmed, E., Belkhiri, A., Heyden, Y. V., Demeyer, K. (2019). Antioxidant activity of *Vitis vinifera, Punica granatum, Citrus aurantium* and *Opuntia ficus indica* fruits cultivated in Algeria. *Heliyon.* 5, 01575.

[53] Afanas'ev, I. B., Dcrozhko, A. I., Brodskii, A., V., Kostyuk, V. A., Potapovitch, A. I. (1989). Chelating and free radical scavenging mechanisms of inhibitory action of rutin and quercetin in lipid peroxidation. *Biochem. Pharmacol.* 38, 1763–1769.

[54] Minatel, I. O., Borges, C. V., Ferreira, M. I., Gomez, H. A. G., Chen, C. Y. O., Lima, G.P . P. Phenolic compounds: functional properties, impact of processing and bioavailability. In: Soto-Hernandez M, Palma-Tenango M, Garcia-Mateos MR, editors. *Phenolic Compounds - Biological Activity.* Intech Open; 2017.

[55] Radušienė, J., Ivanauskas, L., Janulis, V., Jakštas, V. (2010). Phenolic compounds and antioxidant activity of *Origanum vulgare.Acta. Hortic.* 860, 93–96.

[56] Koldaş, S., Demirtas, I., Ozen, T., Demirci, M. A., Behçet, L. (2015). Phytochemical screening, anticancer and antioxidant activities of *Origanum vulgare* L. ssp. *viride* (Boiss.) Hayek, a plant of traditional usage. *J. Sci. Food Agric.* 95,786–798.

[57] Matei, A. O., Gatea, F., Radu, G. L. (2015). Analysis of phenolic compounds in some medicinal herbs by LC–MS. *J. Chromatogr. Sci.* 53, 1147–1154.

[58] Chou, T. H., Ding, H. Y., Lin, R. J., Liang, J. Y., Liang, C. H. (2010). Inhibition of melanogenesis and oxidation by protocatechuic acid from *Origanum vulgare* (Oregano). *J. Nat. Prod.* 73, 1767–1774.

[59] Bodirlau, R., Spiridon, I., Teaca, C. A., Anghel, N., Ichim, M., Colceru, S., Armatu, A. (2009). Anti-inflammatory constituents from different plant species.*Environ. Eng. Manag. J.* 8, 785–792.

[60] Kaki, F. A., Benkiniouar, R., Demirtas, I., Merzoug, A., Touil, A. (2019). Phytochemical study of two algerian plants *Origanum vulgare* L. Sbsp. *glandulosum* (Desf) Ietswaart and *Thymus algeriensis* (Boiss. and Reut). *Asian. J. Chem.* 31, 1105–1109.

[61] Hawas, U. W., El-Desoky, S. K., Kawashty, S. A., Sharaf, M. (2008). Two new flavonoids from *Origanum vulgare*. *Nat. Prod. Res.* 22, 1540–1543.

[62] Martins, N., Barros, L., Santos-Buelga, C., Henriques, M., Silva, S., Ferreira, I. C. F. R. (2014). Decoction, infusion and hydroalcoholic extract of *Origanum vulgare* L.: different performances regarding bioactivity and phenolic compounds. *Food Chem.* 158, 73-80.

[63] Gatea, F., Teodor, E. D., Matei, A. O., Badea, G. I., Radu, G. L. (2015). Capillary Electrophoresis Method for 20 Polyphenols Separation in Propolis and Plant Extracts. *Food Anal. Methods*. 8, 1197–1206.

[64] Koukoulitsa, C., Karioti, A., Bergonzi, M. C., Pescitelli, G., Di Bari, L., Skaltsa, H. (2006). Polar constituents from the aerial parts of *Origanum vulgare* L. Ssp. *hirtum* growing wild in Greece. *J. Agric. Food Chem.* 54, 5388–5392.

[65] Kamran, L., Naquvi, J., Ahamad, J., Salma, A., Ansari, S. H., Najmi, A. K. (2019). A critical review on traditional uses, phytochemistry and pharmacological uses og Origanum vulgare. *Int. Res. J. Pharm.* 10, 7-11.

[66] Nowak, R. (2003). Separation and quantification of tiliroside from plant extracts by SPE/RP-HPLC. *Pharm. Biol.* 41, 627–630.

[67] Sadeghi, Z., Akaberi, M., Valizadeh, J. (2014). *Otostegia persica* (Lamiaceae): A review on its ethnopharmacology, phytochemistry, and pharmacology. *Avicenna. J. Phytomed.* 4,79-88.

[68] Balasundram, N., Sundram, K., Samman, S. (2006). Phenolic compounds in plants and agri-industrial by-products: Antioxidant activity, occurrence, and potential uses. *Food Chem.* 99, 191–203.

[69] Bravo, L. (2009). Polyphenols: Chemistry, dietary sources, metabolism, and nutritional significance. *Nutr. Rev.* 56, 317–333.

[70] Andreasen, M. F., Landbo, A. K., Christensen, L. P., Hansen, Å., Meyer, A. S. (2001). Antioxidant effects of phenolic rye (*Secale cereale* L.) extracts, monomeric hydroxycinnamates, and ferulic acid dehydrodimers on human low-density lipoproteins. *J. Agric. Food Chem.* 49, 4090–4096.

[71] Rice-Evans, C. A., Miller, N. J., Paganga, G. (1996). Structure-antioxidant activity relationships of flavonoids and phenolic acids. *Free Radic. Biol. Med.* 20, 933–956.

[72] Adomako-Bonsu, A. G., Chan, S. L., Pratten, M., Fry, J. R. (2017). Antioxidant activity of rosmarinic acid and its principal metabolites in chemical and cellular systems: Importance of physico-chemical characteristics. *Toxicol. In Vitro.* 40, 248-255.

[73] Nunes, S., Madureira, A. R., Campos, D., Sarmento, B., Gomes, A. M., Pintado, M., Reis, F. (2017). Therapeutic and nutraceutical potential of rosmarinic acid-Cytoprotective properties and pharmacokinetic profil. *Crit. Rev. Food Sci. Nutr.* 57, 1799-1806.

[74] Osakabe, N., Takano, H., Sanbongi, C., Yasuda, A., Yanagisawa, R., Inoue, K., Yoshikawa, T. (2004). Anti-inflammatory and anti-allergic effect of rosmarinic acid (RA); inhibition of seasonal allergic rhinoconjunctivitis (SAR) and its mechanism. *Biofactors.* 21, 127-131.

[75] Robards, K., Prenzler, P. D., Tucker, G., Swatsitang, P., Glover, W. (1999). Phenolic compounds and their role in oxidative processes in fruits. *Food Chem.* 66, 401–436.

[76] Russo, D. (2018). Flavonoids and the structure-antioxidant activity relationship. *J. Pharmacogn. Nat. Prod.* 04, 1–2.

[77] Pietta, P. G. (2000). Flavonoids as antioxidants. *J. Nat. Prod.* 63, 1035–42.

[78] Seeram, N. P., Nair, M. G. (2002) Inhibition of lipid peroxidation and structure−activity-related studies of the dietary constituents anthocyanins, anthocyanidins, and catechins. *J. Agric. Food Chem.* 50, 5308–5312.

[79] Xiao, J., Chen, T., Cao, H. (2014). Flavonoid glycosylation and biological benefits. *Biotechnol. Adv.*

[80] Xiao, J. (2015). Dietary flavonoid aglycones and their glycosides: Which show better biological significance? *Crit. Rev. Food Sci. Nutr.* 57, 1874-1905.

[81] Kaurinovic, B., Vastag, D. (2019). Flavonoids and phenolic acids as potential natural antioxidants. *Antioxidants* [Working Title]. IntechOpen; 2019. *Flavonoids and Phenolic Acids as Potential Natural Antioxidants.* DOI: http://dx.doi.org/10.5772/intechopen.83731 Prof. Emad Ahmed Shalaby

[82] Kaurinovic, B., Popovic, M., Vlaisavljevic, S., Trivic, S. (2011). Antioxidant capacity of *Ocimum basilicum* L. and *Origanum vulgare* L. extracts. *Molecules.* 16, 7401–7414.

[83] Bendary, E., Francis, R. R., Ali, H. M. G., Sarwat, M. I., El Hady, S. (2013). Antioxidant and structure–activity relationships (SARs) of some phenolic and anilines compounds. *Ann. Agric. Sci.* 58, 173–181.

[84] Hmid, I., Elothmani, D., Hanine, H., Oukabli, A., Mehinagic, E. (2017). Comparative study of phenolic compounds and their antioxidant attributes of eighteen pomegranate (*Punica granatum* L.) cultivars grown in Morocco. *Arab. J. Chem.* 10, 2675–2684.

[85] Rebaya, A., Igueld Belghith, S., Cherif J. k., Trabelsi-Ayadi, M. (2016). Total phenolic compounds and zntioxidant potential of Rokrose (*Cistus salviifolius)* leaves and flowers grown in Tunisia. *Int. J. Pharmacogn. Phytochem. Res.* 8, 327-331.

[86] Zielinski, A. A. F., Haminiuk, C. W. I., Alberti, A., Nogueiram A., Demiate, I. M. (2014). A comparative study of the phenolic compounds and the in vitro antioxidant activity of different Brazilian teas using multivariate statistical techniques. *Food Res. Int.* 60, 246–254.

[87] Zlatić, N., Jakovljević, D., Stanković, M. (2019). Temporal, plant part, and interpopulation variability of secondary metabolites and antioxidant activity of *Inula helenium* L. *Plants.* 8, 179.

[88] Valares Masa, C., Sosa Díaz, T., Alías Gallego, J., Chaves Lobón, N. (2016). Quantitative variation of favonoids and diterpenes in leaves and stems of Cistus ladanifer L. at different ages. *Molecules*. 21, 275.

[89] Majewska-Wierzbicka, M., Czeczot, H. (2012). Flavonoids in the prevention and treatment of cardiovascular diseases. *Pol. Merk. Lek.* 32, 50-54.

[90] Kozłowska, A., Dorota Szostak-Węgierek, D. (2014). Flavonoids--food sources and health benefits. *Rocz. Panstw. Zakl. Hig.* 65, 79-85.

[91] McCullough, M. L., Peterson, J. J., Patel, R., Jacques, P. F., Shah, R., Dwyer, J. T. (2012). Flavonoid intake and cardiovascular disease mortality in a prospective cohort of US adults. *Am. J. Clin. Nutr.* 95, 454–464.

[92] Stojiljković, D., Arsić, I., Tadić, V. (2016). Extracts of wild apple fruit (*Malus sylvestris* (L.) Mill., Rosaceae) as a source of antioxidant substances for use in production of nutraceuticals and cosmeceuticals. *Ind. Crops Prod.* 80, 165–176.

[93] Cassidy, A., O'Reilly, É. J., Kay, C., Sampson, L., Franz, M., Forman, J., Curhan, G., Rimm, E. B. (2011). Habitual intake of flavonoid subclasses and incident hypertension in adults. *Am. J. Clin. Nutr.* 93, 338–347.

[94] Macready, A. L., Kennedy, O. B., Ellis, J. A., Williams, C. M., Spencer, J. P. E., Butler, L. T. (2009). Flavonoids and cognitive function: a review of human randomized controlled trial studies and recommendations for future studies. *Genes Nutr.* 4, 227–242.

[95] Prasain, J. K., Carlson, S. H., Wyss, J. M. (2010). Flavonoids and age-related disease: Risk, benefits and critical windows. *Maturitas.* 66, 163–171.

[96] Vauzour, D., Vafeiadou, K., Rodriguez-Mateos, A., Rendeiro, C., Spencer, J. P. E. (2008). The neuroprotective potential of flavonoids: a multiplicity of effects. *Genes Nutr.* 3, 115.

[97] Youdim, K. A., Joseph, J. A. (2001). A possible emerging role of phytochemicals in improving age-related neurological dysfunctions: a multiplicity of effects. *Free Radic. Biol. Med.* 30, 583–594.

[98] Sharma, N., Dobhal, M., Joshi, Y., Chahar, M. (2011). Flavonoids: A versatile source of anticancer drugs. *Pharmacogn. Rev.* 5, 1.

[99] Olejnik, A., Tomczyk, J., Kowalska, K., Grajek, W. (2010). The role of natural dietary compounds in colorectal cancer chemoprevention. *Postępy. Hig. Med. Dosw.* 64, 175- 187.

[100] Majewski, G., Lubecka-Pietruszewska, K., Kaufman-Szymczak, A., Fabianowska-Majewska, K. (2012). Anticarcinogenic capabilities of plant polyphenols: flavonoids and stilbene. *Pol. J. Public. Health.* 122, 434-439.

[101] Zalega, J., Szostak-Węgierek, D. (2013). Nutrition in cancer prevention. Part I. Plant polyphenols, carotenoids, dietary fiber. *Probl. Hig. Epidemiol.* 94, 41-49.

Biographical Sketch

Vanja M. Tadic, PhD

Affiliation: Institute for Medicinal Plant Research, Department of Pharmaceutical Research and Development, Belgrade, Serbia

Education:

2000 – 2002.	Postdoctoral Research Fellow Department of Biochemistry and Molecular Biology, Mayo Clinic and Foundation, Rochester, MN 55905 (USA), Head of Department: Dr. Frank Rusnak
1995 – 1998.	PhD Thesis "Comparative chemical studies on the plants *Anthemis carpatica* and *Anthemis montana*," Faculty of Chemistry, University of Belgrade, Supervisors: Prof.

	Dr. Vlatka Vajs and Prof. Dr. Slobodan Milosavljevic
1990 – 1994.	MSci. Thesis "Flavonoids from Wood Anemone, *Anemone nemorosa* L., *Ranunculaceae*," Faculty of Pharmacy, Department of Pharmacognosy, University of Belgrade, Supervisor: Prof. Dr. Momcilo Gorunovic
1985 – 1990.	Bachelor of science Faculty of Pharmacy, University of Belgrade

Business Address: Institute for Medicinal Plant Research, Department of Pharmaceutical Research and Development, Tadeušsa Košćuška 1, Belgrade, Serbia

Research and Professional Experience:

Natural product chemistry, Chemical and biological evaluation of medicinal plant extracts, Development of new phytopreparations

Vanja M. Tadić works as Science Advisor at Department of Pharmaceutical Research and Development, Institute for Medicinal Plant Research "Dr Josif Pancic," Belgrade. She has been taking part as one of the principal investigator in several research projects, and was leader of the project which resulted in patents winning the Gold Medal and Diploma for the best innovation of the WIPO 2010. Her main research interests are chemistry of natural products (extraction, identification, fractionation and isolation of chemical compounds from natural matrices, especially medicinal plants) and their biological activities. She has the experience in the formulation of high value products in pharmaceuticals, nutraceuticals and food sectors.

Professional Appointments:

Since 2007.	Principal Researcher Institute of Medicinal Plant Research "Dr J. Pancic," Belgrade, R. Serbia

2002-2007.	Senior Research Fellow Institute of Medicinal Plant Research "Dr J. Pancic," Belgrade, R. Serbia
1998-2002.	Research Fellow Institute of Medicinal Plant Research "Dr J. Pancic," Belgrade, R. Serbia
1995-1998.	Research Associate Institute of Medicinal Plant Research "Dr J. Pancic," Belgrade, R. Serbia

Honors:

1. Three phytopreparatipons, "Composition and process for obtaining phytopreparations based on combination of mineral water, medicinal mud and plants for the chronic venous insufficiency";, "Composition and process for obtaining phytopreparations based on combination of mineral water, medicinal mud and plants for skin disease of various etiologies" and "Composition and process for obtaining phytopreparations based on combination of mineral water, medicinal mud and plants for rheumatic diseases," developed by the team which leader was Vanja M. Tadić won:

- Gold Medal and special award of on the World Fair XIV Moscow International Salon of Inventions and Innovation technologies "Archimedes –2011" (5. – 8. April, 2011).
- Gold Medal at the exhibition "Innovation-Belgrade 2010" (30th International Exhibition of Inventions, New Technologies and Industrial Design)
- Gold Medal and Diploma of Inventors Association of Vojvodina at the International Innovation and Knowledge Exhibition TESLA FEST 2010, Novi Sad.
- Gold Medal and Diploma for best innovation of the World Intellectual Property Organisation WIPO (World Intellectual Property Organization) 2010.

- Gold Medal at the 30th International Exhibition MAKINOVA 2010th (exhibition of inventions, technical improvements, new products and the creativity of young people) and international exhibition of technical improvements and innovations of environment EKONOVA 2010, Skopje, under the auspices of the Ministry of Education and Science of the Republic of Macedonia.
- Gold Medal and special award from the organizers at the World Fair in Seoul inventions SIIF 2010th (Seoul International Invention Fair 2010)

Publications Last 3 Years:

1. Tadic V, Zizovic I, Zivkovic J, Arsic I, Rój E (2017): "Chemical Characterization and Antioxidant Potential of Diterpenoids Rich Fractions Isolated from Selected Plant Species," *In* Brandon Jones (Ed): *Diterpenoids: Types, Functions and Research,* pp 75-117, Publisher: NOVA Science Publisher, Inc, New York, ISBN: 9781536106718 (hardcover), ISBN: 0781536106855 (eBook).
2. Tadić V, Arsić I, Stanković M, Maksimović S (2016): "Analysis of supercritical carbon(IV)oxide extracts from selected plants," *In* Roj Edward (Ed): *Supercritical Fluid Applications,* pp 19 – 44, New Chemical Syntheses Institute, Pulawy, Poland, ISBN: 978-83-935354-1-5.
3. Roganović S, Arsić I, Tadić V (2016): "Possibilities and advantages of the application of supercritical extracts as active ingredients in cosmetic products," *In* Roj Edward (Ed): *Supercritical Fluid Applications,* pp 75 - 92, New Chemical Syntheses Institute, Pulawy, Poland, ISBN: 978-83-935354-1-5.
4. Tasić-Kostov M, Arsić I, Pavlović D, Stojanović S, Najman S, Naumović S, Tadić V (2019). Towards a modern approach to traditional use: in vitro and in vivo evaluation of *Alchemilla vulgaris* L. gel wound healing potential, *Journal of Ethnopharmacology*, 238 pp. 111777-111789.
5. Jeremic I, Petricevic S, Tadic V, Petrovic D, Tosic J, Stanojevic Z, Petronijevic M, Vidicevic S, Trajkovic V, Isakovic A (2019). Effects

of *Sideritis scardica* Extract on Glucose Tolerance, Triglyceride Levels and Markers of Oxidative Stress in Ovariectomized Rats, Planta Medıca, 85, 465-472.

6. Nesic I, Stojiljkovic D, Savic S, Tasic-Kostov M, Tadic V (2019). Stability, antioxidant activity, in vivo safety and efficacy of creams with standardized wild apple fruit extract: a comparison of conventional and biodegradable emulsifiers. *Internatıonal Journal of Cosmetıc Scıence,* 41, 300-310.
7. Oliva A, Garzoli S, Sabatino M, Tadic V, Costantini S, Ragno R, Božovic M (2018). Chemical Composition and Antimicrobial Activity of Essential Oil of *Helichrysum italicum* (Roth) G. Don fil. (Asteraceae) from Montenegro, Natural Product Research, doi: 10.1080/14786419.2018.1538218.
8. Radovic M, Adamovic T, Pavlovic J, Rusmirovic J, Tadic V, Brankovic Z, Ivanovic J (2019). Supercritical CO_2 impregnation of gelatin-chitosan films with clove essential oil and characterization thereof, *Chemical Industry & Chemical Engineering Quarterly*, 25, 119-130.
9. Brašanac-Vukanović S, Tadić VM, Blagojević N, Vukašinović-Pešić V, Đurđić S, Stanković M, Mutić J (2019). Element accumulation capacity of *Vaccinium myrtillus* from Montenegro: Comparison of element contents in water and ethanol extracts of bilberry plant parts, *Archives of Biological Sciences,* 71, 145-157.
10. Stankovic M, Savic V, Zivkovic J, Stanojevic Lj, Tadic V, Arsic I (2019). Tyrosinase Inhibitory and Antioxidant Activity of Wild *Rosa Canina* L. and *Sorbus Aucuparia* L. Fruit Extracts. *Acta Polonıae Pharmaceutıca,* 76, 523-533.
11. Pajnik J, Radetić M, Stojanovic DB, Jankovic-Častvan I, Tadic V, Stanković MV, Jovanović DM, Zizovic I (2018). Functionalization of polypropylene, polyamide and cellulose acetate materials with pyrethrum extract as a natural repellent in supercritical carbon dioxide, *The Journal of Supercritical Fluids,* 136, 70–81.
12. Brašanac-Vukanović S, Mutić J, Stanković DM, Arsić I, Blagojević N, Vukašinović-Pešić V, TadićVM (2018). Wild bilberry (*Vaccinium*

myrtillus L., Ericaceae) originated from Montenegro as the source of antioxidants for use in production of nutraceuticals, *Molecules, 23*(8), 1864; https://doi.org/10.3390/molecules23081864.

13. Nikolic I, Lunter DJ, Randjelovic D, Zugic A, Tadic V, Markovic B, Cekic N, Zivkovic L, Topalovic D, Spremo-Potparevic B, Daniels R, Savic S (2018). Curcumin-loaded low-energy nanoemulsions as a prototype of multifunctional vehicles for different administration routes: Physicochemical and in vitro peculiarities important for dermal application, *International Journal of Pharmaceutics*, 550, 333–346.
14. Garzoli S, Božović M, Baldisserotto A, Andreotti E, Pepi F, Tadić V, Manfredini S, Ragno R (2018). *Sideritis romana* L. subsp. *purpurea* (Tal. ex Benth.) Heywood, a new chemotype from Montenegro, *Natural Product Research*, 32, 1056-1061.
15. Maksimović S, Tadić V, Ivanović J, Radmanović T, Milovanović S, Stanković M, Žižović I (2018). Utilization of the integrated process of supercritical extraction and impregnation for incorporation of *Helichrysum italicum* extract into corn starch xerogel, *Chemical Industry and Chemical Engineering Quarterly,* 24, 191-200.
16. Stojiljkovic D, Tadic V, Stankovic M, Roganovic S, Arsic I (2018). Standardized extract of wild apple fruit in alkyl-polyglucoside based cosmetic cream – estimation of stability, safety, antioxidant activity and efficiency, *International Journal of Cosmetic Science,* 40, 285–294.
17. Bojović D, Šoškić M, Tadić V (2018). Comparative study of chemical composition of the essential oils from *Satureja cuneifolia* Ten. and *Satureja montana* L., *Lamiaceae* collected at National park Lovćen, Montenegro, Studia Universitatis Babes-Bolyai. Ser. Chemia, Studia Ubb Chemia, 63, 167-180.
18. Žugić A, Tadić V, Kundaković T, Savić S (2018). Chemical composition and biological activities of the extracts and secondary metabolites of lichens belonging to the genus *Usnea,* Parmeliaceae, Lekovite Sirovine, 38, 68-81.
19. Stanković M, Maksimović S, Tadić V, Arsić I (2018). The oil content of wild fruits from different plant species obtained by conventional

Soxhlet extraction technique. *Acta Facultatis Medicae Naissensis*, 35, 193-200.

20. Maksimović S, Tadić V, Ivanović J, Radmanović T, Milovanović S, Stanković M, Žižović I (2018). Utilization of the integrated process of supercritical extraction and impregnation for incorporation of Helichrysum italicum extract into corn starch xerogel, *Chemical industry & chemical engineering quarterly,* 24, 191-200.
21. Tadić V, Arsić I, Zvezdanovic J, Žugić A, Cvetkovic D, Pavkov S (2017). The estimation of the traditionally used Yarrow (*Achillea millefolium* L. Asteraceae) Oil extracts with anti-inflamatory potential in topical application, *Journal of Ethnopharmacology,* 199, 138 – 148.
22. Maksimovic S, Tadic V, Skala D, Zizovic I (2017). Separation of phytochemicals from *Helichrysum italicum*: An analysis of different isolation techniques and biological activity of prepared extracts, *Phytochemistry,* 138, 9-28.
23. Tadić V, Oliva A, Božović M, Cipolla A, De Angelis M, Vullo V, Garzoli S, Ragno R (2017). Chemical and Antimicrobial Analyses of *Sideritis romana* L. subsp. *Purpurea* (Tal. ex Benth.) Heywood, an Endemic of the Western Balkan, *Molecules,* 22, 1395. doi:10.3390/molecules22091395.
24. Stanković M, Živković J, Tadić V, Arsić I. (2017). Skin protection against solar UV radiation by natural plant products: extracts from elder fruit (*Sambucus nigra* L.). The Fifth International Conference on Radiation and Applications in Various Fields of Research, June 12-16, 2017. Budva, RAD Conference Proceedings, vol. 2, pp. 231-236. ISSN: 2466-4626 (Online) DOI: 10.21175/RadProc.2017.48.
25. Mišan A, Šarić B, Milovanović I, Jovanov P, Sedej I, Tadić V, Mandić A, Sakač M (2017). Phenolic profile and antioxidant properties of dried buckwheat leaf and flower extracts, Chemical Industry and Chemical Engineering Quarterly, Savez hemijskih inzenjera, 23, 39-47.
26. Miletić S, Švarc Gajić J, Milovanović I, Đurendić-Brenesel M, Mandić A, Tadić V, Arsić I, Mišan A (2016). Anti-atherogenic properties of a functional herbal mixture, *Acta periodica technologica*, 47, 205 – 218.

27. Bogdanovic A, Tadic V, Stamenic M, Petrovic S, Skala D (2016). Supercritical carbon dioxide extraction of *Trigonella foenum-graecum* L. seeds: Process optimization using response surface methodology, *Journal of Supercritical Fluids,* 107, 44–50.
28. Bogdanovic A, Tadic V, Arsic I, Milovanovic S, Petrovic S, Skala D (2016). Supercritical and High Pressure Subcritical Fluid Extraction from Lemon Balm (*Melissa officinalis* L., Lamiaceae), *Journal of Supercritical Fluids,* 107, 234-242.
29. Mišan A, Sakač M, Medić Đ, Tadić V, Marković G, Gyura J, Pagano E, Izzo AA, Borrelli F, Šarić B, Milovanović I, Milić N (2016). Antioxidant and Physicochemical Properties of Hydrogen Peroxide-Treated Sugar Beet Dietary Fibre, Phytotherapy research, 30, 855-860.
30. Šoškić M, Bojović D, Tadić V (2016). Comparative chemical analysis of essential oils from lavender of different geographic origins, Studia Ubb Chemia, 61, 127 – 136.
31. Stojiljkovic D, Arsic I, Tadic V (2016). Extracts of Wild apple fruit (*Malus sylvestris* (L.) Mill., Rosaceae) as a source of antioxidant substances for use in production of nutraceuticals and cosmeceuticals. *Industrial Crops & Products,* 80, 165 – 176.

In: The Lamiaceae Family: An Overview
Editor: Alexander Adler
ISBN: 978-1-53617-078-8

Chapter 3

LIPOSOME ENCAPSULATION OF *ORIGANUM COMPACTUM* ESSENTIAL OIL AS THE ACTIVE INGREDIENT: ASSURING THE DRUG EFFECT

Merima Ibišević*[1]*, Vanja M. Tadić*[2,*]*, Snežana Savić*[3]*,
***Jelena Đoković*[3] *and Ivana Nešić*[4]**

[1]Department of Pharmaceutical Technology, Faculty of Pharmacy, University of Tuzla, Tuzla, Bosnia and Herzegovina
[2]Institute for Medicinal Plant Research "Dr. Josif Pančić," Belgrade, Serbia
[3]Department of Pharmaceutical Technology and Cosmetology, Faculty of Pharmacy, University of Belgrade, Belgrade, Serbia
[4]University of Niš, Faculty of Medicine, Department of Pharmaceutical Technology, Niš, Serbia

[*] Corresponding Author's Email: vtadic@mocbilja.rs.

ABSTRACT

Origanum compactum Benth, Lamiaceae, is an endemic Moroccan medicinal herb. Although known as culinary spice in the country of origin Morocco, *O. compactum* has been traditionally used in the treatment of many diseases such as colitis, bronchopulmonary, gastric acidity and gastrointestinal diseases. Numerous studies showed that the *O. compactum* extracts, essential oil and its main components, thymol and carvacrol possess broad spectrum of pharmacological and therapeutic activities such as antibacterial, antifungal, antioxidant and anticancer.

Essential oils generally in pure form are not suitable for administration, because of their instability and susceptibility to the presence of extrinsic factors, like oxygen, light and temperature. In order to protect essential oils from degradation, to improve their stability and bioavailability, different methods of their encapsulation in various colloidal systems such as microcapsules, microspheres, nanoemulsions and liposomes were introduced. In addition, literature data revealed that the incorporation of synthetic as well as herbal drugs into phospholipids carriers such as liposomes tended to improve their efficacy, with the possibility to modulate the pharmacokinetics and/or tissue distribution of the drug in a beneficial way. In order to provide a starting active ingredient that could be incorporated in various pharmaceutical forms, the aim of our investigation was to prepare liposomal dispersion with encapsulated *O. compactum* essential oil with defined chemical and antibacterial properties. Liposomal dispersion was prepared by dissolving Phosal IP 40 (Lipoid, Germany), defined amounts of *O. compactum* essential oil and *Aqua ad injectabilia* applying homogenization. Further, in order to optimize homogenization, the obtained dispersion was homogenized in a high pressure homogenizer. Chemical composition was determined applying GC/MS and GC/FID methods, while the antibacterial activity was evaluated by agar diffusion and microdilution methods on standardized bacterial strains from ATCC collection *Staphylococcus aureus* ATCC 25923, *Enterococcus faecalis* ATCC 51299, *Escherichia coli* ATCC 25922, as well as on clinical strains collected from patients: *Staphylococcus sp., E. faecalis, E. coli and Streptococcus agalactiae*.

Results indicated that *O. compactum* essential oil was rich in thymol and carvacrol, as main components. It possessed wide spectrum of activities against all bacterial strains tested, although in different degrees. The inhibition zones varied from 20.3 ± 0.57 mm to 47.1 ± 1.15 mm, and the values for MIC ranged from 0.195 mg/ml to 0.781 mg/ml for tested bacteria. For determining the degree of liposome encapsulation, GC/MS and GC/FID were applied and it was confirmed that more than 75% of thymol and more than 65% of carvacrol were encapsulated.

The obtained results of microbiological analysis indicated that *O. compactum* essential oil had potent antibacterial properties, what might place it as a potential replacement for conventional antimicrobial products. Undesirable effects can be diminished by effective incorporation into phospholipid carriers such as liposomes.

Keywords: *Origanum compactum*, essential oil, antibacterial activity, liposomes

1. INTRODUCTION

Medicinal plants, as a source of biologically active substances, have been widely used in the production of medicinal products (herbal remedies, traditional herbal remedies), dietary supplements, nutraceuticals and cosmetic products [1, 2]. Applying the modern methods of investigation, the traditional application of a large number of plants has been confirmed and justified [3-6]. Plant extracts, essential and fatty oils, but isolated individual, physiologically active components, as well, have been used most commonly in different products as ingredients with multifunctional and significant biological activities. The extraction methods applied for obtaining extracts might be various, where through the process of optimization the set aim is to maximize the utilization of the plant raw material in respect to its active constituents [7-10], while applying non-conventional, new, green sustainable methods (supercritical carbon-dioxide extraction) additional advantages are the resulting extracts with improved stability, at the same time not containing a range of hazardous, relatively expensive and environmentally damaging organic substances/solvents usually present in conventional extracts [11, 12].

In the Mediterranean region the genus *Origanum* (Lamiaceae) is represented with 39 different species [13]. In Morocco, *Lamiaceae* family is represented by 30 genera and 225 species (more than 90 being endemics), including many species with medicinal and aromatic properties [14]. The genus *Origanum* is represented by five species of which three are endemic, including *Origanum compactum* Benth. It is the sacred plant of

Morocco, known, together with several more species *O. floribundum* Munby, *O. vulgare* subsp. *hirtum* (Link) Ietsw. and *Satureja peltieri* Maire by the name "Zâhtar," although this name is used for all plants with thymol and carvacrol odour, as well. *O. compactum* is a perennial species, has generally pubescent, hairy stems, covered with long hair (3 mm). In the country of origin, Morocco, the area of distribution comprises the region of Quezzane, in the massif of Khatouat, Mamora. This plant occurs in Iberian Peninsula and North Africa, as well [15, 16]. It spontaneously grows in forests, calcareous substrates, bushes and rocky pastures of the plains and low mountains. The flowering time is from May-July (Figure 1). The presence of secretary organ (glandular and non-glandular trichomes) is the principal morphological characteristic of this species. These structures secrete essential oils with a unique flavour, and for that characteristic odour are responsible thymol and carvacrol (Figure 2.) Besides, carvacrol, a phenolic monoterpenoid, has been known to possess a wide range of bioactivities such as antimicrobial, antioxidant, and anticancer. The significant antimicrobial potential of carvacrol might be assigned to presence of phenol moiety and its hydrophobic properties [17]. Thymol, monoterpene phenol, exhibits antioxidant, anti-inflammatory, analgesic, antispasmodic, antibacterial, antifungal, antiseptic and antitumor activities [18].

Figure 1. *O. compactum* Benth.: a) Vegetative stage, b) Flowering stage.

O. compactum has been known as culinary spice in Morocco, the country of origin, and has been traditionally used in the treatment of many diseases such as colitis, bronchopulmonary, gastric acidity and impairments in gastrointestinal function [13, 19]. Numerous studies showed that the *O. compactum* extract and essential oil rich in thymol and carvacrol possessed broad spectrum of pharmacological and therapeutic activities such as antibacterial, antifungal, antioxidant and anticancer [20, 21].

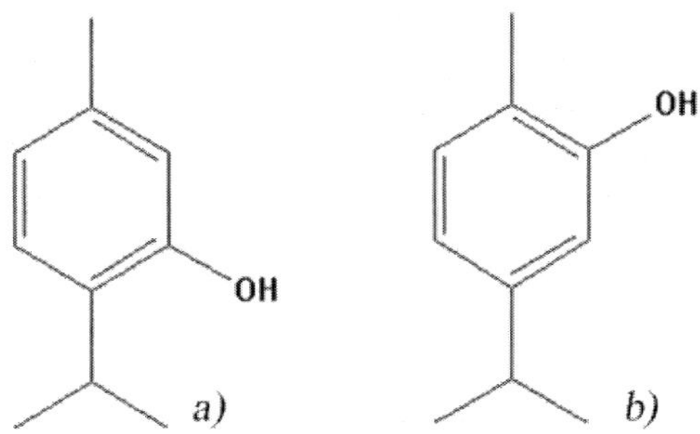

Figure 2. Structures of thymol (a) and carvacrol (b).

Many studies have demonstrated strong antibacterial potential of essential oil and plant extracts against pathogenic bacteria. Because of complex variability in quantity and chemical profiles of the secondary metabolites usually present in plants, several mechanisms have been suggested to provide the sufficient explanation for their antimicrobial activity. As these components have several sites of action at the cellular level, the possible mechanisms of antimicrobial action might include disintegration of cytoplasmic membrane, interaction with membrane proteins (ATPases and others), disturbance of the outer membrane of Gram-negative bacteria with the release of lipopolysaccharides, proton motive force depletion, with leakage of ions, coagulation of the cell content, and inhibition of enzymes synthesis [13].

Although known for their potent antibacterial properties, essential oils have been known for their instability and susceptibility to the presence of extrinsic factors, like oxygen, light and temperature. They can also be unstable, poorly soluble in water and by that poorly delivered to target cells. But, the specifically designated therapeutic systems that might allow the continuous release of the active substance or its specific action at a target place in the body are of great medical importance [22]. Their administration would improve the pharmacokinetic properties of the drug, change its therapeutic index and ensure a constant, desired concentration in the blood or at the site of action over a long period of time. This might significantly reduce the side effects, toxicity and frequency of dosing, making the drug more appropriate for the patient [23]. These effects can be achieved by incorporating the drug substance into suitable carriers such as microcapsules, microspheres, nanospheres, micelles, polymeric prodrugs, tiresome, nanofibrils and in various lipid vesicles, among which liposomes occupy a prominent place [24]. Regarding the essential oils, the incorporation into liposomes might reduce their irritant effect, while at the same time it would lead to better efficacy and prolonged action of the preparation itself. Liposomes are considered as a highly desirable as drug carriers because they can incorporate hydrophilic, hydrophobic and amphipathic drug substances (drugs), and they are also physiologically acceptable due to their similarity with biological membranes and biodegradability. Liposome encapsulation of actives is used for local therapy because of their improved stability and skin bioavailability, as well as the potential of decreasing adverse effects of drugs. In addition, 'empty liposomes' have been shown to improve skin barrier function, due to the fusion of their constituents (e.g., phosphatidylholin) with lipids or proteins of the *stratum corneum*, this being of significance as a preventative in maintaining the normal skin barrier function or in restoring the barrier function, if impaired in some skin diseases [25]. According to the requirements related to a specific field of application, the liposomes might protect a fragile or unstable compound from its surrounding environment, protect the user from the side-effects of the encapsulated compound, trap a compound (aromas, organic solvents, pesticides, essential oils ...), modify the density of a liquid, change a liquid into a solid, enable the coexisting in the same medium two incompatible compounds, control the release of the encapsulated compound. Antimicrobials were among the first substances that were encapsulated in liposomes. Namely, econazole is a broad-spectrum antifungal agent and the first drug to be commercially available as a liposome preparation. Econazole bioavailability studies indicated the significant improvement when econazole liposome therapy was applied compared to conventional cream, gel and lotion. Namely, the liposomal preparation reduced the dose of the drug, shortened the duration of therapy and reduced skin irritation [26].

O. compactum essential oil is one of the strongest antibacterial and antifungal oils. It is known for beneficial effects exhibiting on respiratory, fungal, urogenital and other infections. In recent studies, *O. compactum* essential oil tested against Gram-positive bacteria (*Staphylococcus aureus, Listeria monocytogenes serovar4b, L. innocua* and *Enterococcus faecium*) and Gram-negative bacteria (*Escherichia coli K12, E. coli serovar O157:H7, Proteus mirabilis, Pseudomonas aeruginosa, P. fluorescens* and *Bacillus subtilis*), exhibited strong activity against the mentioned strains [27]. The possible explanation might be that carvacrol and thymol take part in disintegration of the outer membrane of Gram-negative bacteria, releasing lipopolysaccharides (LPS) and inducing the increase in the permeability of the cytoplasmic membrane toward ATP [13].

In addition, the antifungal activity of this essential oil was tested in several studies confirming its significant antifungal effect, pointing out that the exhibited activity might be attributed to the phenolic compounds. Phenolic compounds can disturb the integrity of cell membranes, causing leakage of cellular materials and ultimately the microorganism death [28]. Some studies also have demonstrated the antitumor activity of *O. compactum* against human cancer cell lines [29]. It was discovered that thymol can be associated with cell death of many types of cell line of melanoma [30], osteosarcoma [31], glioblastoma [32], breast cancer [33], leukaemia [18], hepatoma, lung cancer and colon cancer [34].

In spite the shown significant and numerous biological activities, *O. compactum* essential oils has been known as extremely irritating to the skin and mucous membranes. It is allowed for oral use only in special oleocapsules. Therefore, it is also very important to adhere to the dosage recommendations. The recommended dose ranges from 75 mg to a maximum of 450 mg per day. Maximum daily doses of 450 mg are not recommended for long periods of time and should always be consulted with a pharmacist or doctor. This essential oil has a strong taste that can stimulate stronger secretion of saliva, so it should always dilute before use. It is not recommended for children, pregnant and lactating women. It should not come in contact with eyes or genitals, and special caution should be taken before use on skin. External use is possible with dilution in vegetable oil (sesame, hazelnut).

The aim of our investigation was to prepare liposomal dispersion with encapsulated *O. compactum* essential oil with defined chemical and antibacterial properties.

2. About Our Research

2.1. Materials and Methods

O. compactum essential oil was purchased at Pranarôm International (Ghislenghien, Belgique) (Figure 3.). It was produced by hydrodistillation at a low pressure and stored at 4°C until the experiments started. Gas chromatography was used for determining its chemical composition.

2.1.1. Gas Chromatography (GC-FID)

Gas chromatography analysis of the essential oils were carried out on an HP-5890 Series II GC apparatus [Hewlett-Packard, Waldbronn (Germany)], equipped with the split–splitless, injector and automatic liquid sampler (ALS), attached to HP-5 column (25 m × 0.32 mm i.d. and 0.52 μm film thickness) and fitted with a flame ionization detector (FID). Carrier gas was H2 (1 ml/min), with a split ratio of 1:30, injector temperature was 250°C, detector temperature 300°C, while column temperature was linearly programmed from 40 to 260°C with a rate of change of the 4°C/min, and then kept isothermally at 260°C for 10 min. Solution of essential oil in alcohol (10 mg/ml) were consecutively injected in an amount of 1μl. Area percent reports, obtained as result of standard processing of chromatograms, were used as a base for the quantification analysis. Namely, the relative proportion of the essential oil constituents were expressed as percentages obtained by peak area normalization, all relative response factors being taking as one.

2.1.2. Gas Chromatography/Mass Spectrometry (GC-MS)

The same analytical conditions as those mentioned for GC-FID were employed for GC/MS analysis, along with column HP-5MS (30 m × 0.25 mm i.d. and 0.25 μm film thickness), using HP G 1800C Series II GCD system [Hewlett-Packard, Palo Alto, CA (USA)]. As the carrier gas used helium. The transfer line was heated to 260°C. The mass spectra were obtained in EI mode, with an ionisation voltage of 70 electron volt (eV); in the range from 40 to 450 m/z. The amount of the injected sample, dissolved in alcohol (10 mg/ml) was 0.2 μl. The components of the essential oil were identified by comparison of their mass spectra to those from Wiley 275 and NIST/NBS (NIST–National Institute of Standards and Technology/NBS-National Bureau of Standards) libraries, using different search engines. Identification of the compounds was achieved by comparing their Kovats' retention indices and mass spectra with those reported in the literature [35] and supplemented by the Automated Mass Spectral Deconvolution and Identification System software (AMDIS ver. 2.1), GC-MS library [36]. The experimental values for Kovats' retention indices were determined by the use of calibrated Automated Mass Spectral Deconvolution and Identification System Software (AMDIS ver. 2.1), GCMS library [36], and also compared to those from available literature (Adams 2007) [35] and used as additional tool to approve MS findings. The relative proportion of the essential oil constituents were expressed as percentages obtained by peak area normalization, all the relative response factors were entered as one.

2.1.3. Antibacterial Activity of O. compactum Essential Oil

Antibacterial activity of *O. compactum* essential oil was tested on standard and clinical bacterial strains. Standard bacterial strains were from ATCC collection: *Staphylococcus aureus* ATCC 25923, *Enterococcus faecalis* ATCC 51299, *Escherichia coli* ATCC 25922, as well as on clinical strains *Staphylococcus sp.*, *E. faecalis*, *E. coli* and *Streptococcus agalactiae* collected from vaginal and cervical swabs of patients located at the University Clinical Centre Tuzla (BIH), according to EUCAST Disk Diffusion Method for Antimicrobial Susceptibility Testing Version 7.0 (January 2019) [37].

After isolation, strains were conserved at -20 ^{0}C in BHI (Brain Hart Infusion broth, HiMedia, India) and grown on Luria-Bertani medium and Plate Count Agar (PCA, Biokar) for 24h prior to antimicrobial testing. Mueller Hinton Agar (HiMedia, India) and Mueller Hinton Broth (Liofilchem, Italy) were used for antibacterial testing. Antibacterial activity was first tested against three standard strains (*S. aureus* ATCC 25923, *E. coli* ATCC 25922, *E. faecalis* ATCC 51299) and then on clinical strains, using agar diffusion method according to CLSI guidelines [38] with some modifications.

To obtain individual colonies, bacteria were cultured on Luria-Bertani medium for 24h and on Plate Count Agar (PCA, Biokar) overnight at 37°C. Then the colonies were suspended in 0.9% sterile saline to achieve turbidity equal to 0.5 McFarland standard (1.5 x 10^8 CFU/ml). After that, Mueller Hinton agar plates (HiMedia, India) were inoculated with bacterial suspension. Essential oil was added on each plate (50μL). Incubation was at 37°C for 24 h and after incubation, the size of the inhibition zones was measured.

The minimum inhibitory concentration (MIC) for *O. compactum* essential oil was tested using 96 well broth microdilution method in Muller Hinton Broth (MHB) as per the guidelines given by Clinical and Laboratory Standards Institute (SAD) [39]. The suspensions of overnight fresh bacterial cultures were adjusted at 0.5 McFarland turbidity. *O. compactum* essential oil was dissolved in dimethyl sulfoxide (DMSO) to render the proper dissolution with MHB [40]. After that, a series of double dilutions were made and introduced in a microtiter plates with 96 wells, 10 μl at 90 μl inoculated MHB. The final volume in each well was 100 μl, final density of bacterial cells 10^6 CFU/ml, and concentrations of the examined oil were in the range of 0.098 to 12.5 mg/ml. Incubation of microtiter plates was 24 hours at 37° C. For detection of bacterial growth was added 20 μl of 0.5% aqueous triphenyltetrazolium chloride solution (TTC).

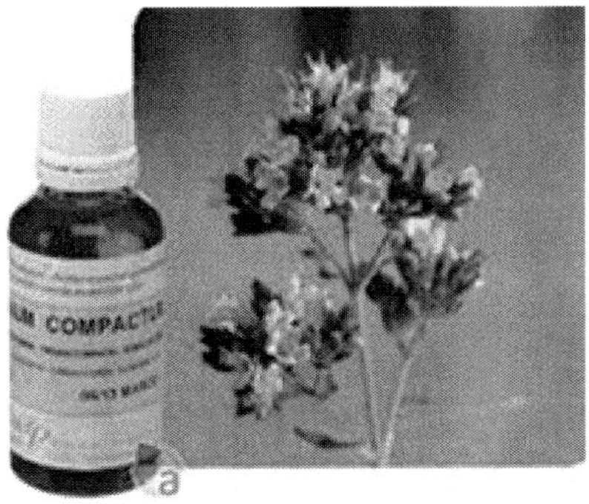

Figure 3. *O. compactum* essential oil used in experimental section.

MIC was defined as the lowest concentration of investigated essential oil in which there is no visible growth bacteria, red coloured colonies at the bottom of the recess microtiter plates after adding TTC.

2.1.4. Preparation of Liposome with O. compactum Essential Oil

To obtain the liposomal dispersion with encapsulated *O. compactum* essential oil with defined chemical composition, Phosal IP 40 (Lipoid, Germany, being the combination of Phospholipids of natural origin (Soya lecithin): Phosphatidylcholine + Lysophosphatidylcholine; Non-polar lipids; Ethanol; Tocopherols) was used.

Two liposomal dispersion were prepared, with different essential oil concentrations, namely with 1.0 and 9.3% (Table 1).

Table 1. Composition of liposomal dispersion I and II

	Formulation of liposomal dispersion I	**Formulation of liposomal dispersion II**
Phosal IP 40	10.0 g	12.0 g
Essential oil	1.0 g	9.3 g
Aqua ad injectabilia	89.0 g	79.7 g

Liposomal dispersion was prepared applying homogenisation in a high pressure homogenizer (Emulsiflex-C3, Avestin, Canada) at 500 bar in 5 cycles, which followed the process of dissolving precise amount of Phosal IP 40 and *O. compactum* essential oil in *Aqua ad injectabilia*, a phase of mixing and then process of homogenization (at atmospheric pressure), performed by Ultra Turrax Ika T25-Digital, 15 minutes at 5000 rpm.

2.1.5. Characterization of Liposome Size, Polydispersity Index and Zeta Potential

The structure of liposomal dispersions was evaluated with optical microscope (BA310 Motic optical microscope, Speed Fair Co. Ltd., Hong Kong, China) at a magnification of 1000 x. Microphotograph was taken with Optika Pro 3LT camera processed with Optika Vision Pro Software (Optika Microscopes, Ponteranica, Italy).

Analysis of liposome size and polydispersity index (PdI) was carried out by using Zetasizer (Nano series) ZS 90, Malvern Instruments using dynamic light scattering technique (Dynamic Light Scattering, DLS). Samples were analysed 24 h after preparation and diluted with *Aqua purficata* in a ratio of 1: 1000 (to reduce the influence of the density and viscosity of the dispersions on the measurement results, the samples were diluted prior to the measurement). The measurement was performed in disposable polystyrene cuvettes at a temperature of 25°C at a light scattering angle of 90°. Each sample was measured three times.

Zeta potential measurement was determined using Zetasizer (Nano series) ZS 90, Malvern Instruments. Sample was diluted with physiological solution (conductivity at 50 μS/cm). The dilution was 1: 1000.

2.1.6. Essential Oil Encapsulation Efficiency (%EE)

For determining the degree of liposome encapsulation, the precisely weight of dispersion (5 x 2.0000 g) was centrifuged in ultracentrifuge (Allegra 64R Centrifuge, Beckman Coulter, USA) at 30,000 rpm/min for 30 minutes. The supernatant was then separated and re-extracted with an organic solvent which was evaporated. Sediment was dissolved in an organic solvent (giving the final concentration of 5.2 mg/ml) and analysed for the content of thymol and carvacrol with GC/MS and GC/FID.

2.2. Results and Discussion

2.2.1. Chemical Composition of O. compactum

The main components of *O. compactum* essential oil, as reported in up to now published literature, are phenolic compounds, carvacrol and thymol, which define the biological and pharmacological properties of this oil [41]. Main compounds of *O. compactum* essential oil belong to three categories of compounds: terpenes (monoterpene hydrocarbons and sesquiterpene hydrocarbons), terpenoides (oxygenated monoterpenes and oxygenated sesquiterpenes) and phenylpropanoides. The yield of essential oil of plants is generally dependant on plant species, geographical distribution, storage organ, collection period and method of extraction [13, 42]. The chemical composition of essential oil depends on the geo-climatic location and growing conditions, like concentration of nutrients, temperature, soil type, day length, climate, altitude, amount of available water, season or vegetative period of plant [43]. The intraspecies variability of chemical composition defines different chemotypes of the specific species within genera, and therefore their essential oils might exhibit therapeutic activities that differentiate from each other.

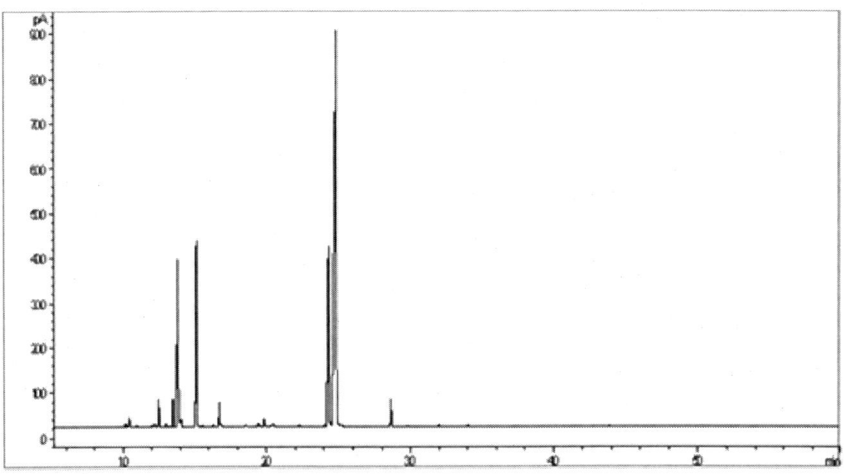

Figure 4. Gas chromatogram of the investigated *O. compactum* essential oil.

Table 2. Quantitative and qualitative composition of *O. compactum* essential oil

	COMPOUND	KI	PA (%)
1	α –thujene	924	0.23
2	α – pinene	932	0.59
3	camphene	946	0.09
4	1-octen-3-ol	974	0.09
5	β-pinene	974	0.16
6	3- octanone	979	0.08
7	myrcene	988	1.41
8	3-octanol	988	t
9	α – phellandrene	1002	0.18
10	Δ-3-carene	1008	0.07
11	α-terpinene	1014	1.54
12	*p*- cymene	1021	10.84
13	limonene	1024	0.30
14	β – phellandrene	1025	0.14
15	1,8-cineole	1026	0.40
16	(*E*)-β-ocimene	1044	0.05
17	γ-terpinene	1054	12.89
18	α-terpinolene	1079	0.11
19	*p*-cymenene	1089	0.12
20	linalol	1095	1.42
21	*trans*-sabinene hydrate	1098	0.09
22	*cis*-thujone	1101	t
23	*cis-p*-menth-2-en-1-ol	1118	t
24	camphor	1141	0.14
25	borneol	1165	0.26
26	terpinene-4-ol	1174	0.59
27	*p*-cymene -8-ol	1179	0.20
28	α-terpineol	1186	t
29	*trans*-chrysanthenyl acetate	1235	t
30	carvacrol, methyl ether	1241	0.11
31	*iso*-isopulegyl acetate	1281	0.09
32	thymol	1289	14.62
33	carvacrol	1298	50.87
34	piperitenone oxide	1352	0.07
35	carvacrol acetate	1370	t
36	α-copaene	1374	t
37	α-gurjunene	1409	t
38	β-caryophyllene	1417	1.71
39	aromadendrene	1432	t
40	α – humulene	1452	0.08
41	allo-aromadendrene	1458	t
42	γ-muurolene	1478	0.06

	COMPOUND	KI	PA (%)
43	β-bisabolene	1505	0.09
44	γ-cadinene	1513	0.05
45	δ-cadinene	1522	0.12
46	caryophyllene oxide	1582	0.12
47	(*2E,6Z*)-farnesol	1714	t
	Total		99.98

KI = Kovats' indices; PA = peak area percentage; t = Chemical compounds with percentage less of 0.05%;).

The compounds present in the investigated essential oil in significant quantities were carvacrol, thymol, γ-terpinene and *p*-cymene (50.87, 14.62, 12.89 and 10.84%, respectively, Table 2, Figure 4). Our investigation revealed the results slightly different from those obtained by Bouhdid et al. [27]. Namely, while the amount of thymol was almost four times less in comparison to carvacrol in investigated essential oil, Bouhdid at al. determined similar contents of both phenolic compounds in their sample (being 27.5 and 30.59%, respectively).

2.2.2. Antibacterial Activity of O. compactum Essential Oil

Among the isolated strains, *E. faecalis* was the most common bacteria and showed the widest range of zones. The least common isolated was *S. aureus*. *O. compactum* essential oil had a strong antibacterial effect on all isolated strains (Figure 5).

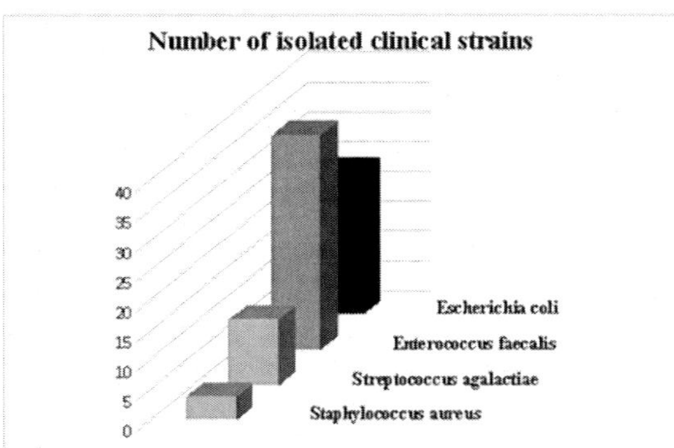

Figure 5. The number of isolated clinical bacterial strains from vaginal and cervical swabs.

Table 3. Antibacterial activity of *O. compactum* EO tested on standard and clinical strains

Bacterial strain	Inhibition zones of *O. compactum* EO (mm)	Ampicillin inhibition zones (mm)	MIC of *O. compactum* EO (mg/ml)
Investigated standard bacterial strains			
E. faecalis (ATCC 51299)	31.00 ± 0.57	18	0.195
E. coli (ATCC 25922)	35.00 ± 0.50	20	0.390
S. aureus (ATCC 25923)	35.00 ± 1.15	19	0.195
Investigated clinical bacterial strains found in cervical and vaginal swabs (isolate number)			
Streptococcus agalactiae (17953)	28.30	18	0.781
S. agalactiae (19600)	21.60	20	0.781
S. agalactiae (39491)	25.00	17	0.781
S. agalactiae (37110)	24.10	19	0.781
S. agalactiae (38135)	26.00	16	1.562
S. agalactiae (17935)	21.00	18	1.562
S. agalactiae (19325)	27.30	21	0.781
S. agalactiae (17769)	27.00	20	0.781
S. agalactiae (19608)	28.00	18	0.390
S. agalactiae (19966)	26.00	20	0.781
S. agalactiae (20299)	21.60	15	1.562
Staphylococcus aureus (17798)	24.00	16	0.781
S. aureus (39169)	22.00	11	0.781
S. aureus (18295	23.30	20	0.781
S. aureus(19965)	25.00	23	0.781
Enterococcus faecalis (19328)	47.10	20	0.195
E. faecalis (17398)	40.00	27	0.195
E. faecalis (17404)	30.00	21	0.390
E. faecalis (17923)	37.30	25	0.390
E. faecalis (17951)	38.00	26	0.390
E. faecalis (18077)	31.60	19	0.781
E. faecalis (39155)	26.30	16	0.781
E. faecalis (40913)	21.30	18	0.781
E. faecalis (40911)	20.30	19	0.781
Escherichia coli (4322)	40.30	25	0.195
E. coli (4737)	35.10	15	0.195
E. coli (5607)	30.30	18	0.195
E. coli (5563)	37.10	17	0.195
E. coli (5421)	27.00	15	0.390
E. coli (6357)	36.00	24	0.195
E. coli (2087)	28.60	15	0.390
E. coli (2101)	28.00	16	0.390
E. coli (40918)	32.00	17	0.195

Bacterial strain	**Inhibition zones of *O. compactum* EO (mm)**	**Ampicillin inhibition zones (mm)**	**MIC of *O. compactum* EO (mg/ml)**
E. coli (40905)	38.00	19	0.195
E. coli (20952)	28.30	20	0.781
E. coli (17013)	23.00	19	0.781
E. coli (21686)	35.00	19	0.195
E. coli (21299)	25.30	15	0.390
E. coli (20734)	30.00	16	0.195
E. coli (40925)	29.00	17	0.390
E. coli (21078)	23.60	19	0.781
E. coli (19329)	34.00	21	0.390
E. coli (17948)	35.00	20	0.195
E. coli (17723)	25.10	18	0.781
E. coli (41652)	26.00	17	0.781
E. coli (3792)	28.00	21	0.781
E. coli (39316)	28.30	20	0.781

Based on the results obtained applying the agar diffusion and broth dilution methods, *O. compactum* essential oil showed a strong antibacterial activity (Table 3, Figures 6). For standard bacteria, the inhibition zones (IZ) ranged from 31.0 ± 0.57 mm to 35.0 ± 1.15 mm and the values for MIC were from 0.195 mg/mL to 0.390 mg/ml. For bacteria isolated from clinical strains, the inhibition zones (IZ) varied from 20.3 ± 0.57 mm to 47.1 ± 1.15 mm, and the values for MIC ranged from 0.195 mg/mL to 0.781 mg/ml.

We chose the mentioned bacterial strains as the most common causative agents in vaginal infections, especially in women in reproductive age. Although non-death threats, the infections are connected with high level of anxiety and decrease in a life quality. Recent investigations revealed that more than 70% of women in the risk group had the vaginal infections caused, at least, by one of the investigated strains [44]. Interestingly, in most cases satisfactory results in the treatment were achieved administrating the medicinal preparations locally; in fact the topical application of the drugs was as effective as oral ones. Topical administration exhibits several advantages over oral, not causing numerous adverse and undesirable effects, interactions with other medicines, and the problem of resistance almost is not existent.

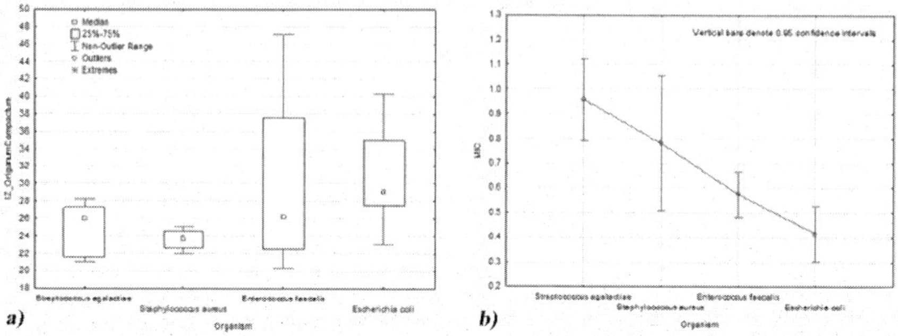

Figure 6. Inhibition zones (IZ) of *O. compactum* essential oil (a) and the minimum inhibitory concentration (MIC) (b), tested on investigated vaginal and cervical clinical strains.

Antibacterial activity of *O. compactum* essential oil was assumed to depend on two main components (carvacrol and thymol). Namely, the presence of phenol groups and a system of delocalized electrons in their chemical structures might take a main role in exhibited antibacterial effects [45]. Also an important characteristic of the investigated essential oil and its constituents is hydrophobicity which could increase the permeability of the bacterial cell membrane and allow easier passage of components through its lipid layer.

Results obtained by disc diffusion method mostly depended on the solubility of components and their diffusion through agar. It is interesting that bacteria with a higher inhibition zone did not have always a lower MIC value.

2.2.3. Characterization of Liposome Size, Polydispersity Index and Zeta Potential in Prepared Liposomal Dispersion

Liposomal dispersion of prepared *O. compactum* essential oil was of milky appearance caused by multiple light scattering, with bluish reflection and a characteristic *O. compactum* essential oil scent. The presence of liposomes was demonstrated by observation under a microscope at a magnification of 100 x (Figure 7).

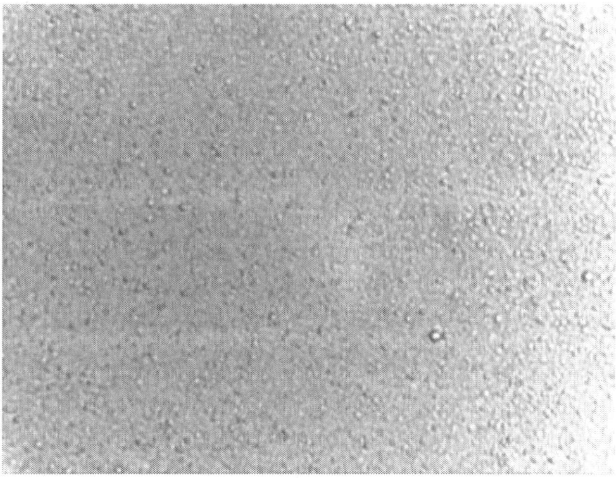

Figure 7. Liposomal dispersion under microscope at a magnification of 100 x (small round features represent actually formed liposomes of different sizes).

Phospholipids and other soybean fractions, constituting the liposomes, have been used in the manufacture of medicinal preparations (parenteral, dermal and cosmetic) and are considered as safe, confirmed by existence of extensive documentation for pharmacological and toxicological properties and implies their production under conditions of good manufacturing practice (GMP). Also, Soya lecithin is a rich source of essential fatty acids, being constituents of liposomes, as well, and due to the high content of unsaturated fatty acids at room temperature, they are in a state of increased fluidity, what might facilitates their application. The chemical composition of liposomes might greatly contribute to the improved stability of encapsulated components. Increase in the phosphatidylcholine content decreases the hydrolysis constant of some encapsulated components and, such as, takes part in increasing their stability [46].

Lately, the production of pharmaceutical products with liposomes has been very attractive. Liposomes can be formulated in liquid (suspension), solid (dry powder) or semi-solid (gel, cream) forms. *In vivo*, they can be used topically or via parenteral route [47]. In addition, liposomes have the ability to increase the antimicrobial activities of encapsulated essential oils. Liposomal incorporation of carvacrol and thymol isolated from the essential oil of *O. dictamnus* and *in vitro* tested antimicrobial activity revealed that antimicrobial activity was significantly increased after the encapsulation in liposomes [48].

Taking into account the abovementioned facts, the liposomal dispersions of *O. compactum* essential oil could be used in gynaecology as the active ingredient of numbers of gynaecological preparations intended for treatment of bacterial and fungal infections. In addition, those dispersions could be used in dermal preparations like ointments for wound healing, having in mind strong antibacterial, antifungal and anti-inflammatory properties of the investigated essential oil. *O. compactum* essential oil could be used in dentistry as well, for the treatment of periodontitis [20].

Results of the characterization studies (size, polydisperzity index (PDI), zeta potential) of liposomal dispersion were shown in Tables 4 and 5. The results were obtained after 5 cycles of high pressure homogenization.

Table 4. Analysis of size and polydispersity index

Liposomal dispersion	Size (nm)	PdI	PdI width
Phosal IP 40 (1:1000) with 1.0% essential oil	260.7 ± 0.568	0.212 ± 0.019	149.80 ± 1.84
Phosal IP 40 (1:1000) with 9.3% essential oil	298.4 ± 0.850	0.231 ± 0.006	143.10 ± 2.25

Table 5. Analysis of zeta potential

Liposomal dispersion	Zeta potential (mV); 25°C
Phosal IP 40 with 1.0% essential oil	67.60 ± 3.99
Phosal IP 40 with 9.3% essential oil	55.60 ± 1.31

Average size and polydispersity index are important parameters, especially when the liposomes are intended for therapeutic use by inhalation or parenteral route [47]. The applied techniques of liposome size and polydispersity index analysis enabled the measurement of the degree of fluctuation in the intensity of scattered light due to the brown motion of particles/drops in a sample and translate it into the size of the drops by the

Stokes – Einstein equation [49]. The width of the size distribution was indicated by the polydispersity index (PdI). PdI is a measure of the size distribution of the droplets and indicates the quality or homogeneity of the dispersion [50]. It ranges from 0 (monodisperse) to 0.5 (relatively wide distribution) [51]. The polydispersity index for liposomal dispersion should be up to 0.3 to be considered homogeneous.

The analysis of particle size of prepared liposomes revealed that they could be used in preparation of dermal products, as their size was estimated to be ≤300 nm. Namely, the liposomes with size ≤ 300 nm are capable to deliver their contents to some extent into the deeper layers of the skin [52].

Essential oils have the ability to decrease the size of liposomes and this effect was explained by the capability of essential oils to cause higher cohesion packing among the apolar chains in the membrane vesicles [53]. Also, it was concluded that monoterpenes might be able to decrease the sizes of liposomes by forcing soybean phosphatidylcholine vesicles to increase its surface curvature.

Polydispersity index (PdI) was also up to 0.3 and liposomal dispersion could be considered homogeneous. The value less than 0.1 for this index shows a homogeneous system and a value greater than 0.3 indicates a high degree of heterogeneity [54]. With each cycle the size of the liposomes decreased and PdI value was reduced.

Zeta potential (surface charge) is important parameter that determines the stability of liposomal dispersion [55]. Particles with a zeta potential of +30 mV and above, or -30 mV or below are considered stable (because they have a charge) (Table 6). However, if the zeta potential is from 0 to +5 mV, the particles tend to coagulate or destabilize [55, 56] (Figure 8.).

Zeta potential is important in the process of making different formulations, in selecting raw materials that would improve stability and increase the shelf life of formulations. Also with the knowledge of the zeta potential, one can predict the particle interaction with the surface.

Table 6. Influence of certain zeta potential value on colloid stability

Zeta Potential (mV)	Stability and Liposome Behavior
0 - ±5	Rapid coagulation or flocculation
±10 - ±30	Initial instability
±30 - ±40	Mean stability
±40 - ±60	Good stability
More than ±60	Excellent stability

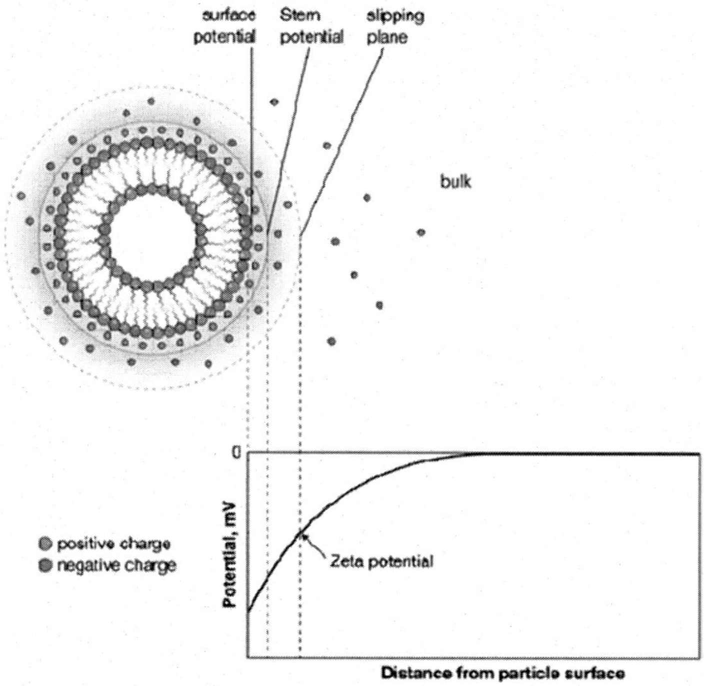

Figure 8. Zeta potential and surface charge.

Zeta potential is an essential parameter of liposome stability. Liposomes made with Phosal IP 40 might be considered stable because liposomal dispersion has a zeta potential – 67.6 mV. Particles with a potential of +30 mV and more, or -30 mV or lower are considered stable [55]. The zeta potential of the liposomes with *O. compactum* essential oil could be attributed to the interaction between the components of essential oil and phosphatidylcholine.

2.2.4. Essential Oil Encapsulation Efficiency (%EE)

The degree of encapsulation of thymol and carvacrol was determined by GC/MS and GC/FID (Figure 9).

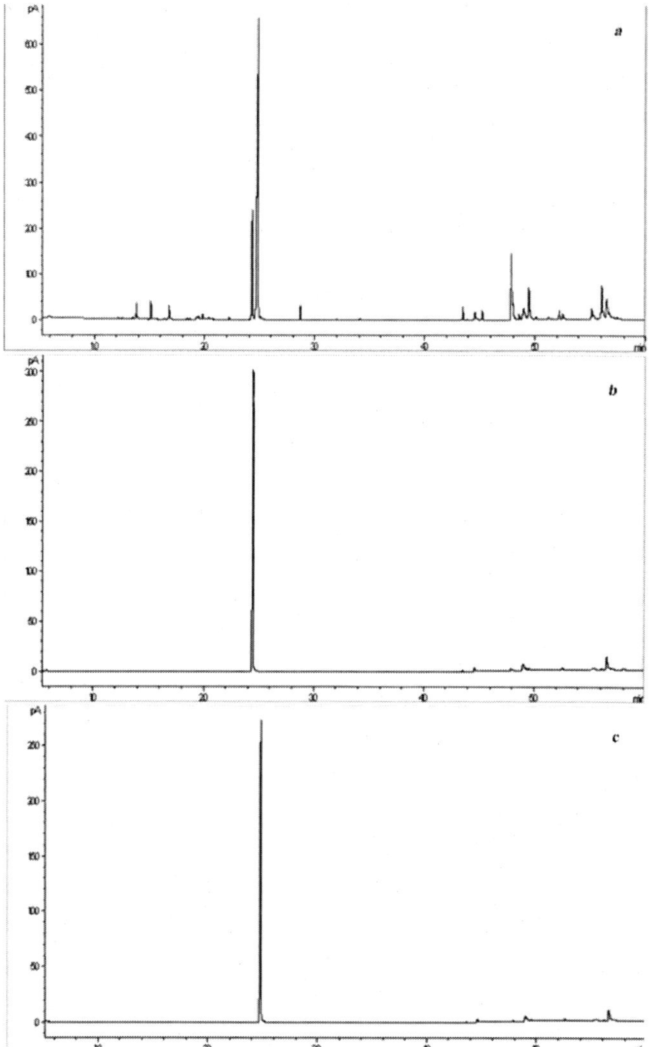

Figure 9. Chromatograms of prepared *O. compactum* essential oil liposomal dispersion (a) with the chromatograms of the standards thymol (b) and carvacrol (c) used in the determination of encapsulation efficiency.

In 1% dispersion, it is confirmed that 75.94% of thymol and 68.09% of carvacrol were encapsulated. For 9.3% liposomal dispersion it was encapsulated 80.75% of thymol and 84.43% of carvacrol. More thymol and carvacrol were encapsulated in 9.3% liposomal dispersion than in 1% liposomal dispersion, with the slightly different trend in efficacy of incorporation for thymol and carvacrol, being more in the case of carvacrol (68.09 and 84.43%, in 1% and 9.3% dispersion, respectively) (Figure 10).

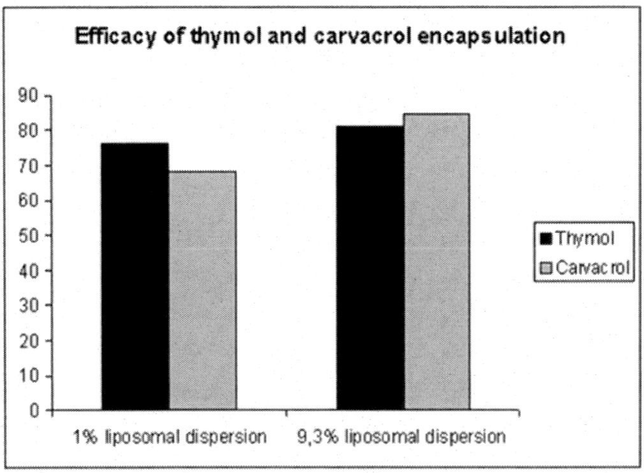

Figure 10. Encapsulation efficiency (EE%) of thymol and carvacrol in the two types of *O. compactum* essential oil liposomal dispersion.

Dispersion prepared with 1% *O. compactum* essential oil might be incorporated non-diluted into the preparations for dermatological and vaginal application. However, the dispersion with higher concentration could potentially serve as a concentrated dispersion that might be diluted just before incorporation into the final product. As highly concentrated dispersions of active substances incorporated into liposome carriers have been frequently present on the market, in our study, we evaluated the efficiency of encapsulation of *O. compactum* essential oil achieved with almost ten times higher concentration (being 9.3%) incorporated into liposomal dispersion. The obtained satisfactory results of concentrated liposome dispersion stability might be a good starting point in further investigation of *O. compactum* essential oil clinical application.

Conclusion

Applying the homogenization under high pressure, using Phosal IP 40 and purified water, different concentrations *O. compactum* essential oil were successfully encapsulated within liposomal dispersions.

Based on obtained results, it can be concluded that *O. compactum* essential oil really has a powerful antibacterial effect on all tested bacteria, and therefore could play an important role in the treatment of various bacterial infectious diseases. Prepared liposomes have proven to be quite stable and at the same time easy to prepare.

Vaginal infections caused by various bacteria and fungi are very common, especially among women of reproductive age. As *O. compactum* essential oil showed a strong antibacterial effect on vaginal infections caused by the investigated bacterial strains, incorporating essential oil into phospholipid carriers might solve the problem of the irritant potential of the essential oil itself, and at the same time lead to a prolonged effect of the preparation, and thus a better efficiency.

Acknowledgments

Serbian Ministry of Education, Science and Technological Development supported this study through the Research Project No. TR 31029 and Eureka E!12689 SCIMPLANT.

References

[1] Tadić, V., Dobrić, S., Marković, G., Djordjević, S., Arsić, I., Menković, N., Stević, T. (2008). Anti-inflammatory,

Gastroprotective, Free-Radical-Scavenging and Antimicrobial Activities of Hawthorn Berries Ethanol Extract. *J. Agr. Food Chem.* 56, 7700-7709.

[2] Cupara, S., Šobajić, S., Tadić, V., Arsić, I., Djordjević, S., Runjaić-Antić, D., Djordjević, B. (2010). Dry sea buckthorn berries (*Hyppophae rhamnoides* L.) – fatty acid and carotene contetnt in pericarp and seed oil, *HealthMED*, 4,788-791.

[3] Tasić-Kostov, M., Arsić, I., Pavlović, D., Stojanović, S., Najman, S., Naumović, S., Tadić, V. (2019). Towards a modern approach to traditional use: *in vitro* and *in vivo* evaluation of *Alchemilla vulgaris* L. gel wound healing potential. *J. Ethnopharmacol.* 238, 111777-111789.

[4] Tadić, V., Arsić, I., Zvezdanović, J., Žugić, A., Cvetković, D., Pavkov, S. (2017). The estimation of the traditionally used yarrow (*Achillea millefolium* L. Asteraceae) oil extracts with anti-inflamatory potential in topical application. *J. Ethnopharmacol.* 199, 138–148.

[5] Zugic, A., Jeremic, I., Isakovic, A., Arsic, I., Savic, S., Tadic, V. (2016). Evaluation of anticancer and antioxidant activity of a commercially available CO_2 supercritical extract of old man's beard (*Usnea barbata*). *PlosONE.* 11, e0146342.

[6] Arsic, I., Zugic, A., Tadic, V., Tasic-Kostov, M., Misic, D., Primorac, M., Runjajic Antic, D. (2012). Estimation of dermatological application of creams with St. John's Wort oil extracts. *Molecules*, 17, 275-294.

[7] Tadic, V., Jeremic, I., Dobric, S., Isakovic, A., Markovic, I., Trajkovic, V., Bojovic, D., Arsic, I. (2012). Anti-inflammatory, Gastroprotective and Cytotoxic Effects of *Sideritis scardica* Extracts. *Planta Med.* 78, 415 – 427.

[8] Stojiljković, D., Arsić, I., Tadić, V. (2016). Extracts of Wild apple fruit (*Malus sylvestris* (L.) Mill., Rosaceae) as a source of antioxidant substances for use in production of nutraceuticals and cosmeceuticals. *Ind. Crops. Prod.* 60, 165-176.

[9] A Arsić, I. A., Tadić, V. M., Đorđević, S. M., Žugić, A. R., Vujić, Z. B., Petrović, S. D. (2014). Optimization of extraction of antioxidant components from Yarrow herb. *Hem. Ind.* 68, 511-517.

[10] Zugic, A., Djordjevic, S., Arsic, I., Markovic, G., Zivkovic, J., Jovanovic, S., Tadic, V. (2014). Antioxidant activity and phenolic compounds in 10 selected herbs from Vrujci Spa, Serbia, *Ind. Crops. Prod.* 52, 519–527.

[11] Tadić, V., Bojović, D.,Arsić, I., Djordjević, S., Aksentijević, K., Stamenić, M., Janković, S. (2012). Chemical and Antimicrobial Evaluation of Supercritical and Conventional *Sideritis scardica* Griseb., Lamiaceae Extracts. *Molecules*, 17, 2683-2703.

[12] Bogdanović, A., Tadić, V., Arsić, I., Petrović, S., Skala, D. (2016). Supercritical and high pressure subcritical fluid extraction from Lemon Balm (*Melissa officinalis* L. *Lamiaceae*). *J. Supercrit. Fluid.* 107, 234-242.

[13] Bouyahya, A., Jamal, A., Edaoudi, F., Et-Touys1, A., Bakri, Y., Dakka, N. (2016). Origanum compactum Benth: A Review on Phytochemistry and Pharmacological Properties. *Med. Aromat. Plants*. 5, 252. doi:10.4172/2167-0412.1000252.

[14] Benabid, A. (2000). *Flore et écosystème du Maroc: évaluation et préservation de la biodiversité* [*Flora and ecosystem of Morocco: evaluation and preservation of biodiversity*]. Ibis Press, Paris, p: 357.

[15] Jahandiez, E., Maire, R. (1934). *Catalogue des plantes du Maroc - Spermaphytes et ptéridophytes* [*Catalog of plants from Morocco - Spermaphytes and pteridophytes*]. 3: 630-632.

[16] Tutin, T. G., Heywood, V. H., Burges, N. A., Moore, D. M., Valintine, D. H., Walters, S. M., Webb, D. A. (1972). *Flora Europaea*. University Press, Cambridge. 3: 171-172.

[17] Sharifi-Rad, M., Varoni, E. M., Iriti, M.,Martorell, M.,Setzer, W. N., Contreras, M. D. M.,Salehi, B., Soltani-Nejad, A.,Rajabi, S., Tajbakhsh, M. et al.(2018). Carvacrol and human health: A comprenhesive review. *Phytother. Resear.* 32, 1675-1687.

[18] Deb, D. D., Parimala, G., Saravana Devi S., Chakraborty, T. (2011). Effect of thymol on peripheral blood mononuclear cell PBMC and

acute promyelotic cancer cell line HL-60. *Chem-Biol. Interact.* 193, 97-106.

[19] Bouyahya, A., Abrini, J., Et-Touys, A. et al. (2017). Indigenous knowledge of the use of medicinal plants in the North-West of Morocco and their biological activities. *Eur. J. Integr. Med.* 13, 9–25.

[20] Bouyahya, A., Guaouguaou, F. E., Dakka, N., Bakri, Y. (2017). Pharmacological activities and medicinal properties of endemic Moroccan medicinal plant *Origanum compactum* (Benth) and their main compounds. *Asian Pacific Journal of Tropical Disease.* 7, 628-640.

[21] Bouyahya, A. et al. Essential oils of *Origanum compactum* increase membrane permeability, disturb cell membrane integrity, and suppress quorum-sensing phenotype in bacteria. (2019). *J. Pharm. Anal.* doi.org/10.1016/j.jpha.2019.03.001.

[22] Vuuren, S., C. du Toit, L., Parry, A., Pillay, V., Choonara, YE. (2010). Encapsulation of essential oils within apolymeric liposomal formulation for enhancement of antimicrobial efficacy. *Nat. Prod. Commun.* 5, 1401- 1408.

[23] Vanić, Ž. (2012). Liposomes as drug carriers: structural properties and classification, *Pharmaceutical Gazette* 68, 6.

[24] Torchilin VP. (2006). *Nanoparticles as drug carriers.* London: Imperial College Press.

[25] Morganti, P., Fabrizi, G., Zhong Feng, X., Yan, S. (2002). A new delivery system to improve acne therapy. *Euro-Cosmetics.* 5, 33-37.

[26] Raab, W. (1988,). Liposomes - a new form of dermatological drug carrier. *Arztl. Kosmetol.* 18, 213-224.

[27] Bouhdid, S., Skali, S. N., Idaomar, M., Zhiri, A., Baudoux et al. (2008). Antibacterial and antioxidant activities of *Origanum compactum* essential oil. *Afr. J. Biotechnol.* 7, 1563-1570.

[28] Chaouki, W., Leger, D. Y., Eljastimi, J., Beneytout, J. L., Hmamouchi, M. (2010). Antiproliferative effect of extracts from *Aristolochia baetica* and *Origanum compactum* on human breast cancer cell line MCF-7. *Pharm. Biol.* 48, 269-274.

[29] Chaouki, W., Meddah, B., Hmamouchi, M. (2015). Antiproliferative activity of *Origanum compactum* extract on lung cancer and hepatoma cells. *Arab. J. Med. Arom. Plant.* 1: 1 doi: 10.13140/RG.2.1.2852.5285.

[30] Satooka, H., Kubo, I. (2012). Effects of thymol on B16-F10 melanoma cells. *J. Agr. Food Chem.* 60, 2746-2752.

[31] Chang, H. T., Hsu, S. S., Chou, C. T., Cheng, J. S., Wang, J. L. et al. (2011). Effect of thymol on $Ca2^+$ homeostasis and viability in MG63 human osteosarcoma cells. *Pharmacol.* 88, 201-212.

[32] Hsu, S. S., Lin, K. L., Chou, C. T., Chiang, A. J., Liang, W. Z. et al. (2011). Effect of thymol on Ca2+ homeostasis and viability in human glioblastoma cells. *Eur. J. Pharmacol.* 670, 85-91.

[33] El Babili, F., Bouajila, J., Souchard, J. P., Bertrand, C., Bellvert, F.(2011) Oregano: chemical analysis and evaluation of its antimalarial, antioxidant, and cytotoxic activities. *J. Food Sci.* 76, C512-518.

[34] Slamenová, D., Horváthová, E., Sramková, M., Marsálková, L. (2007). DNA-protective effects of two components of essential plant oils carvacrol and thymol on mammalian cells cultured in vitro. *Neoplasma.* 54, 108-112.

[35] Adams, P. (2007). *Identification of Essential Oil Components by Gas Chromatography/Mass Spectrometry*, 4th Ed., Allured Publishing Corporation: Carol Stream, Illinois.

[36] *Automated Mass Spectral Deconvolution and Identification System software (AMDIS ver. 2.1).* (2005). National Institute of Standards and Technology (NIST), Standard Reference Data Program, Gaithersburg, MD.

[37] European Committee on Antimicrobial Susceptibility Testing. (2019). *Disk Diffusion Method for Antimicrobial Susceptibility Testing*, Version 7.0.

[38] CLSI MO2-A11 (2012). *Performance Standards for Antimicrobial Disk Susceptibility Tests, Approved Standard*, eleventh edition. Clinical and Laboratory Standards Institute (CLSI), ISBN 1-56238-986-6.

[39] CLSI M7-A7 (2006). *Methods for dilution, antimicrobial susceptibility tests for bacteria that grow aerobically, approved standard.* Clinical and Laboratory Standards Institute (CLSI), 7th ed.

[40] Hajlaoui, H., Mighri, H., Noumi, E., Snoussi, M., Trabelsi, N., Ksouri, R., Bakhrouf, A. (2010). Chemical composition and biological activities of Tunisian *Cuminum cyminum L.* essential oil: A high effectiveness against *Vibrio spp.* Strains. *Food Chem. Toxicol.* 48, 2186-2192.

[41] Aboukhalid, K., Lamiri, A., Agacka-Mołdoch, M. et al. (2016). Chemical polymorphism of *Origanum compactum* grown in all natural habitats in Morocco. *Chem. Biodivers.* 13, 1126 – 1139.

[42] Bojović D, Šoškić M, Tadić V (2018). Comparative study of chemical composition of the essential oils from *Satureja cuneifolia* Ten. and *Satureja montana* L., *Lamiaceae* collected at National park Lovćen, Montenegro, *STUD U BABES-BOL CHE.* 63, 167-180.

[43] Belkamel, A., Bammi, J., Belkamel, A., Douira, A. (2013). Study of essential oil chemical composition of an endemic Ibero-moroccan: *Origanum compactum* (Benth.). *J. Anim. Plant Sci.* 19, 2880-2887.

[44] Palmeira-de-Oliveira, R., Palmeira-de-Oliveira, A., Martinez-de-Oliveira, J. (2015). New strategies for local treatment of vaginal infections. *Adv. Drug Deliver. Rev.* 19, 12810-12818.

[45] Nazzaro, F., Fratianni, F., De Martino, L., Copola, R., De Feo, V. (2013). Effect of essential oils on pathogenic bacteria. *Pharmaceuticals.* 6, 1451-1474.

[46] Šiler Marinković, S., Mojović, Lj., Zahariev, J. (1997). Investigation of the Oxidative and Hydrolytic Stability of Liposomes. *Arch. Pharm.* 3-4, 197-204.

[47] Laouini, A., Jaafar-Maalej, C., Limayem. Blouza, I., Sfar, S., Charcosset, C., Fessi, H. (2012). Preparation, characterization and applications of liposomes: State of the art. *J. Colloid. Sci. Biotechnol.* 1, 147 – 168.

[48] Liolios, C., Gortzi, O., Lalas, S., Tsaknis, J., Chinou, I. (2009). Liposomal incorporation of carvacrol and thymol isolated from the

essential oil of *Origanum dictamnus* L. and in vitro antimicrobial activity. *Food Chem.* 112, 77–83.

[49] Rupp, C., Steckel, H.,Müller, B. W. (2010). Mixed micelle formation with phosphatidylcholines: The influence of surfactants with different molecule structures. *Int. J. Pharmaceut.* 387, 120–128.

[50] Anton, N., Gayet, P., Benoit, J. P., Saulnier, P.(2007). Nano-emulsions and nanocapsules by the PIT method: an investigation on the role of the temperature cycling on the emulsion phase inversion. *Int. J. Pharmaceut.* 344, 44–52.

[51] Müller, R. H., Schmidt, S., Buttle, I., Akkar, A., Schmitt, J., Brömer, S. (2004). SolEmuls® – novel technology for the formulation of i.v. emulsions with poorly soluble drugs. *Int. J. Pharmaceut.* 269, 293–302.

[52] Verma, D. D., Verma, S., Blume, G., Fahr, A. (2003). Particle size of liposomes influences dermal delivery of substances into skin. *Int. J. Pharmaceut.* 258, pp.141–151.

[53] Valenti, D., De Logu, A., Loy, G. (2001). Liposome- incorporated Santolinainsularis essential oil: preparation, characterization, and in vitro antiviral activity. *J. Liposome Res.* 11, 73–90.

[54] Khatibi, S. A., Misaghi, A., Moosavy, M.-H., Amoabediny, G., Basti, A. A. (2015). Effect of preparation methods on the properties of Zataria multiflora boiss. essential oil loaded nanoliposomes: Characterization of size, encapsulation efficiency and stability. *Pharm. Sci.* 20, 141 – 148.

[55] Hunter, R. J. (1981). *Zeta potential in Colloid Science*, Academic Press, NY

[56] Poudel, A., Gachumi, G., Wasan, K. M., Bashi, Z. D., El-Aneed, A., Badea, I. (2019). Development and Characterization of Liposomal Formulations Containing Phytosterols Extracted from Canola Oil Deodorizer Distillate along with Tocopherols as Food Additives. *Pharmaceutics.* 11, 185.

BIOGRAPHICAL SKETCH

Vanja M. Tadic, PhD

Affiliation: Institute for Medicinal Plant Research, Department of Pharmaceutical Research and Development, Belgrade, Serbia

Education:

2000 – 2002. Postdoctoral Research Fellow
Department of Biochemistry and Molecular Biology, Mayo Clinic and Foundation, Rochester, MN 55905 (USA), Head of Department: Dr. Frank Rusnak

1995 – 1998. PhD Thesis
"Comparative chemical studies on the plants *Anthemis carpatica* and *Anthemis montana*" Faculty of Chemistry, University of Belgrade, Supervisors: Prof. Dr.Vlatka Vajs and Prof. Dr. Slobodan Milosavljevic

1990 – 1994. M. Sci. Thesis
"Flavonoids from Wood Anemone, *Anemone nemorosa* L., Ranunculaceae,"
Faculty of Pharmacy, Department of Pharmacognosy, University of Belgrade, Supervisor: Prof. Dr. Momcilo Gorunovic

1985 – 1990. Bachelor of Science, Faculty of Pharmacy, University of Belgrade

Business Address: Institute for Medicinal Plant Research, Department of Pharmaceutical Research and Development, Tadeušsa Košćuška 1, Belgrade, Serbia

Research and Professional Experience:

Natural product chemistry, Chemical and biological evaluation of medicinal plant extracts, Development of new phytopreparations

Vanja M. Tadić works as Science Advisor at Department of Pharmaceutical Research and Development, Institute for Medicinal Plant Research “Dr Josif Pancic,” Belgrade. She has been taking part as one of the principal investigator in several research projects, and was leader of the project which resulted in patents winning the Gold Medal and Diploma for the best innovation of the WIPO 2010. Her main research interests are chemistry of natural products (extraction, identification, fractionation and isolation of chemical compounds from natural matrices, especially medicinal plants) and their biological activities. She has the experience in the formulation of high value products in pharmaceuticals, nutraceuticals and food sectors.

Professional Appointments:

Since 2007.	Principal Science Fellow Institute of Medicinal Plant Research “Dr J. Pancic,” Belgrade, R. Serbia
2002-2007.	Senior Research Fellow Institute of Medicinal Plant Research “Dr J. Pancic,” Belgrade, R. Serbia
1998-2002.	Research Fellow Institute of Medicinal Plant Research “Dr J. Pancic,” Belgrade, R. Serbia
1995-1998.	Research Associate Institute of Medicinal Plant Research “Dr J. Pancic,” Belgrade, R. Serbia

Honours:

Three phytopreparatipons, “Composition and process for obtaining phytopreparations based on combination of mineral water, medicinal mud and plants for the chronic venous insufficiency”;, “Composition and process for obtaining phytopreparations based on combination of mineral water, medicinal mud and plants for skin disease of various etiologies" and "Composition and process for obtaining phytopreparations based on combination of mineral water, medicinal mud and plants for rheumatic diseases," developed by the team which leader was Vanja M. Tadić won:

Gold Medal and special award of on the World Fair XIV Moscow International Salon of Inventions and Innovation technologies "Archimedes –2011" (5. – 8. April, 2011).

Gold Medal at the exhibition "Innovation-Belgrade 2010" (30th International Exhibition of Inventions, New Technologies and Industrial Design)

Gold Medal and Diploma of Inventors Association of Vojvodina at the International Innovation and Knowledge Exhibition TESLA FEST 2010, Novi Sad.

Gold Medal and Diploma for best innovation of the World Intellectual Property Organisation WIPO (World Intellectual Property Organization) 2010.

Gold Medal at the 30th International Exhibition MAKINOVA 2010th (exhibition of inventions, technical improvements, new products and the creativity of young people) and international exhibition of technical improvements and innovations of environment EKONOVA 2010, Skopje, under the auspices of the Ministry of Education and Science of the Republic of Macedonia.

Gold Medal and special award from the organizers at the World Fair in Seoul inventions SIIF 2010th (Seoul International Invention Fair 2010)

Publications Last 3 Years:

1. Tadic V, Zizovic I, Zivkovic J, Arsic I, Rój E (2017): "Chemical Characterization and Antioxidant Potential of Diterpenoids Rich Fractions Isolated from Selected Plant Species," *In* Brandon Jones (Ed): *Diterpenoids: Types, Functions and Research*, pp 75-117, Publisher: NOVA Science Publisher, Inc, New York, ISBN: 9781536106718 (hardcover), ISBN: 0781536106855 (eBook).

2. Tadić V, Arsić I, Stanković M, Maksimović S (2016): "Analysis of supercritical carbon(IV)oxide extracts from selected plants," *In* Roj Edward (Ed): *Supercritical Fluid Applications*, pp 19 – 44, New Chemical Syntheses Institute, Pulawy, Poland, ISBN: 978-83-935354-1-5.
3. Roganović S, Arsić I, Tadić V (2016): "Possibilities and advantages of the application of supercritical extracts as active ingredients in cosmetic products," *In* Roj Edward (Ed): *Supercritical Fluid Applications*, pp 75 - 92, New Chemical Syntheses Institute, Pulawy, Poland, ISBN: 978-83-935354-1-5.
4. Tasić-Kostov M, Arsić I, Pavlović D, Stojanović S, Najman S, Naumović S, Tadić V (2019). Towards a modern approach to traditional use: in vitro and in vivo evaluation of *Alchemilla vulgaris* L. gel wound healing potential, *Journal of Ethnopharmacology*, 238 pp. 111777-111789.
5. Jeremic I, Petricevic S, Tadic V, Petrovic D, Tosic J, Stanojevic Z, Petronijevic M, Vidicevic S, Trajkovic V, Isakovic A (2019). Effects of *Sideritis scardica* Extract on Glucose Tolerance, Triglyceride Levels and Markers of Oxidative Stress in Ovariectomized Rats, *Planta Medıca*, 85, 465-472.
6. Nesic I, Stojiljkovic D, Savic S, Tasic-Kostov M, Tadic V (2019). Stability, antioxidant activity, in vivo safety and efficacy of creams with standardized wild apple fruit extract: a comparison of conventional and biodegradable emulsifiers. *Internatıonal Journal of Cosmetıc Scıence*, 41, 300-310.
7. Oliva A, Garzoli S, Sabatino M, Tadic V, Costantini S, Ragno R, Božovic M (2018). Chemical Composition and Antimicrobial Activity of Essential Oil of *Helichrysum italicum* (Roth) G. Don fil. (Asteraceae) from Montenegro, *Natural Product Research*, DOI: 10.1080/14786419.2018.1538218.
8. Radovic M, Adamovic T, Pavlovic J, Rusmirovic J, Tadic V, Brankovic Z, Ivanovic J (2019). Supercritical CO_2 impregnation of gelatin-chitosan films with clove essential oil and characterization

thereof, *Chemical Industry & Chemical Engineering Quarterly*, 25, 119-130.
9. Brašanac-Vukanović S, Tadić VM, Blagojević N, Vukašinović-Pešić V, Đurđić S, Stanković M, Mutić J (2019). Element accumulation capacity of *Vaccinium myrtillus* from Montenegro: Comparison of element contents in water and ethanol extracts of bilberry plant parts, *Archives of Biological Sciences*, 71, 145-157.
10. Stankovic M, Savic V, Zivkovic J, Stanojevic Lj, Tadic V, Arsic I (2019). Tyrosinase Inhibitory and Antioxidant Activity of Wild *Rosa Canina* L. and *Sorbus Aucuparia* L. Fruit Extracts. *Acta Polonıae Pharmaceutıca, 76, 523-533.*
11. Pajnik J, Radetić M, Stojanovic DB, Jankovic-Častvan I, Tadic V, Stanković MV, Jovanović DM, Zizovic I (2018). Functionalization of polypropylene, polyamide and cellulose acetate materials with pyrethrum extract as a natural repellent in supercritical carbon dioxide, *The Journal of Supercritical Fluids*, 136, 70–81.
12. Brašanac-Vukanović S, Mutić J, Stanković DM, Arsić I, Blagojević N, Vukašinović-Pešić V, TadićVM (2018). Wild bilberry (*Vaccinium myrtillus* L., Ericaceae) originated from Montenegro as the source of antioxidants for use in production of nutraceuticals, *Molecules, 23*(8), 1864; https://doi.org/10.3390/molecules23081864.
13. Nikolic I, Lunter DJ, Randjelovic D, Zugic A, Tadic V, Markovic B, Cekic N, Zivkovic L, Topalovic D, Spremo-Potparevic B, Daniels R, Savic S (2018). Curcumin-loaded low-energy nanoemulsions as a prototype of multifunctional vehicles for different administration routes: Physicochemical and in vitro peculiarities important for dermal application, *International Journal of Pharmaceutics*, 550, 333–346.
14. Garzoli S, Božović M, Baldisserotto A, Andreotti E, Pepi F, Tadić V, Manfredini S, Ragno R (2018). *Sideritis romana* L. subsp. *purpurea* (Tal. ex Benth.) Heywood, a new chemotype from Montenegro, *Natural Product Research*, 32, 1056-1061.
15. Maksimović S, Tadić V, Ivanović J, Radmanović T, Milovanović S, Stanković M, Žižović I (2018). Utilization of the integrated process of supercritical extraction and impregnation for incorporation of

Helichrysum italicum extract into corn starch xerogel, *Chemical Industry and Chemical Engineering Quarterly,* 24, 191-200.

16. Stojiljkovic D, Tadic V, Stankovic M, Roganovic S, Arsic I (2018). Standardized extract of wild apple fruit in alkyl-polyglucoside based cosmetic cream – estimation of stability, safety, antioxidant activity and efficiency, *International Journal of Cosmetic Science*, 40, 285–294.
17. Bojović D, Šoškić M, Tadić V (2018). Comparative study of chemical composition of the essential oils from *Satureja cuneifolia* Ten. and *Satureja montana* L., *Lamiaceae* collected at National park Lovćen, Montenegro, Studia Universitatis Babes-Bolyai. Ser. Chemia, *STUDIA UBB CHEMIA*, 63, 167-180.
18. Žugić A, Tadić V, Kundaković T, Savić S (2018). Chemical composition and biological activities of the extracts and secondary metabolites of lichens belonging to the genus *Usnea,* Parmeliaceae, Lekovite Sirovine, 38, 68-81.
19. Stanković M, Maksimović S, Tadić V, Arsić I (2018). The oil content of wild fruits from different plant species obtained by conventional Soxhlet extraction technique. *Acta Facultatis Medicae Naissensis*, 35, 193-200.
20. Maksimović S, Tadić V, Ivanović J, Radmanović T, Milovanović S, Stanković M, Žižović I (2018). Utilization of the integrated process of supercritical extraction and impregnation for incorporation of Helichrysum italicum extract into corn starch xerogel, *Chemical industry & chemical engineering quarterly*, 24, 191-200.
21. Tadić V, Arsić I, Zvezdanovic J, Žugić A, Cvetkovic D, Pavkov S (2017). The estimation of the traditionally used Yarrow (*Achillea millefolium* L. Asteraceae) Oil extracts with anti-inflamatory potential in topical application, *Journal of Ethnopharmacology*, 199, 138 – 148.
22. Maksimovic S, Tadic V, Skala D, Zizovic I (2017). Separation of phytochemicals from *Helichrysum italicum*: An analysis of different isolation techniques and biological activity of prepared extracts, *Phytochemistry*, 138, 9-28.

23. Tadić V, Oliva A, Božović M, Cipolla A, De Angelis M, Vullo V, Garzoli S, Ragno R (2017). Chemical and Antimicrobial Analyses of *Sideritis romana* L. subsp. *purpurea*(Tal. ex Benth.) Heywood, an Endemic of the Western Balkan, *Molecules*, 22, 1395. doi:10.3390/molecules22091395.
24. Stanković M, Živković J, Tadić V, Arsić I. (2017). Skin protection against solar UV radiation by natural plant products: extracts from elder fruit (*Sambucus nigra* L.). The Fifth International Conference on Radiation and Applications in Various Fields of Research, June 12-16, 2017. Budva, *RAD Conference Proceedings*, vol. 2, pp. 231-236. ISSN: 2466-4626 (Online) DOI: 10.21175/RadProc.2017.48.
25. Mišan A, Šarić B, Milovanović I, Jovanov P, Sedej I, Tadić V, Mandić A, Sakač M (2017). Phenolic profile and antioxidant properties of dried buckwheat leaf and flower extracts, *Chemical Industry and Chemical Engineering Quarterly*, Savez hemijskih inzenjera, 23, 39-47.
26. Miletić S, Švarc Gajić J, Milovanović I, Đurendić-Brenesel M, Mandić A, Tadić V, Arsić I, Mišan A (2016). Anti-atherogenic properties of a functional herbal mixture, *Acta periodica technologica*, 47, 205 – 218.
27. Bogdanovic A, Tadic V, Stamenic M, Petrovic S, Skala D (2016). Supercritical carbon dioxide extraction of *Trigonella foenum-graecum* L. seeds: Process optimization using response surface methodology, *Journal of Supercritical Fluids*, 107, 44–50.
28. Bogdanovic A, Tadic V, Arsic I, Milovanovic S, Petrovic S, Skala D (2016). Supercritical and High Pressure Subcritical Fluid Extraction from Lemon Balm (*Melissa officinalis* L., Lamiaceae), *Journal of Supercritical Fluids*, 107, 234-242.
29. Mišan A, Sakač M, Medić Đ, Tadić V, Marković G, Gyura J, Pagano E, Izzo AA, Borrelli F, Šarić B, Milovanović I, Milić N (2016). *Antioxidant and Physicochemical Properties of Hydrogen Peroxide-Treated Sugar Beet Dietary Fibre, Phytotherapy research*, 30, 855-860.

30. Šoškić M, Bojović D, Tadić V (2016). Comparative chemical analysis of essential oils from lavender of different geographic origins, *STUDIA UBB CHEMIA*, 61, 127 – 136.
31. Stojiljkovic D, Arsic I, Tadic V (2016). Extracts of Wild apple fruit (*Malus sylvestris* (L.) Mill., Rosaceae) as a source of antioxidant substances for use in production of nutraceuticals and cosmeceuticals. *Industrial Crops & Products*, 80, 165 – 176.

In: The Lamiaceae Family: An Overview
Editor: Alexander Adler
ISBN: 978-1-53617-078-8

Chapter 4

IMPACT OF GREEN SYNTHESISED SILVER NANOPARTICLES ON *IN VITRO*-GROWN *LAVANDULA OFFICINALIS* PLANTLETS

Pınar Nartop and Züleyha Günbeldek
Department of Biomedical Engineering, Faculty of Engineering, Namık Kemal University, Tekirdağ, Turkey

ABSTRACT

Metal nanoparticles have been continuously used in many applications in nanotechnology. Silver nanoparticles are the most used ones amongst the other metal nanoparticles in nanobiotechnology. They have strong antiseptic, antibacterial, antifungal and antiviral effects depending on their sizes and shapes. Moreover, they are eco-friendly elements which have relatively low toxicity towards the environment and human. These nanoparticles can also be used in plant biotechnology applications. Our previous studies showed that green synthesized silver nanoparticles can be used safely and effectively as a surface sterilization agent for *Lamiaceae* seeds. In this study, *in vitro*-grown plantlets of *Lavandula officinalis* were subjected to green-synthesized silver nanoparticles at 1, 2 and 4 mg/L concentrations. Growth parameters such as elongation, shoot and root formation and biomass accumulations were investigated to understand the effects of silver nanoparticles on micropropagation of *L. officinalis*. The results showed that silver nanoparticles enhanced single shoot formations and average node length in *L. officinalis* node cultures.

Keywords: *Lavandula officinalis*, node culture, *in vitro* biomass production, silver nanoparticles, *Syzygium aromaticum*, green synthesis

1. INTRODUCTION

Higher plants can produce a wide variety of compounds. These bioproducts can be chemically synthesized via pathways of plant metabolism. Generally, plant metabolites can be categorized into two classes; primary and secondary metabolites. Primary metabolites are the main nutritional source of plants and consist of proteins, lipids and carbohydrates. Secondary metabolites are synthesized from primary metabolites and have more complex pathways than them. Although they are not essential compounds for plant survival like primary metabolites, they should be synthesized for higher life quality of plants. They mostly act as an immune system for plants; protect and defend plants against biotic and abiotic stress factors. These compounds generally synthesized at very low concentrations (not more than 1% dw). Plant secondary metabolites can only be found in plants. They can not be produced by other organisms. They mostly have very complex structures. As a result of this, their chemical synthesis can rarely be achieved by researchers. This complexity makes plants very unique because these metabolites are also very important and beneficial for human and animals. Most of the plant secondary metabolites have antibacterial, antioxidant, antifungal, anticarcinogenic, antifungal, antiviral and antibiotic activities. Because of these biological effects, they are used as pharmaceuticals (anticarcinogenic, cardiac glycosides, allergens, immunochemical), agrochemicals (hormones, biopesticides, plant virus inhibitors), food additives (spices, sweeteners, emulsifiers), flavors, coloring agents (dyes) and fragrances (Scott and Dougall, 1987; Rao and Ravishankar, 2002; Nartop and Gürel, 2014).

About 100,000 secondary metabolites are present in the plant kingdom confined to specific taxonomic groups. There are three major groups of secondary metabolites in plants based on their biosynthetic pathway; nitrogen-containing compounds (cyanogenic glycosides, alkaloids, and glucosinolates), phenolic compounds (flavonoids and phenylpropanoids) and terpenes (isoprenoids) (Fang et al., 2011; Ashraf et al., 2018).

Biomass is an organic matter and a total mass of living plants. It is composed of organized living cells such as tissues, organs and biological systems, which have biologic catalysts such as enzymes, that regulate plant primary and secondary metabolisms. Therefore, plant biomass is the "biofactory" that produce plant metabolites. The quantity of biofactory is the first bottleneck of metabolite production which specifies the quality of production. If biomass accumulation cannot be achieved, plant metabolites cannot be produced in sufficient amounts (Nartop, 2018a).

The commercial importance of plant secondary metabolites has caused a great interest in research and finding alternative ways for their production. Plant biotechnology is one of the most utilized branches of biotechnological bioprocesses. *In vitro* production of plants, also known as plant tissue cultures, is the most popular alternative way of biomass production. For this purpose, some biotechnological applications such as cell, tissue, organ, root and hairy root cultures have been implemented.

In vitro production systems have some advantages over traditional agricultural techniques: (i) bioactive secondary metabolites can be synthesized under controlled conditions independently from geographical, seasonal variations and environmental factors, (ii) biotechnological procedures can be identified for continuous production of uniform quality and yield, (iii) high-yielding cell lines can be selected for better quality, (iv) rapid production is possible, (v) harmful herbicides and pesticides are not used, (vi) process cost can be decreased by growth control and process regulations, (vii) novel compounds which are not normally found in nature can be synthesized from low-cost precursors in culture conditions

(Adamse, 1990; Chattopadhyay et al., 2002; Rao and Ravishankar, 2002; Mulabagal and Tsay, 2004; Cai et al., 2012; Nartop and Gürel, 2014).

These culture types can produce secondary metabolites in their bodies. However, as mentioned above, a high-quality biomass accumulation is essential for obtaining a high secondary metabolite production. At this point, the optimization of media compositions and culture conditions are required. Media should be supplemented with optimum concentrations of growth regulators, vitamins, microelements and macroelements. Optimum light intensity (1000-4000 lux), temperature (22-24°C) and pH are also very important factors in *in vitro* conditions. Without optimization, biofactories cannot be obtained and secondary metabolites cannot be produced in sufficient amounts (Nartop, 2018a).

Nanotechnological applications are being currently used in many areas of nanobiotechnology. "Nanobiotechnology" term is the combination of nanotechnology and biotechnology terms and is the intersection of nanotechnology, biomedical sciences and molecular biology. In both nanotechnology and nanobiotechnology, the main area of interest is nanoparticles. Because of their unique properties, silver nanoparticles became the most popular nanoparticle type amongst the other nanoparticle types such as gold, zinc, copper, etc. They have broad-spectrum bioactivities (antiseptic, antibacterial, antifungal, antiprotozoal, anti-inflammatory, antiviral and anticancer effects) depending on their sizes and shapes. Despite the biological effects on microorganisms, they show relatively low cytotoxic effects on human (Lansdown, 2007; dos Santos et al., 2012; Babapour et al., 2012; Perello, 2013; Tran et al., 2013; Nartop, 2019).

Silver ions are highly reactive and can form silver nanoparticles in different shapes and sizes. Smaller and spherical nanoparticles enhance the extent of bacterial elimination and are more effective than bigger ones, due to their larger contact surface area (Pal et al., 2007). This feature gives a great opportunity to synthesize selectively antimicrobial or selectively toxic silver nanoparticles (Cao, 2017). Raffi et al. (2008) reported that the bacterial membrane was completely ruptured by silver nanoparticles in a few minutes after the contact that indicates the high efficacy of silver nanoparticles owing to the unique property of large surface area to volume ratio. Studies showed that silver nanoparticles affect growth inhibition, membrane distraction and cause the death of bacterial cells. Although the precise mechanism of action has not been fully understood yet, three mechanisms are revealed; (i) bacteria can uptake free silver ions which leads to the interruption of ATP synthesis and DNA replication, (ii) generation of radical oxygen species by silver nanoparticles or silver ions, (iii) silver nanoparticles may direct damage to cell membranes (Jones and Hoek, 2010; Anjum et al., 2016).

Silver nanoparticles can be synthesized through biological, physical, chemical and photochemical procedures (Tran et al., 2013; Sintubin et al., 2012). In biological synthesis, also known as "green synthesis" or "green bioreduction," silver nanoparticles have been synthesized using plant extracts. Plant extracts contain plant molecules (mostly phenolics) which have –OH groups which act as bioreductors for silver ions. These metabolites convert Ag^+ to Ag^0 and lead to the formation of colloidal silver nanoparticles (Dhoondia and Chakraborty, 2012).

The green synthesis method is more beneficial, advantageous and rapid than other methods. It is cost-effective, energy-efficient, environmentally friendly and can easily be scaled up for large-scale synthesis (Yasin et al., 2013). Green-synthesized AgNPs have various applications in many sectors such as electronics, clothing, food industry, paints, sunscreens, cosmetics, biosensing, medicines, drug delivery and medical devices (Anjum et al., 2016).

The only materials used in this method are plant water extracts (stock extract; source of bioreductor molecules) and silver nitrate solution (source of silver ions). Plant water extract can be prepared from any plant material. In the most common method, 5-10 g of pulverized or crumbled plant biomass (leaves, flowers, roots, rhizomes etc.) is added into 100 ml distilled water and kept in a water bath at 80°C for 1 h. In some reports, instead of a water bath, a hot plate is utilized and plant biomaterial is boiled for 5-20 mins. To obtain a stock extract, the mixture is filtered through filter paper. For the best results, it should be kept in glass bottles for one day at room temperature (22 ± 2°C) before use. This stock extract can be used in two weeks, and it should also be kept at freezer at +4°C.

To obtain a colloidal silver nanoparticles solution, the plant extract is added into 1-5 mM silver nitrate solution at 1:20 or 1:10 ratios. Bioreduction starts as soon as the plant extract added to silver nitrate solution. Unless stated otherwise, there is no need to apply heat or light. Generally, the bioreduction process can be carried out at room conditions. Depending on the plant extract, bioreduction lasts a few minutes or hours. In some cases, it may continue for 24 hours. Silver nanoparticles formation can physically be observed by the color change as a result of the biochemical reduction.

Silver nanoparticle synthesis can be confirmed by pH changes of colloidal solutions and UV–visible spectrophotometry. Not only does temperature, the concentration of plant extract and silver nitrate solution, pH, incubation time affect silver nanoparticles formation, but also their shapes and sizes (Rai and Yadav, 2013).

Many factors affect secondary metabolite production in intact plants and *in vitro* cultures. These factors are called elicitors and can be classified into two groups; abiotic and biotic elicitors. Abiotic elicitors are non-biological factors such as injuries, light, osmotic stress, salinity, drought, thermal stress, metal ions, heavy metals, mineral salts, gases and hormonal elicitors (jasmonates, salicylates, ethylene, cytokinins, gibberellic acid). Biotic elicitors comprise biological factors including bacterial, viral and fungal infections. Carbohydrates (polysaccharides, oligosaccharides, peptides; oligogalacturonic acid, agropectin, chitosan, chitin, mannuronate, guluronate, mannan, galacturonides, glutathione, hepta-β-glucoside, β-glucan), proteins (cellulase, oligandrin), lipids (lipopolysaccharides), yeast extracts, fungal elicitors (necrotrophic pathogens, biotrophic pathogens, *Agrobacterium rhizogenes*, *Agrobacterium tumefaciens*, *Bacillus cereus*, *Bacillus aminovorans*, *Rhizobacterium*, *Pseudomonas syringae*) are classified as biotic elicitors (Naik and Al-Khayri, 2016; Baenas et al., 2014; Nartop, 2018a).

Elicitors have been frequently used in plant cell and tissue cultures to enhance secondary metabolite production. Most commonly used elicitors are methyl jasmonate, jasmonic acid, salicylic acid, dextran, pectin, alginate, chitin, chitosan. However, these molecules also cause changes in the primary metabolism of plants. They affect photosynthesis, growth cycle, respiration, nutrient uptake, homeostasis, tissue differentiation, biosynthesis of carbohydrates, lipids, and protein; and gene expressions (Baenas et al., 2014; Nartop, 2018a). As a result of the decrease in biomass accumulation, the mass of biofactory diminishes. In order to annihilate these negative impacts, researchers are still trying to find new solutions.

In this perspective, nanoparticles such as gold, zinc, copper and silver have been used to induce both secondary metabolite and biomass accumulations. The increasing rates of primary metabolism such as cell proliferation, total protein contents, dry weight, root length, leaf area, shoot production rates, enzyme activities (superoxide dismutase, peroxide, catalase) have been observed in different types of cell and tissue cultures when silver nanoparticles were used as elicitors (Aghdaei et al., 2012; Zhang et al., 2013; Mohammed Al-Oubaidi and Mohammed-Ameen, 2014; Raei et al., 2014; Homaee and Ehsanpour, 2015; Al-Sowaidi and Al-Oubaidi, 2015; Gorczyca et al., 2015; Fazal et al., 2016; Bhat and Bhat, 2016; Ghanati and Bakhtiarian, 2016; Zafar et al., 2016; Zaka and Abbasi, 2016; Nartop, 2018a). In these studies, it was also concluded that phenolics, flavonoids, glycosides and essential oils productions were triggered, which means silver nanoparticles did not suppress primary metabolism while inducing secondary metabolism.

Silver nanoparticles have also been used as surface sterilization agents in plant tissue culture studies. Green synthesized silver nanoparticles were used in surface sterilization of *Rosmarinus officinalis* and *Pyracantha coccinea* stem explants and some seeds (Salvia farinecae, *Ocimum basilicum* – Large Leaf Italian, *Thymus vulgaris*, *Ocimum basilicum* var. *Purpurascens*), *Centaurea cyanus* L., *Digitalis purpurea* L., *Calendula officinalis* L., *Lavandula officinalis* L. and *Melissa officinalis* L.). In these studies, it was found that silver nanoparticles showed no adverse effects; no browning, prevention of cell division, growth and callus formation on stem explants of *R. officinalis* and *P. coccinea*; no adverse effects on seed

Table 1. Biosynthetic AgNPs were added at 1, 2 and 4 mg/L concentrations to media. Basal WPM medium was used as control. The pH of the media was adjusted to 5.8. The media were autoclaved for 15 minutes at 121ºC and a pressure of 1.2 kg/cm3. Nodal explants were incubated for four weeks at 16 h photoperiod (4000 lux) and 24 ± 1°C temperature.

Table 1. Media compositions used for examining AgNP effect on *L. officinalis*

Medium Code	Basal Medium	Sugar Content (g/L)	pH	Agar Concentration (g/L)	AgNP Concentration (mg/L)
WPM (Control)	WPM	30	5.8	6	0
NP1	WPM	30	5.8	6	1
NP2	WPM	30	5.8	6	2
NP4	WPM	30	5.8	6	4

2.5. Growth Parameters of *L. officinalis* Cultures

The following growth parameters were investigated to determine the effects of AgNP on *L. officinalis*: shoot elongation; single and multiple shoot formations; root formations; shoot, root and node numbers; shoot, root and node numbers per explant; average node length; shoot, root and total fresh weights; shoot, root and total dry weights; relative shoot, root and total dry weights.

Observations were made with regards to shoot, root and node regenerations. To calculate shoot, root and node numbers per explant, the total number of shoots, roots and nodes were divided by the total number of explants. Single and multiple shoot formation percentages (SSF % and MSF %) were calculated using the following formulas (Nartop, 2018d; Nartop et al., 2015):

SSF % = (Total SSF/ Total explant number) x 100

MSF % = (Total MSF/ Total explant number) x 100

After plantlets were taken out from culture vessels, their roots and bottom parts were cleared from agar with the help of distilled water and a napkin. The shoot and root parts of plantlets were separated from each other and their FWs were recorded. They were dried for seven days at room temperature (24 ± 1°C) and their DWs were recorded. In order to calculate relative dry weight (RDW) percentages and biomass production rates per explant (BPR), the following formulas were used (Nartop, 2018d; Nartop et al., 2015);

RDW% = [DW(g)/FW(g)] x 100

BPR (g/month/explant) = FW or DW/explant number

2.6. Statistical Analyses

The study was implemented in a factorial randomized plot design and each experiment was repeated thrice. Fifteen explants were used for each replicate. Data were analyzed with one way ANOVA and the post hoc tests were performed using TUKEY test ($p \leq 0.05$).

3. Results and Discussion

Formation of silver nanoparticles was observed 2h after the addition of plant extract into silver nitrate solution. The clear and transparent mixture turned into a very dark brown and thick suspension. *S. aromaticum* silver nanoparticles showed characteristic bands in the UV-Visible region at 388 nm (Figure 1). Moreover, the pH changes in the mixture which is another sing of bioreduction of silver ions (Maria *et al.*, 2015) were detected and the results were summarized in Figure 2. pH values of colloidal solutions decreased gradually from 4,92 to 3,78. Similar results were reported in green-synthesized silver nanoparticle studies (Amin *et al.*, 2012; Sinha and Paul, 2014; Kumar *et al.*, 2017; Nartop, 2017; Nartop 2018b). Formation of silver nanoparticles was also confirmed by SEM analysis. As shown in Figure 3, spherical silver nanoparticles can be observed in the colloidal solution.

Effects of these spherical silver nanoparticles were observed in *in vitro*-grown *L. officinalis* plantlets and data revealed that there are differences in the effect of the different concentrations of silver nanoparticles.

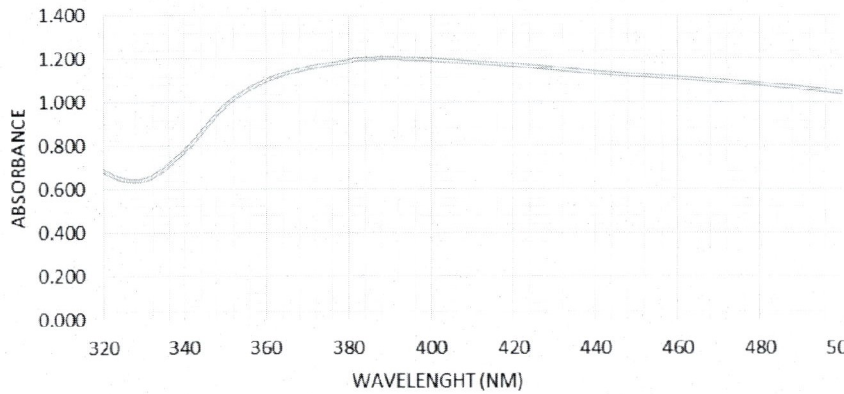

Figure 1. UV-Vis chromatograms of AgNPs.

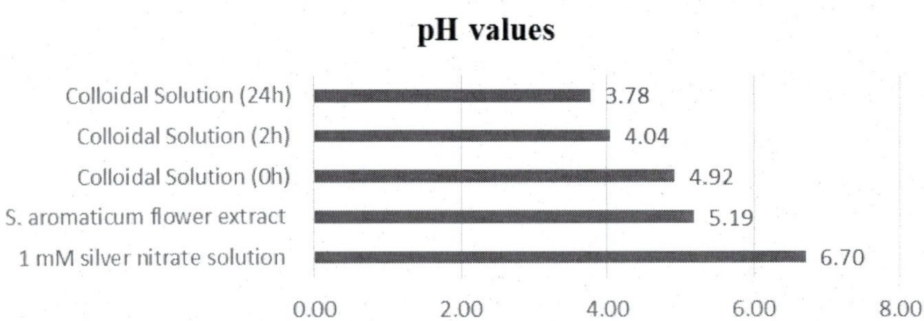

Figure 2. pH values of $AgNO_3$ solution, *S. aromaticum* flower extract and colloidal solution.

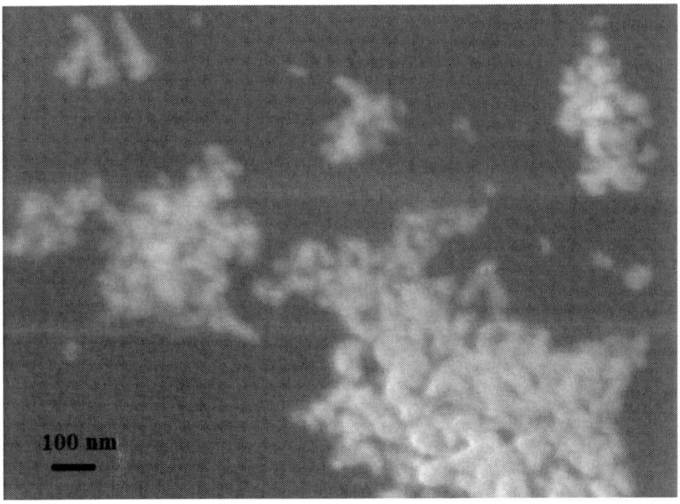

Figure 3. SEM images of *S. aromaticum* flower AgNP samples (24h samples).

Shoot elongation rates (Figure 4) were lower at NP1 and NP4 media. Interestingly, 2 mg/L concentration (NP2) did not affect plantlets and the rates were very close (1,55 cm) to WPM (1,50 cm).

Single shoot formation percentages did not affect statistically by the addition of silver nanoparticles. However, the percentages of silver nanoparticle-added media (NP1, NP2 and NP4) were all higher than the control group (WPM). NP1 medium had the highest percentage (62,22%) and it was more than two times higher than the control group (28.89%). Conversely, multiple shoot formation percentages were negatively affected by silver nanoparticles. The highest percentage (51,11%) was obtained in NP2 medium, which was 76,66% lower than WPM medium (Figure 5).

Numbers of shoots and shoots/explant rates were negatively affected by silver nanoparticles (Figure 6 and 7). The numbers were gradually decreased as the silver nanoparticles concentrations increased and there was no statistical significance between NP1, NP2 and NP4 media.

Root formations of plantlets were also affected negatively by silver nanoparticles (Figure 8). Although the results were found statistically significant, the percentages were decreased gradually as the concentration increased, as observed in multiple shoot formation and shoot numbers.

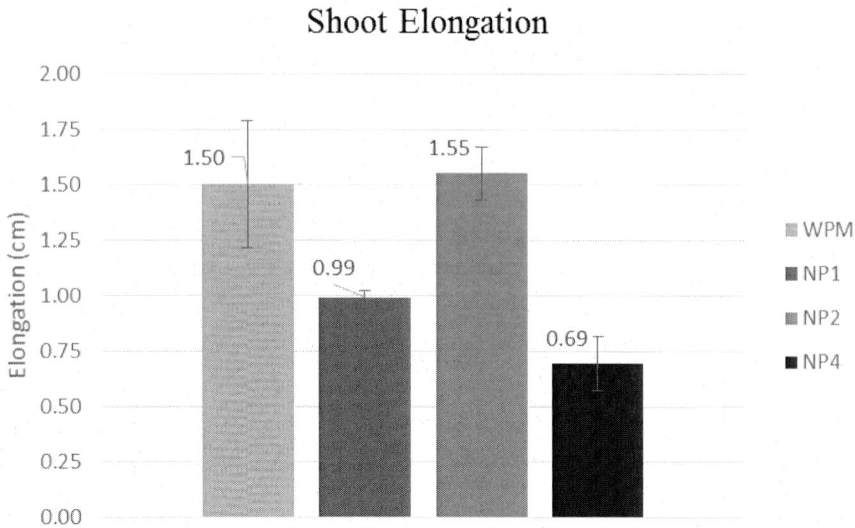

F = 6,08; MSE = 0,08464.

Figure 4. Shoot elongations in WPM, NP1, NP2 and NP4 media.

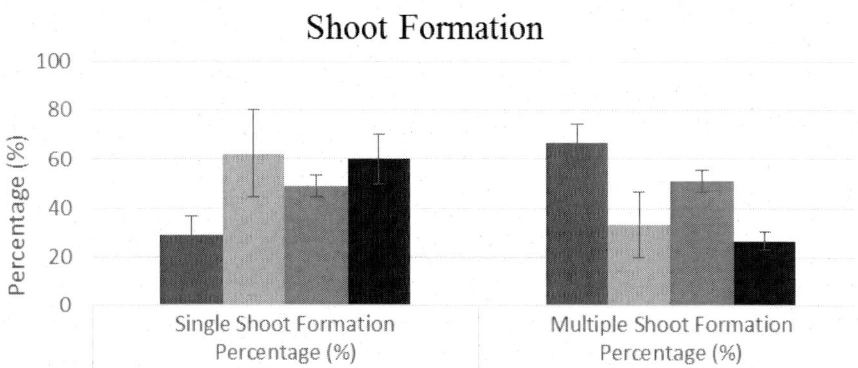

	Single Shoot Formation Percentage (%)	Multiple Shoot Formation Percentage (%)
WPM	28.89	66.67
NP1	62.22	33.33
NP2	48.89	51.11
NP4	60.00	26.67

F = 1,84; MSE = 337,8.

Figure 5. Percentages of single and multiple shoot formations in WPM, NP1, NP2 and NP4 media.

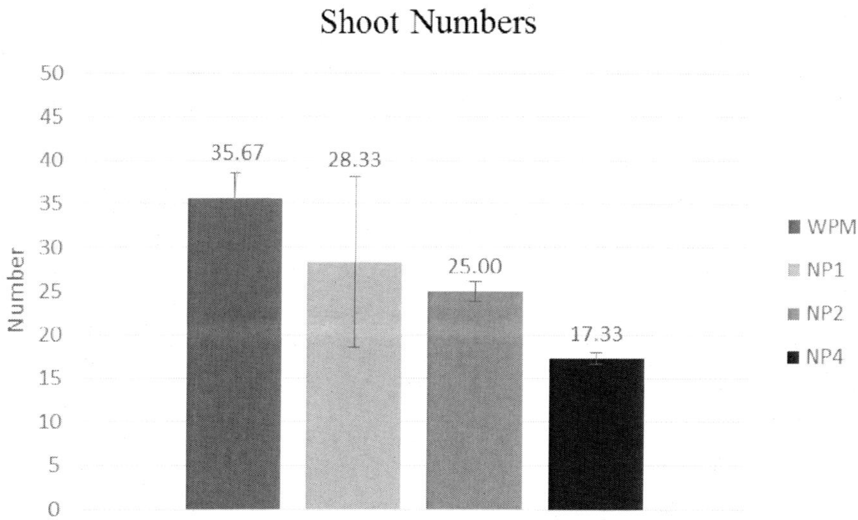

F = 2,16; MSE = 80,50.

Figure 6. Shoot numbers obtained in WPM, NP1, NP2 and NP4 media.

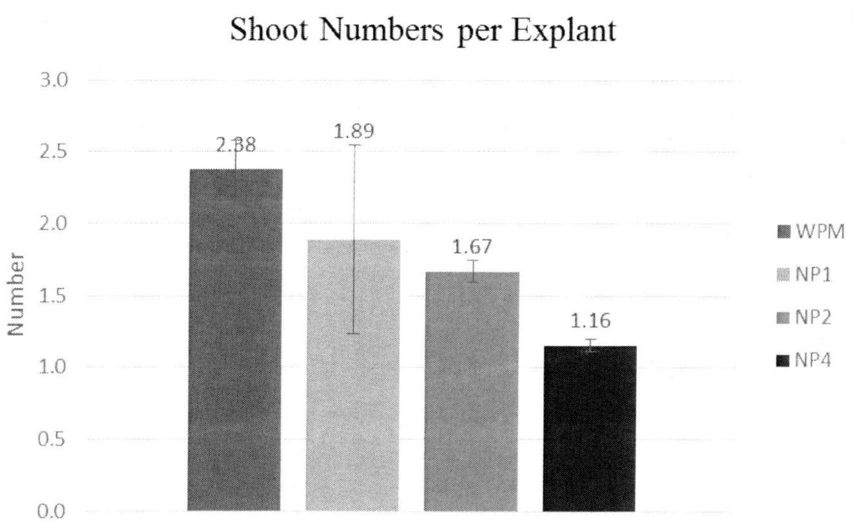

F = 2,16; MSE = 0,3578.

Figure 7. Shoot numbers per explant obtained in WPM, NP1, NP2 and NP4 media.

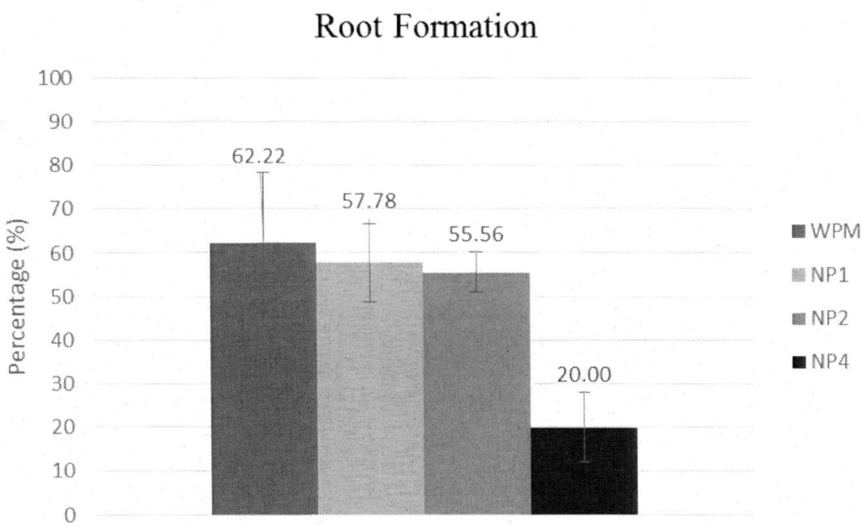

F = 2,84; MSE = 314,8.

Figure 8. Percentages of root formations in WPM, NP1, NP2 and NP4 media.

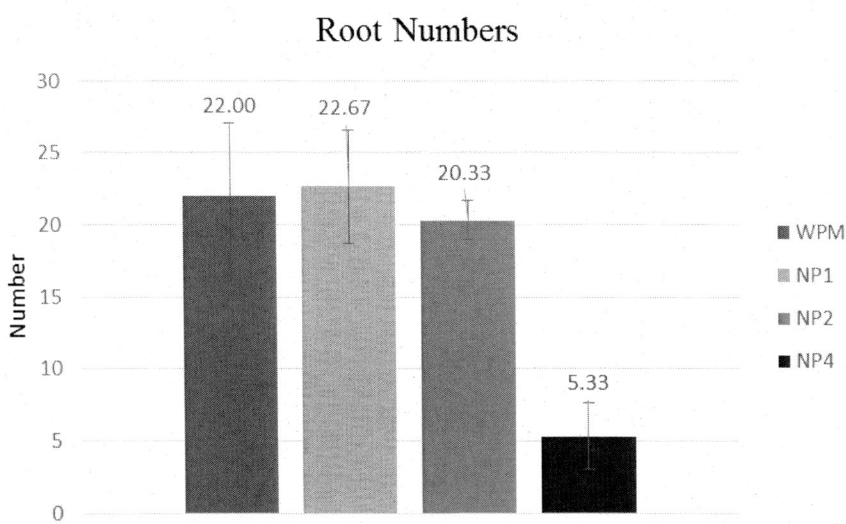

F = 5,64; MSE = 36,00.

Figure 9. Root numbers obtained in WPM, NP1, NP2 and NP4 media.

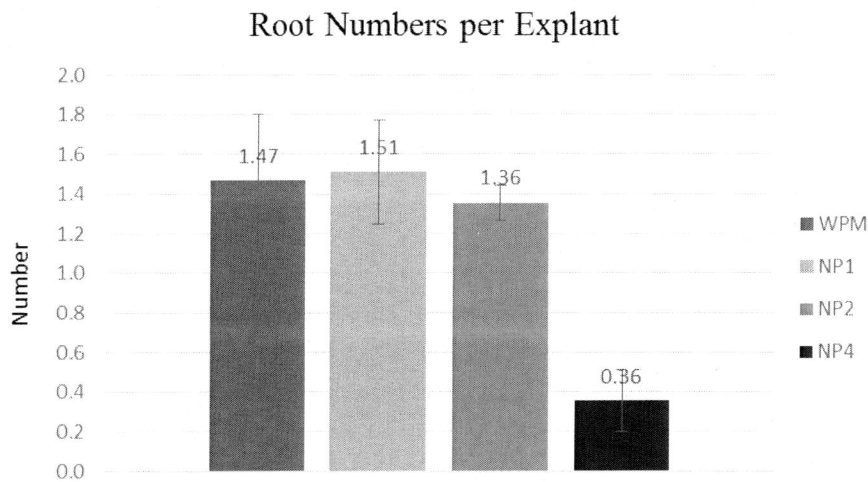

F = 5,64; MSE = 0,16.

Figure 10. Root numbers per explant obtained in WPM, NP1, NP2 and NP4 media.

Effects of silver nanoparticles on root numbers and root numbers/explant rates were found significant at $p \leq 0.05$. The highest root number was obtained in NP1 medium (22,67%). In NP2 medium, root number decreased to 20,33%. However, in NP4 medium, root numbers (5,33%) were found almost four times less than WPM medium (Figure 9). Root numbers/explant values also showed similar results as root number values (Figure 10).

Although there was no statistical significance, (Figure 11 and 12) silver nanoparticles negatively affected the formation of *L. officinalis* nodes in *in vitro* cultures. The highest node number (145) was detected in the WPM medium. Amongst NP1, NP2 and NP4 media, the highest number was obtained in NP1, but these results were almost 80% lower than WPM media.

Interestingly, the average node length was significantly affected by the addition of 2 mg/L silver nanoparticles (Figure 13). NP1 and NP4 media values were found lower than WPM medium. However, in NP2 medium, average node length showed a 74% increase and enhanced to 0,80 cm from 0,46 cm.

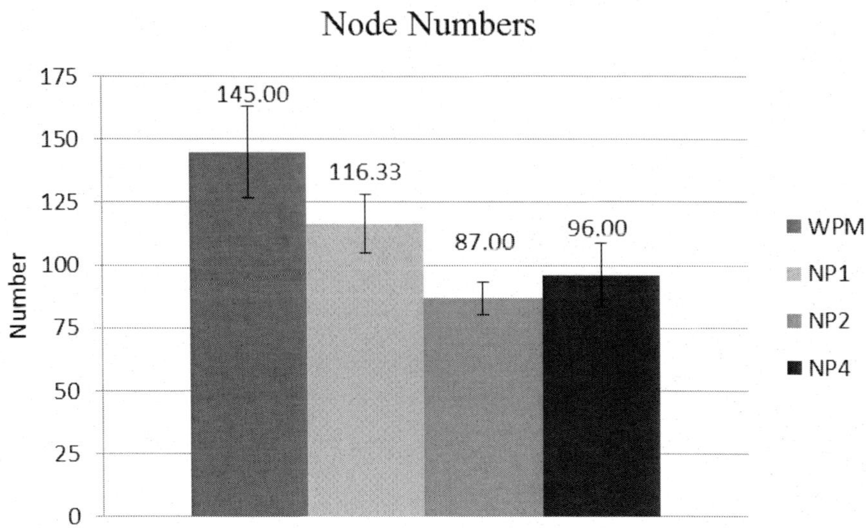

F = 3,92; MSE = 507,1.

Figure 11. Root numbers obtained in WPM, NP1, NP2 and NP4 media.

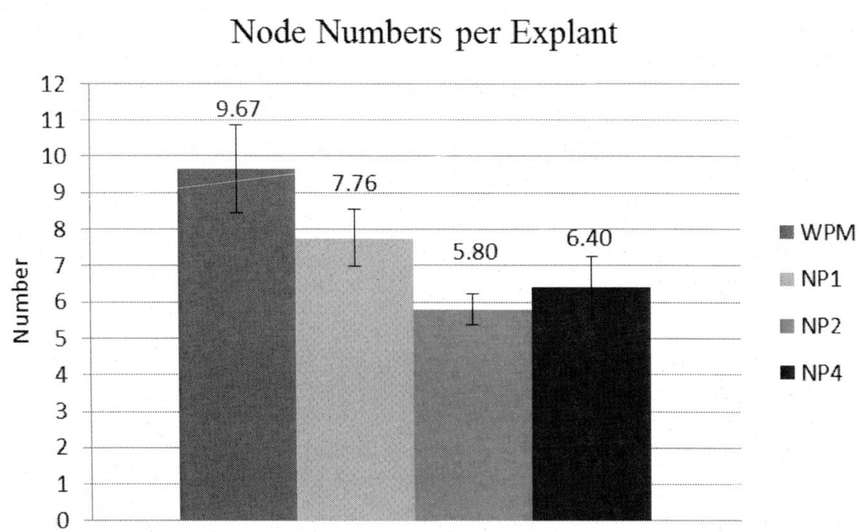

F = 3,92; MSE = 2,254.

Figure 12. Root numbers per explant obtained in WPM, NP1, NP2 and NP4 media.

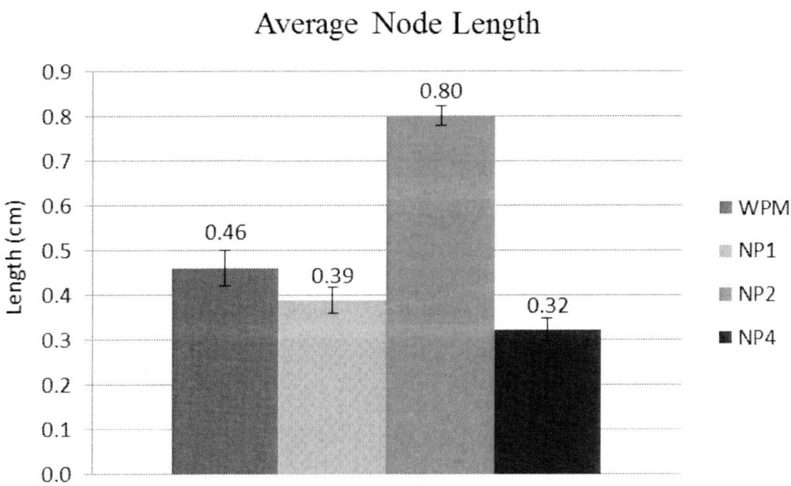

F = 51,15; MSE = 0,002660.

Figure 13. Average node length obtained in WPM, NP1, NP2 and NP4 media.

Fresh weight accumulations were decreased by the addition of silver nanoparticles (Figure 14). Although fresh weight accumulations of *L. officinalis* node cultures were not significantly affected by silver nanoparticles, the highest fresh weights were obtained in control medium. Amongst NP1, NP2 and NP4 media, the highest shoot fresh weights were detected in NP4 medium, whereas root and total fresh weights were higher in NP1 medium.

Dry weights of *L. officinalis* node cultures showed similarities as fresh weight results. There was no statistical significance between the media tested. However, the highest accumulations were detected in WPM medium. Amongst NP1, NP2 and NP4 media, the highest dry weight accumulations were observed in NP1 medium (Figure 15).

In relative dry weight percentages, no statistical significance was observed. Interestingly, relative shoot and total dry weight percentages had the highest values in NP2 medium, whereas relative root dry weight percentage was higher in NP1 medium. The results obtained in NP1, NP2 and NP4 media were higher than WPM medium, which means silver nanoparticles enhanced relative dry weight percentages in *L. officinalis* node cultures (Figure 16).

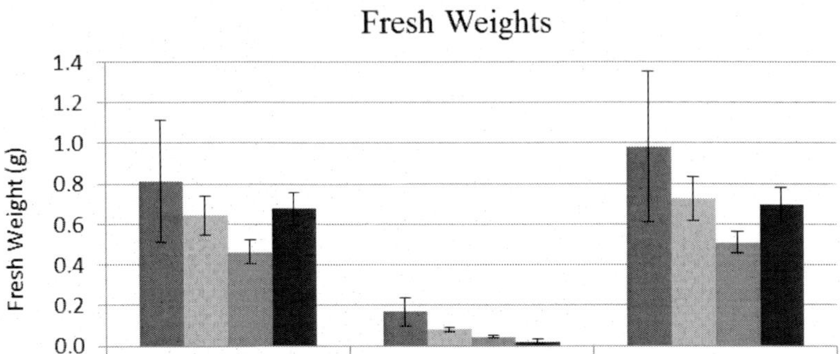

	Shoot Fresh Weight (g)	Root Fresh Weight (g)	Total Fresh Weight (g)
WPM	0.81	0.17	0.98
NP1	0.64	0.08	0.73
NP2	0.47	0.04	0.51
NP4	0.676	0.020	0.696

Shoot FW: F = 0,75; MSE = 0,08197) (Root FW: F = 3,40; MSE = 0,003757. Total FW: F = 0,96; MSE = 0,1182.

Figure 14. Shoot, root and total fresh weights obtained in WPM, NP1, NP2 and NP4 media.

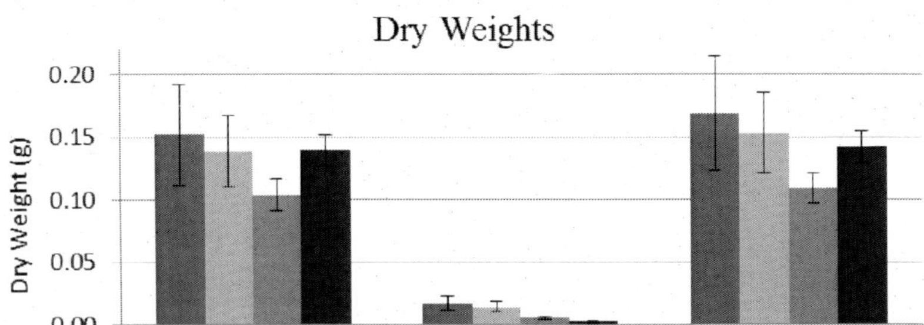

	Shoot Dry Weight (g)	Root Dry Weight (g)	Total Dry Weight (g)
WPM	0.15	0.02	0.17
NP1	0.14	0.01	0.15
NP2	0.10	0.01	0.11
NP4	0.139	0.002	0.142

(Shoot DW: F = 0,63; MSE = 0,002070) (Root DW: F = 3,78; MSE = 0,000039). (Total DW: F = 0,76; MSE = 0,002566).

Figure 15. Shoot, root and total dry weights obtained in WPM, NP1, NP2 and NP4 media.

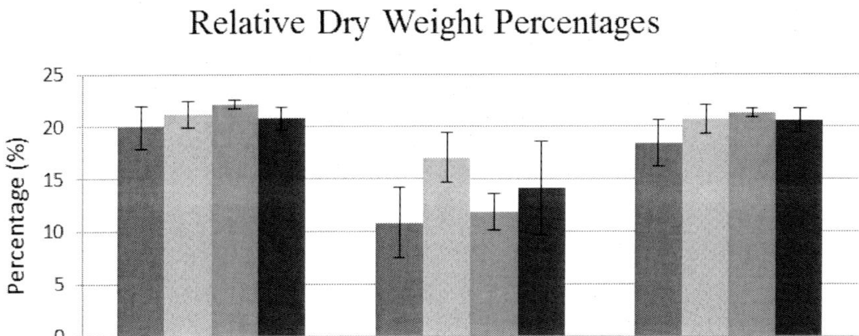

	Relative Shoot Dry Weight Percentage (%)	Relative Root Dry Weight Percentage (%)	Relative Total Dry Weight Percentage (%)
WPM	19.95	10.84	18.43
NP1	21.17	17.05	20.70
NP2	22.20	11.83	21.32
NP4	20.82	14.13	20.60

Shoot RDW: F = 0,49; MSE = 5,332) (Root RDW: F = 0,76; MSE = 29,97.
Total RDW: F = 0,80; MSE = 6,020.

Figure 16. Relative shoot, root and total dry weights percentages obtained in WPM, NP1, NP2 and NP4 media.

Addition of silver nanoparticles caused a reduction in both fresh and dry biomass production rates per explant (*BPR - g/month/explant*). The highest BPRs were obtained in WPM medium (Figure 17 and 18). However, the effects of silver nanoparticles were different between shoot, root and total fresh BPRs: the highest shoot BPR was detected in NP4 medium; the root and total BPRs were higher in NP1 media (Figure 17).

In dry BPRs, WPM and NP1 media results were very close to each other (Figure 17). NP2 medium had the lowest values in shoot BPR (0,007 g) and total BPR (0.007 g). In root BPR, the lowest value was obtained in NP4 medium, which means roots are not as resistant as shoots to silver nanoparticles.

Application of 1,2 and 4 mg/L silver nanoparticle concentrations did not affect plantlets morphologically. As seen in Figure 19, no visible morphological deformation was observed both on plantlets and their roots formations in *in vitro* conditions.

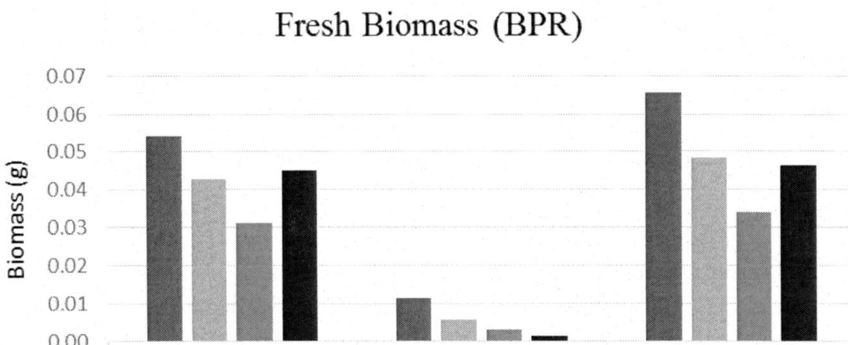

	Shoot Biomass (FW)	Root Biomass (FW)	Total Biomass (FW)
WPM	0.054	0.011	0.065
NP1	0.043	0.005	0.048
NP2	0.031	0.003	0.034
NP4	0.045	0.001	0.046

Figure 17. Fresh biomass accumulations obtained in WPM, NP1, NP2 and NP4 media.

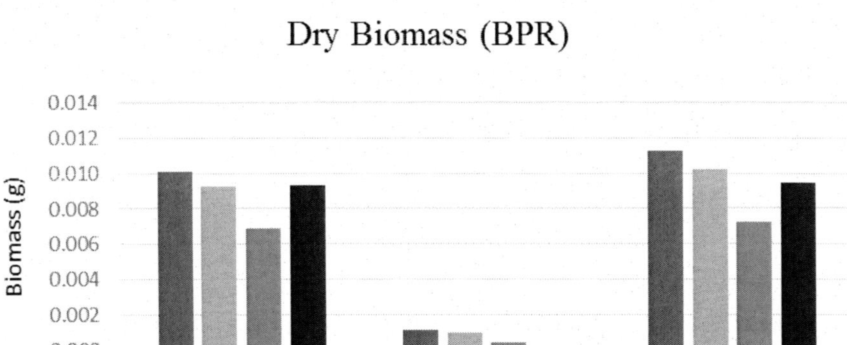

	Shoot Biomass (DW)	Root Biomass (DW)	Total Biomass (DW)
WPM	0.010	0.001	0.011
NP1	0.009	0.0010	0.010
NP2	0.007	0.0004	0.007
NP4	0.009	0.0002	0.009

Figure 18. Dry biomass accumulations obtained in WPM, NP1, NP2 and NP4 media.

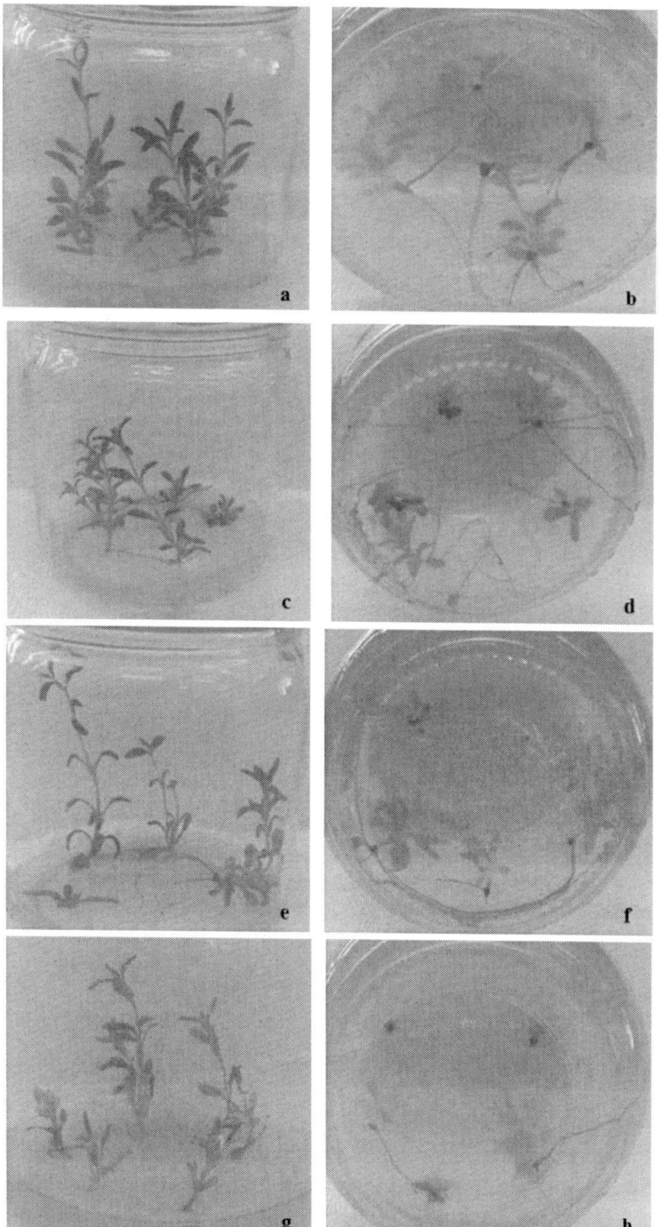

Figure 19. *In vitro*-grown *L. officinalis* plantlets and their root formations; (a, c, e, g) plantlet growth, (b, d, f, h) root formations in WPM, NP1, NP2, NP4 media, respectively.

Conclusion

Effects of nanoparticles have been investigated in many studies about plants. However, there is no information about the effects of biosynthetic silver nanoparticles in *in vitro* conditions. There are some studies about the effects of nanoparticles such as TiO_2 and CuO nanoparticles; Mandeh et al. (2012) reported that TiO2 nanoparticles increased callugenesis and the size of calli in barley cultures; Anwaar et al. (2016) reported that callus induction was enhanced by CuO nanoparticles in *Oryza sativa* cultures. In our study, it was observed that silver nanoparticles enhanced single shoot formations and average node length in *L. officinalis* node cultures. In our previous studies, it was also demonstrated that the use of silver nanoparticles had some beneficial effects in *in vitro* cultures.

In our opinion, silver nanoparticles obtained from different sources should be integrated into plant tissue culture studies more frequently. Especially, green-synthesized silver nanoparticles may be a low-cost option for more productive plant tissue culture studies.

References

Adamse, P. (1990). Selection of high-yielding cell lines of *Tagetes* using flow cytometry, Progress in plant cellular and molecular biology, *Proceedings of the VIIth International Congress on Plant Tissue Culture*, Amsterdam, The Netherlands, 24-29 June, (Nijkamp, HJJ, Van der Plas, LHW and Van Aartrijk, J. Eds.) Kluwer Academic Publishers, Dordrecht/Boston/London, pp. 726-731.

Aghdaei, M., Salehi, H. and Sarmast, M. K. (2012). Effects of silver nanoparticles on *Tecomella undulata* (Roxb) Seem, micropropagation. *Advances in Horticulture and Science*, 26 (1), 21–24.

Al-Sowaidi, W. M. M. and Al-Oubaidi, H. K. M. (2015). Increasing (glycosides compounds) of *Olea europaea* L. from shoot tips using

AgNO3 nanoparticle *in vitro*. *International Journal of Phytopharmacy*, 6 (1), 31–35.

Amin, M., Anwar, F., Janjua, M. R. S. A., Iqbal, M. A., and U. Rashid (2012). Green synthesis of silver nanoparticles through reduction with *Solanum xanthocarpum* L. berry extract: Characterization, antimicrobial and urease inhibitory activities against *Helicobacter pylori*. *International Journal of Molecular Science*, 13, 9923 - 9941.

Anjum, S., Abbasi, B. H. and Shinwari, Z. K. (2016). Plant-mediated green synthesis of silver nanoparticles for biomedical applications: Challenges and opportunities, *Pakistan Journal of Botany*, 48 (4): 1731 - 1760.

Anwaar, S., Maqbool, Q., Jabeen, N., Nazar, M., Abbas, F., Nawaz, B., Hussain, T. and Hussain, S. Z. (2016). The effect of green synthesized CuO nanoparticles on callogenesis and regeneration of *Oryza sativa* L., *Frontiers of Plant Sciences*, 7: 1-9.

Ashraf, M. A., Iqbal, M., Rasheed, R., Hussain, I., Riaz, M. and Arif, M. S. (2018). *Environmental Stress and Secondary Metabolites in Plants: An Overview. Plant Metabolites and Regulation Under Environmental Stress*, (eds.) Parvaiz Ahmad, Mohammad Abass Ahanger, Vijay Pratap Singh, Chapter 8, Elsevier, pp. 153-167, ISBN: 978-0-12-812689-9.

Babapour, A., Yang, B., Bahang, S. and Cao, W. (2012). Biocide silver nanoparticles in two different silica-based coating. International conference on fundamental and applied sciences 2012, American Institute of Physics, *AIP Conference Proceedings*, 1482, 681–684.

Baenas, N., García-Viguera, C. and Moreno, D. A. (2014). Elicitation: A tool for enriching the bioactive composition of foods. *Molecules*, 19, 13541–13563.

Bhat, P. and Bhat, A. (2016). Silver nanoparticles for the enhancement of accumulation of capsaicin in suspension culture of *Capsicum* sp. *Journal of Experimental Science*, 7 (2), 1–6.

Cai, Z., Kastell, A., Knorr, D. and Smetanska, I. (2012). Exudation: An expanding technique for continuous product and release of secondary metabolites from plant cell suspension and hairy root cultures, *Plant Cell Reports*, 31, 461 - 477.

Cao, H. (2017). Toward selectively toxic silver nanoparticles,' in Cao, H. (Ed.): '*Silver nanoparticles for antibacterial devices: biocompatibility and toxicity*,' Chapter 12 (RCR Press, Taylor & Francis Group, Boca Raton, 2017), ISBN 9781498725323.

Chattopadhyay, S., Farkya, S., Srivastava, A. K. and Bisaria, V. S. (2002). Bioprocess considerations for production of secondary metabolites by plant cell suspension cultures. *Biotechnology and Bioprocess Engineering*, 7: 138 - 149.

Dhoondia, Z. H. and Chakraborty, H. (2012). Lactobacillus mediated synthesis of silver oxide nanoparticles. *Nanomaterials and Nanotechnology*, 15 (2), 1–7.

dos Santos, C. A., Jozala, A. F., Pessoa, A. Jr. and Seckler, M. M. (2012). Antimicrobial effectiveness of silver nanoparticles co-stabilized by the bioactive copolymer pluronic F68. *Journal of Nanobiotechnology*, 10: 43.

Fang, X., Yang, C-Q, Wei, Y-K, Ma, Q-X, Yang, L. and Chen, X-Y. (2011). Genomics grand for diversified plant secondary metabolites. *Plant Diversity and Resources*, 33(1), 53–64.

Fazal, H., Abbasi, B. H., Ahmad, N. and Ali, M. (2016). Elicitation of medicinally important antioxidant secondary metabolites with silver and gold nanoparticles in callus cultures of *Prunella vulgaris* L. *Applied Biochemistry and Biotechnology*, 180 (6), 1076–1092.

Ghanati, F. and Bakhtiarian, S. (2016). Changes of natural compounds of *Artemisia annua* L. by methyl jasmonate and silver nanoparticles. *Advanced Environmental Biology*, 2251.

Gorczyca, A., Pociecha, E., Kasprowicz, M. and Niemiec, M. (2015). Effect of nanosilver in wheat seedlings and *Fusarium culmorum* culture systems. *European Journal of Plant Pathology*, 142 (2), 251–261.

Homaee, M. B. and Ehsanpour, A. A. (2015). Physiological and biochemical responses of potato (*Solanum tuberosum*) to silver nanoparticles and silver nitrate treatments under *in vitro* conditions. *Indian Journal of Plant Physiology*, 20 (4), 353–359.

Jones, M. C. and Hoek E. M. (2010). A review of the antibacterial effects of silver nanomaterials and potential implications for human health and the environment. *Journal of Nanoparticle Research*, 12: 1531-1551.

Kumar, V., Singh D. K., Mohan, S., Bano, D., Gundampati, R. K. and Hasan S. H. (2017). Green synthesis of silver nanoparticle for the selective and sensitive colorimetric detection of mercury (II) ion. *Journal of Photochemistry & Photobiology B: Biology*. 168: 67–77.

Lansdown, A. B. G. (2007). Critical observations on the neurotoxicity of silver. *Critical Reviews in Toxicology*, 37 (3), 237–250.

Lloyd, G. B. and McCown, B. H. (1980). Commercial-feasible micropropagation of mountain laurel- *Kalmia latifolia* by use of shoot-tip culture. *Proceedings - International Plant Propagators' Society*, 30, 421–427.

Mandeh, M., Omidi, M. And Rahaie, M., 2012. *In vitro* influences of TiO^2 nanoparticles on Barley (*Hordeum vulgare* L.) tissue culture, *Biological Trace Element Research*, 150 (1-3): 376-380.

Maria, B. S., Devadiga, A., Kodialbail, V. S. and M. B. Saidutta (2015). Synthesis of silver nanoparticles using medicinal *Zizyphus xylopyrus* bark extract. *Applied Nanoscience*, 5, 755–762.

Mohammed Al-Oubaidi, H. K. and Mohammed-Ameen, A. S. (2014). The effect of (AgNO3) NPs on increasing of secondary metabolites of *Calendula officinalis* L. *in vitro*. *International Journal of Pharmaceutical Therapeutics*, 5 (4), 267–272.

Mulabagal, V. and Tsay, H-S. (2004). Plant cell cultures - An alternative and efficient source for the production of biological important secondary metabolites. *International Journal of Applied Science and Engineering*, 2 (1), 29–48.

Naik, P. M. and Al-Khayri, J. M. (2016). Impact of abiotic elicitors on *in vitro* production of plant secondary metabolites: A review. *Journal of Advanced Resources and Biotechnology*, 1 (1), 7.

Nartop, P. (2017). Use of biosynthetic silver nanoparticles in the surface sterilization of Pyracantha coccinea stem explants, *Pamukkale University Journal of Engineering Sciences*, 23(6), 759-761.

Nartop P, (2018a). Engineering of Biomass Accumulation and Secondary Metabolite Production in Plant Cell and Tissue Cultures, *Plant Metabolites and Regulation under Environmental Stress*, (eds.) Parvaiz Ahmad, Mohammad Abass Ahanger, Vijay Pratap Singh, Chapter 9, Elsevier, pp. 169-194, ISBN: 978-0-12-812689-9.

Nartop, P. (201b). Effects of surface sterilization with green synthesized silver nanoparticles on *Lamiaceae* seeds, *IET Nanobiotechnology*, 12(5), 663 - 668.

Nartop, P. (2018c). Green sterilization of *Rosmarinus officinalis* L. stem surfaces with silver nanoparticles synthesized using *Rubia tinctorum* L. cell culture extracts, *Iranian Journal of Science and Technology, Transactions A: Science*, 42 (2), 411 - 414.

Nartop, P. (2018d). *In vitro* biomass production of *Astragalus gilvus*, *Iranian Journal of Science and Technology, Transactions A: Science*, 42 (3), 1157-1161.

Nartop, P. (2019). Silver nanoparticles: Ecofriendly surface sterilization of plant seeds in different shapes and sizes, *Journal of Animal and Plant Sciences*, 29 (2), 453 - 460.

Nartop, P. and Gürel, A. (2014). Immobilization of Plant Cells and Its Effects on Secondary Metabolite and Biomass Production, *Biotechnology and Genetic Engineering II of Recent Progress in Medicinal Plants Series, Multi-Volume Set*, Vol. 39: 1-933699-99-X Series, ISBN: 0-9656038-5-7 (Chapter), Studium Press.

Nartop, P., Gurel, A., Akgun, I. H. and Bedir, E. (2015). Astragaloside IV and cycloastragenol production of *Astragalus trojanus* Stev. *in vitro* conditions. *Indian Journal of Biotechnology*, 14, 540–546.

Pal, S., Tak, Y. K. and Song, J. M. (2007). Does the antibacterial activity of silver nanoparticles depend on the shape of the nanoparticle? A study of the gram negative bacterium *Escherichia coli*. *Applied Environmental Microbiology*, 73 (6), 1712–1720.

Perello, M. (2013). Effects of direct and dietary exposure to silver nanoparticles on a tritrophic System, *Proceedings of the National Conference on Undergraduate Research* (NCUR), University of Wisconsin La Crosse, WI, April 11–13.

Raei, M., Angaji, S. A., Omidi, M. and Khodayari, M. (2014). Effect of abiotic elicitors on tissue culture of *Aloe vera. International Journal of Bioscience*, 5 (1), 74–81.

Raffi, M., Hussain, F., Bhatti, T., Akhter, J., Hameed, A. and Hasan, M. (2008). Antibacterial characterization of silvernanoparticles against *E. coli* ATCC-15224. *Journal of Materials Science and Technology*, 24, 192 - 196.

Rai, M. and Yadav, A. (2013). Plants as potential synthesiser of precious metal nanoparticles: Progress and prospects. *IET Nanobiotechnology*, 7, 117 - 124.

Rao, R. S. and Ravishankar, G. A. (2002). Plant cell cultures: Chemical factories of secondary metabolites. *Biotechological Advances,* 20, 101–153.

Scott, C. D. and Dougall, D. K. (1987). Plant cell tissue culture: A potential source of chemicals. Martin Marielta Energy Systems, INC for the US Department of Energy.

Sinha, S. N., and Paul D. (2014). Eco-friendly green synthesis and spectrophotometric characterization of silver nanoparticles synthesized using some common Indian spices. *International Journal of Green Herbal Chemistry*, 3 (2), 401 - 408.

Sintubin, L., Verstraete, W. and Boon, N. (2012). Biologically produced nanosilver: current state and future perspectives. *Biotechnology and Bioengineering,* 109 (10), 2422–2436.

Tran, Q. H., Nguyen, V. Q. and Le, A-T. (2013). Silver nanoparticles: synthesis, properties, toxicology, applications and perspectives. *Advances in Natural Science: Nanoscience and Nanotechnology*, 4, 033001 (20 pp).

Yasin, S., Liu, L. and Yao, J. (2013). Biosynthesis of silver nanoparticles by bamboo leaves extract and their antimicrobial activity. *Journal of Fiber Bioenginerring and Informatics*, 6 (1), 77–84.

Zafar, H., Ali, A., Ali, J. S., Ihsan, U., Haq, I. U. and Zia, M. (2016). Effect of ZnO nanoparticles on *Brassica nigra* seedlings and stem explants: growth dynamics and antioxidative response. *Frontiers in Plant Sciences*, 7, 535.

Zaka, M. and Abbasi, B. H. (2016). Effects of bimetallic nanoparticles on seed germination frequency and biochemical characterisation of *Eruca sativa*. *IET Nanobiotechnology*, http://dx.doi.org/10.1049/iet-nbt.2016.0004.

Zhang, B., Zheng, L. P., Ping, Y. L. and Wen Wang, J. (2013). Stimulation of artemisinin production in *Artemisia annua* hairy roots by Ag-SiO_2 core-shell nanoparticles. *Current Nanoscience*, 9 (3), 363–370 (8).

In: The Lamiaceae Family: An Overview ISBN: 978-1-53617-078-8
Editor: Alexander Adler

Chapter 5

VARIATION OF SECONDARY METABOLITES IN *LAMIACEAE* HERBS WATER EXTRACTS

Hasib Ahmad[1,*] and Yoichi Matsubara[2*]
[1]The United Graduate School of Agricultural Science, Gifu University, Gifu, Japan
[2]Faculty of Applied Biological Sciences, Gifu University, Gifu, Japan

ABSTRACT

Lamiaceae, also known as Mint family is well known for including many medicinally important plants used in flavouring, preservative and especially medicinal purpose due to having curative and preventive properties. They have also found their use in combating crop pest and diseases owing to containing a complex mixture of secondary metabolites. However, information about the variation of these compounds in different *Lamiaceae* herbs especially in water extracts is a less ventured area. As such, in the present study, the variation of major secondary metabolites in water extracts of 13 different *Lamiaceae* herbs was investigated. After raising for 8 weeks in the greenhouse, the herbs

* Corresponding Author's Email: ymatsu@gifu-u.ac.jp

were harvested and cryopreserved using liquid nitrogen. Afterward, sample extracts (20% w/v) were prepared from the leaf using ultrapure water and analyzed using UPLC-MS/MS. The top three peaks observed in the chromatogram of each herb and their corresponding m/z values were cross-checked with the mass bank and the identified compounds were considered to be the major components of the extract. About six flavonoids (apigenin, camphenol, luteolin, epigallocatechin, oenin and limocitrin) and two compounds involved in flavonoid synthesis (shikimic acid and coumaric acid) were identified. Furthermore, some other phenolic compounds like geniposide, rosmarininc acid and camphene were also identified. Among the 13 herbs, presence of apigenin in 9, kaempferol in 7, geniposide in 6, shikimic acid in 5, luteolin in 5 and coumaric acid in 2 were determined. Oenin and camphene were only found in hyssop whereas epigallocatechin was only found in rosemary. Rosmarinic acid and limocitrin were also found individually only in catnip and thyme respectively. From these results, it can be stated that the herbs under *Lamiaceae* family contain a variety of secondary metabolites which undoubtedly plays an important role in their characteristics as well as potential use.

INTRODUCTION

Plants have been the primary food source for humans since the beginning of time and will continue to be so. Besides being the means of sustenance, plants have also been utilized for overcoming various other obstacles that hurdled the progress of human race. One of such hurdles are the diseases that plagued humans both directly and indirectly. Plants have been the source for medicinal treatments for both as curative and preventive measures for thousands of years (Carović-Stanko et al., 2016). According to the World Health Organization (WHO), about 80% of the world population still relies mainly on plant-based drugs (Bahmani et al., 2014), thus lowering at the same time the impact of self-medication side effects (Alexa et al., 2014). Plants contains a vast array of natural compounds that are responsible for these actions which pave ways to novel remedies with less side effects compared to synthetics.

Among the various plants used for this purpose, the *Lamiaceae* is one of the most important herbal families that has been used in folk medicine since ancient times. It contains about 236 genera and more than 6000 species and the largest genera are *Salvia* (900), *Scutellaria* (360), *Stachys* (300), *Plectranthus* (300), *Hyptis* (280), *Teucrium* (250), *Vitex* (250), *Thymus* (220) and *Nepeta* (200) (Raja, 2012). The most popular members of the family are thyme, mint, oregano, basil, sage, savory, rosemary, hyssop, lemon balm etc due to their aroma and flavor. They grow in different agroclimatic range and many of these are cultivated for use in cosmetics, flavoring, fragrance, perfumery, pesticides and pharmaceuticals industries (Özkan, 2008). The essential oils as well as other secondary metabolites present in these herbs are used in these industries.

Plants generally being incapable of movement and lacking an immune system like animals, are prey to different external sources like herbivors and microbes. To protect against these external factors, plants have developed their own security system through production of several secondary metabolites in the evolutionary process. Secondary metabolites are chemical compounds that are produced in a plant cell through the primary metabolic pathway. They are shown to have various biological effects like antiallergenic, antidiabetic, antioxidant, antimutagenic, anticarcinogenic, antimicrobial, and anti-inflammatory agents, enhancers of the gastrointestinal function, immune-modulators and stimulators as well as blood pressure and cholesterol reducing agents (Vaishali et al., 2013), which makes the basis of their use in traditional medicine. Beside these, antioxidative and antimicrobial activities of the *Lamiaceae* herb extracts against plant pathogen in vitro have also been reported (Gomes et al., 2014). However, these attributes of the secondary metabolites were derived from the organic solutions of the herb extracts as well as from essential oils (EOs). Although in case of human disease control these organic extractions and essential oils are quite applicable, for use in large scale agricultural system, it becomes impractical due to certain innate characteristics (volatility) and high cost of organic extraction process. Conversely, a suitable substitute for it could be the use of water extracts of herbs. The process of extraction with water is relatively easy and inexpensive as well as they do not have the problem of spontaneous volatility that EOs have. However, information about the secondary metabolites present in water extracts of *Lamiaceae* herbs is a very less ventured area. As such, in the present study, we investigated the variation of major secondary metabolites present in 13 different *Lamiaceae* herb water extracts.

MATERIALS AND METHODS

Growing Herbs and Preparation of Its Water Extract

Seeds of the 13 *Lamiaceae* herbs i.e., oregano (*Origanum vulgare* L.), Catnip (*Nepeta cataria* L.), sage (*Salvia officinalis* L.), dark opal (*Ocimum* spp.), thyme (*Thymus vulgaris* L.), basil (*Ocimum basilicum* L.), hyssop (*Hyssopus officinalis* L.), peppermint (*Mentha piperita* L.), lamb's ear (*Stachys byzantina* K.), lemon balm (*Melissa officinalis* L.), rosemary (*Rosmarinus officinalis* L.), sweet marjoram (*Origanum majorana* L.) and summer savory (*Satureja hortensis* L.) were sown in separate plastic containers (31.9 cm × 26.4 cm × 15.3 cm) containing commercial soil (Supermix A, Sakata Co. Ltd., Japan) and grown in a greenhouse. Eight weeks after sowing, the plants were uprooted and the shoots were cryopreserved using liquid nitrogen.

Analysis of Herb Water Extracts Using UPLC-MS/MS and Identification of Major Compounds

From the cryopreserved samples of five plants of each herb, 0.6 g of leaves were pulverized separately in a mortar with liquid nitrogen to fine powder and mixed with 3 ml ultrapure water to prepare sample extract solution of each herb (20%, w/v). The sample solutions were then centrifuged (13,000 rpm, 4°C, 15 min) and the supernatants were filtered through a sterilizing filter (0.45 μm, ADVANTECH Co. Ltd., Japan). The sample was centrifuged (13,000 rpm, 4°C, 15 min) using Nanosep 10K (Nihon Pall Ltd. Tokyo, Japan) to remove proteins in each extract.

The samples were analyzed using UPLC-MS/MS (Waters Corporation, Milford, USA). A reversed-phase column (ACQUITY UPLC BEH C18, 1.7 μm, 2.1 × 100 mm, Waters Corporation, Milford, USA) with thermostation at 25°C was used for the analysis. The mobile phases comprised 0.1% formic acid in water (A) and 0.1% formic acid 2-propanol-acetonitrile (30:70, v/v) (B) at a flow rate of 4 ml/min. The gradient profile was as follows: 0-6 min, 95% A; 6-12 min, 75% A; 12-30 min, 65% A; 30-32.5 min, 5% A and 32.5-35 min, 95% A. The sample injection volume was 10 μl. The mass spectrometer (Xevo Q Tof MS, Waters Corporation, Milford, USA) analyzed the mass range of electrospray ionization in negative mode at 100-1000 m/z; MS/MS collision was performed at 30V. A mass chromatogram of the m/z value of each component in the extract was prepared from the measurement results obtained using retention time.

From the resulting chromatogram, the three major peaks were selected based on the retention time and the corresponding m/z values were identified from the spectrum graph. Then, these values were cross referenced with MassBank database (www.massbank.jp) to identify the compounds.

RESULTS

The analysis of the different herb extracts was conducted by liquid chromatography-mass spectrometry (LC-MS) and represented as chromatogram and mass spectrum. All the herbs expressed different level of compounds present in the water extract at various retention times. From the chromatogram, the most promising region of major compounds in these herbs were observed within retention times ranging from 9.00 min to 20 min except for Lamb's ear where highest was observed at 31.34 min. Interestingly, the peak sizes of the lamb's ear extract compared to the other herbs were considerably smaller. The top three peaks within this range was selected for each herb and considered to be the major components of the respective herb extracts (Figure 1). The m/z values corresponding to the top three retention time of each herbs were identified from the mass spectrum graph (Figure 2) and these values were then cross reference with the MassBank database and the compounds were identified. The presence of Apigenin (270.05 m/z) in 9 types, kaempferol (286.05 m/z) in 7 types, geniposide (388.14 m/z) in 6 types, shikimic acid (174.05 m/z) in 5 types, luteolin (286.05 m/z) in 5 types and coumaric acid (164.05 m/z) in 2 types out of the 13 herbs was detected. In addition, components found solely such as oenin (493.13 m/z) and camphene (136.13 m/z) in hyssop, epigallocatechin (306.07 m/z) in rosemary, rosmarinic acid (360.08) in catnip, and limocitrin (346.07 m/z) in thyme were also detected. Table 1 illustrates the three major compounds present in each herb.

Among the 11 compounds that were identified, 6 were under flavonoid class (apigenin, kaempferol, luteolin, epigallocatechin, oenin and limocitrin), 2 were under prenol lipid class (geniposide and camphene), 2 were under cinnamic acid derivative class (coumaric acid, rosmarinic acid) and 1 under organo oxygen compound class (shikimic acid).

Table 1. Major compounds present in water extract of each herb

Herb	**Apigenin**	**Kampherol**	**Geniposide**	**Shikimic acid**	**Luteolin**	**Coumaric acid**	**Epigallocatechin**	**Rosmarinic acid**	**Oenin**	**Camphene**	**Limocitrin**
Oregano	✓		✓	✓							
Sage	✓	✓	✓								
Hyssop		✓							✓	✓	
Lemon balm		✓	✓		✓						
Basil	✓			✓		✓					
Peppermint	✓	✓		✓							
Dark opal	✓				✓	✓					
Thyme		✓			✓						✓
Rosemary			✓		✓		✓				
Catnip	✓			✓				✓			
Lamb's ear	✓	✓		✓							
Sweet marjoram	✓	✓	✓								
Summer savory	✓		✓		✓						

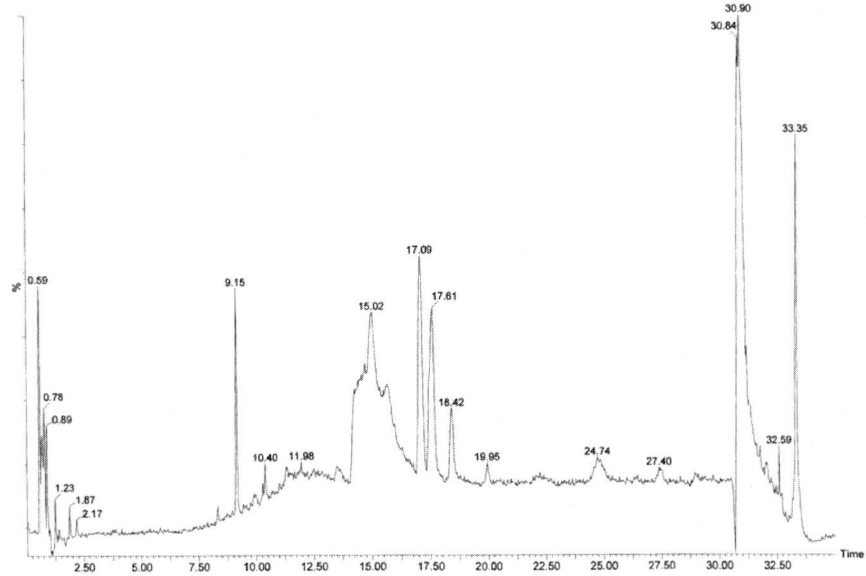

(a)

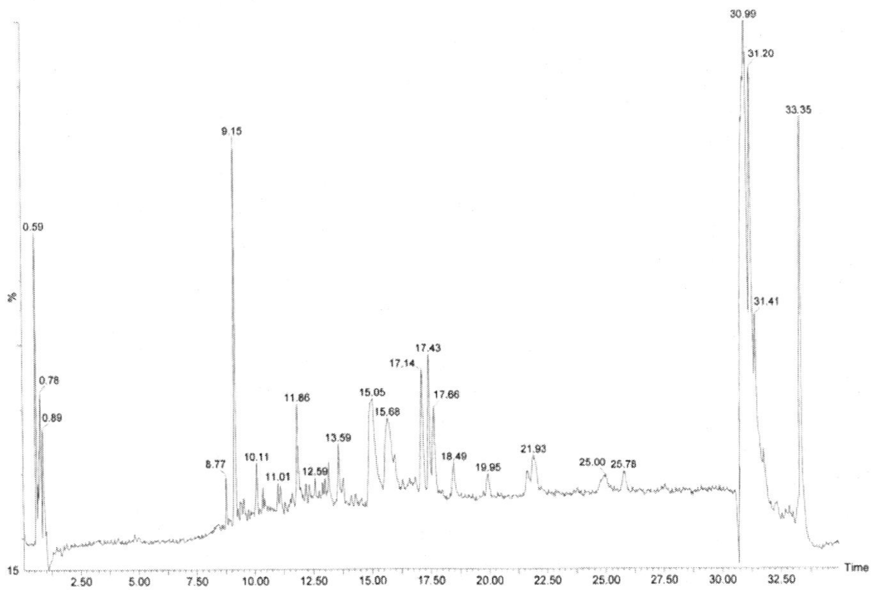

(b)

Figure 1. (Continued).

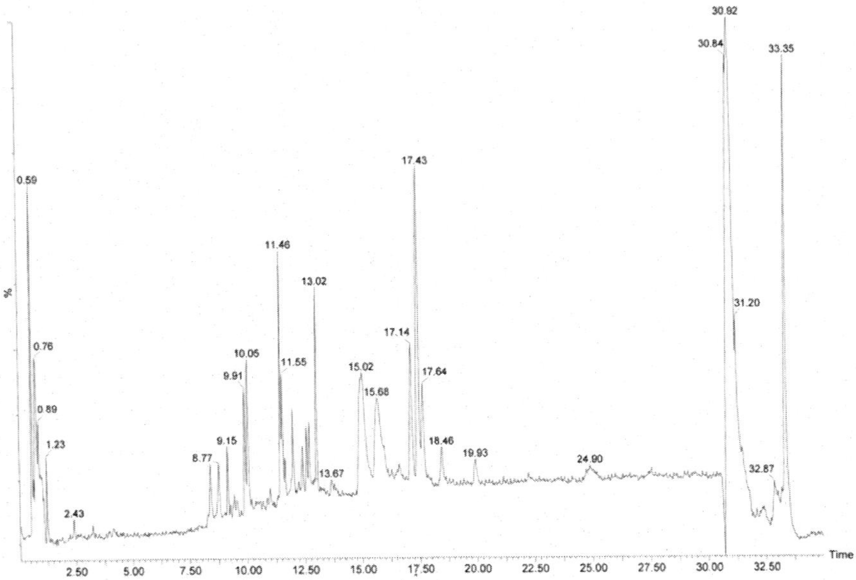

(c)

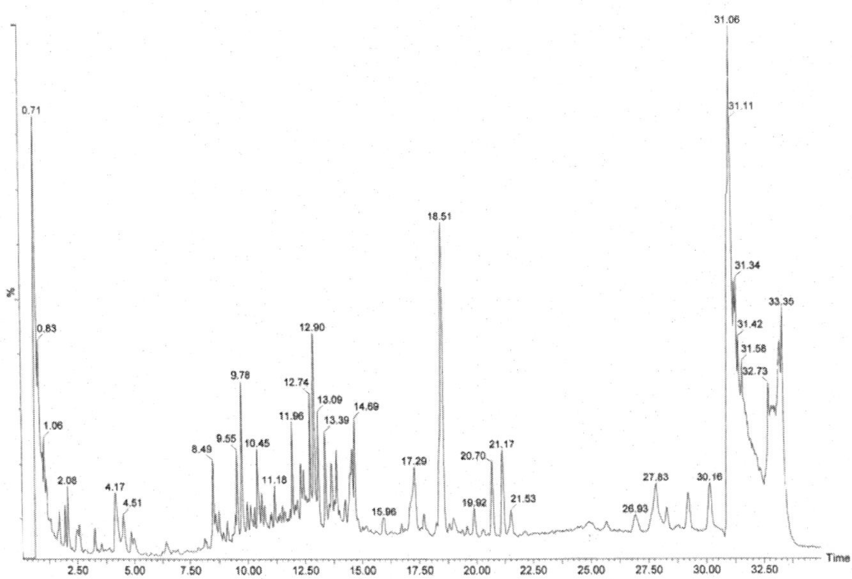

(d)

Figure 1. (Continued).

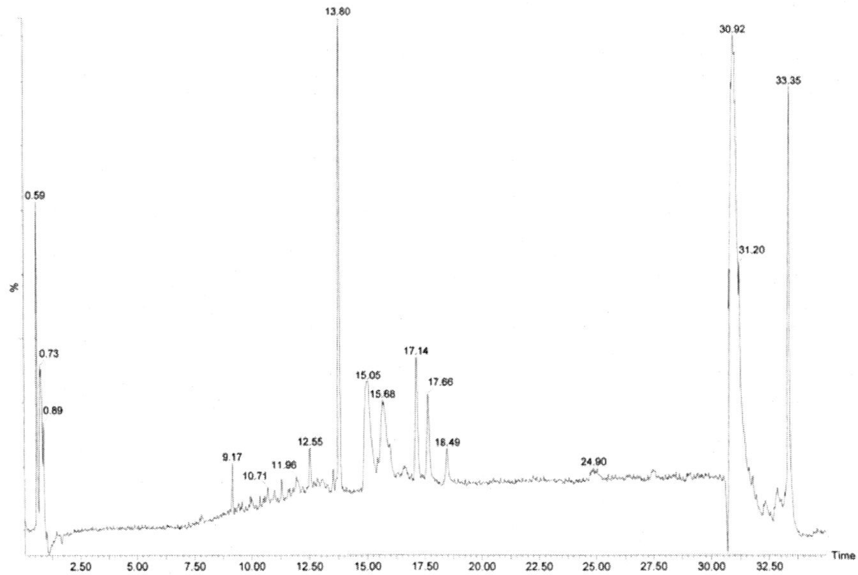

(e)

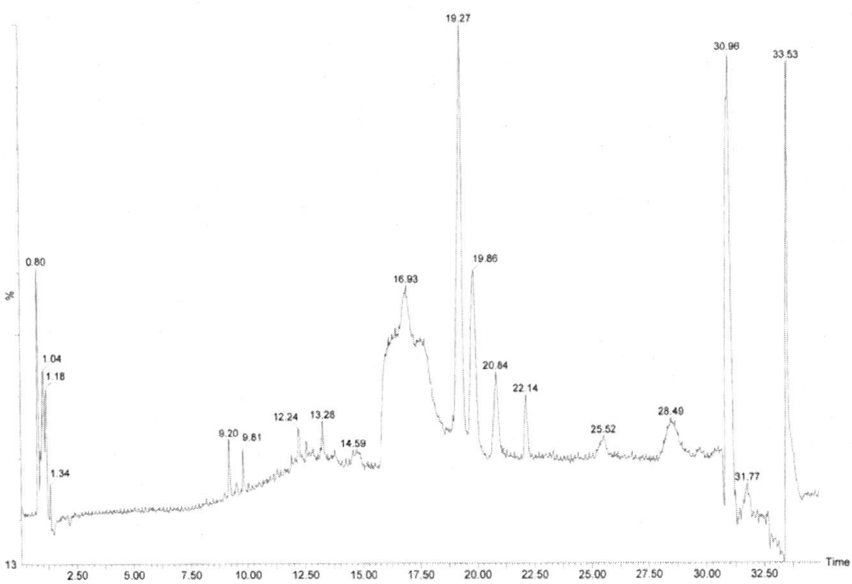

(f)

Figure 1. (Continued).

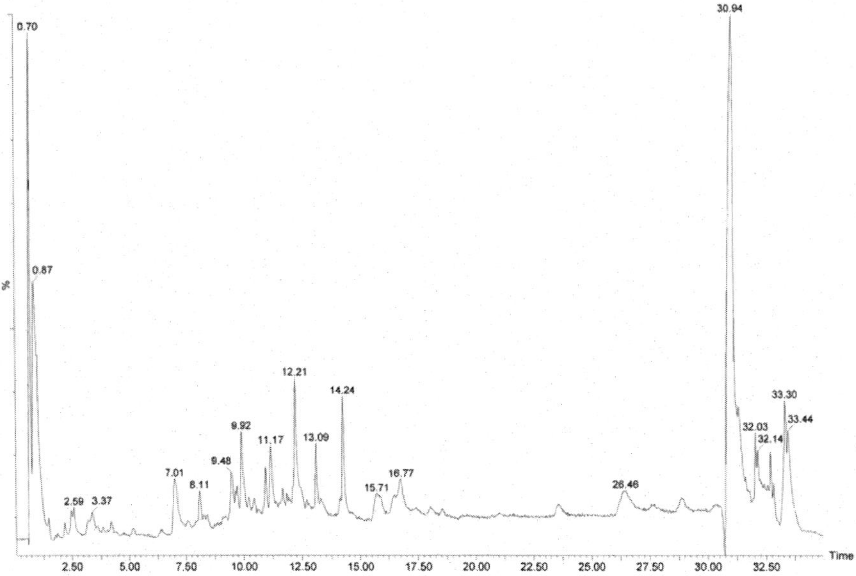

(g)

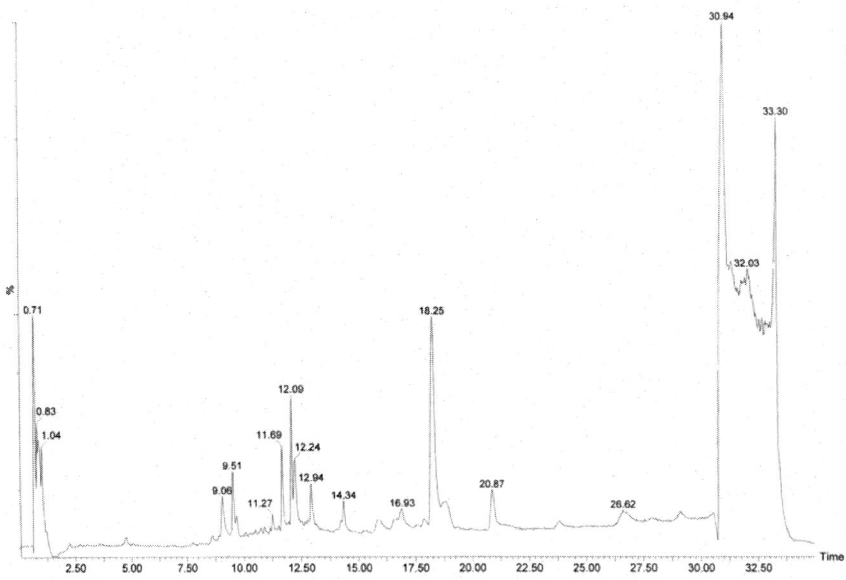

(h)

Figure 1. (Continued).

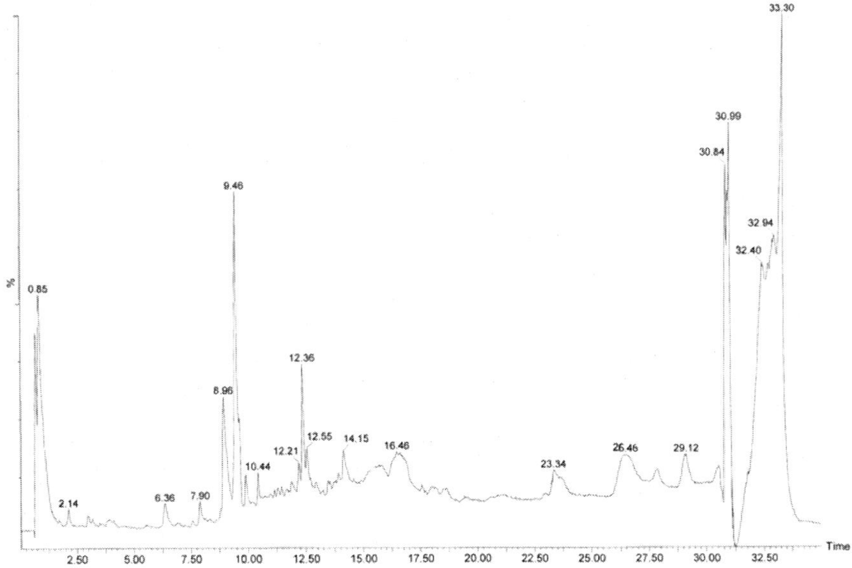

(i)

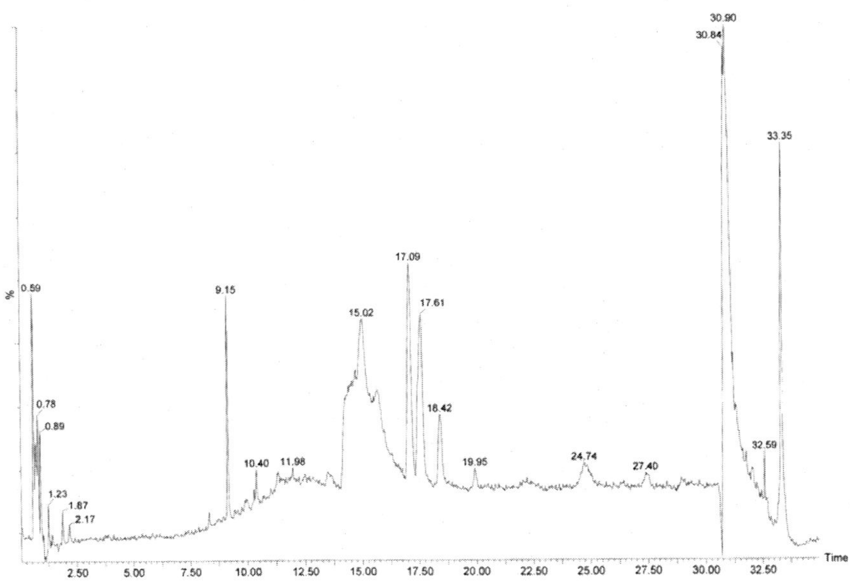

(j)

Figure 1. (Continued).

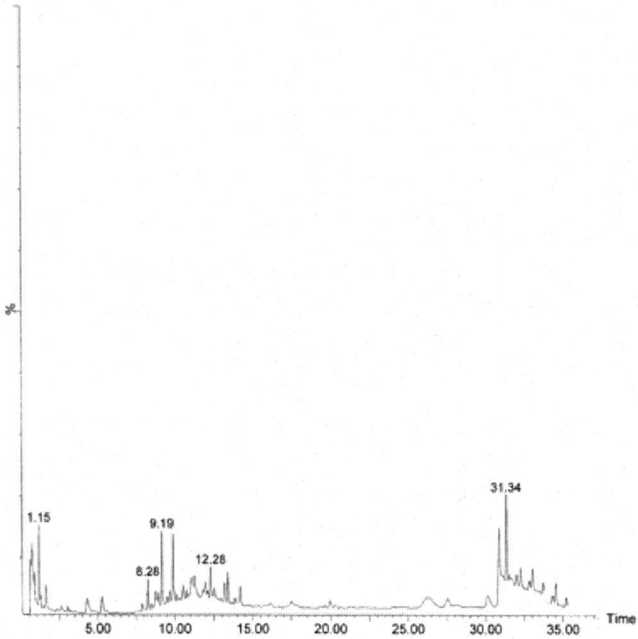

(k)

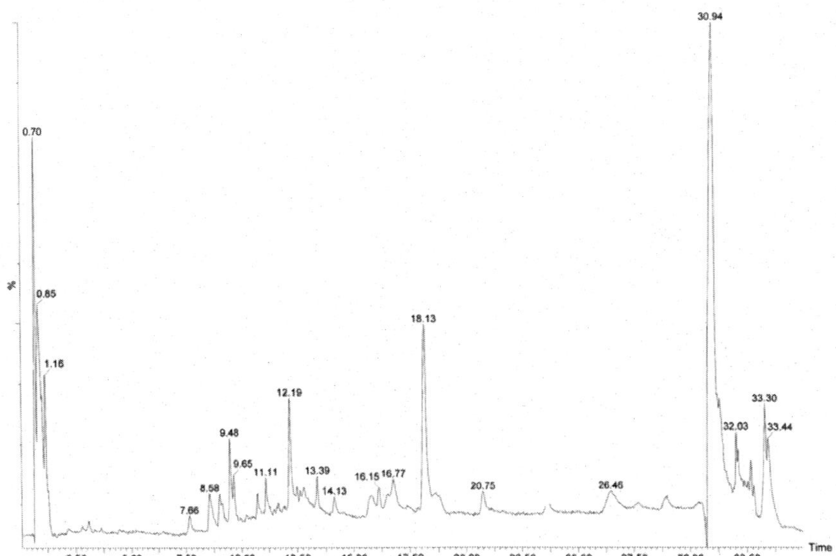

(l)

Figure 1. (Continued).

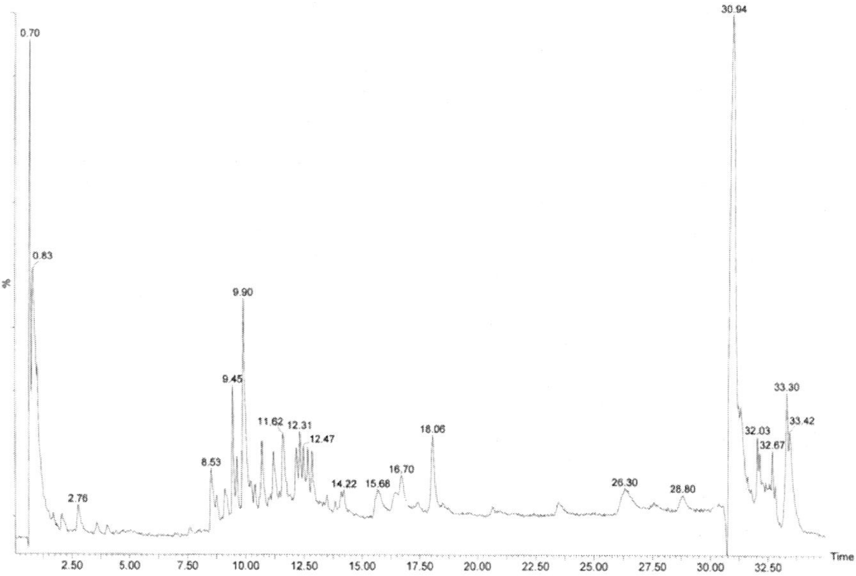

(m)

Figure 1. Chromatogram of 13 *Lamiaceae* herbs water extract derived by UPLC-MS. Here, a, oregano; b, sage; c, hyssop; d, lemon balm; e, basil; f, peppermint; g, dark opal; h, thyme; i, rosemary; j, catnip; k, lamb's ear, l, sweet marjoram; m, summer savory.

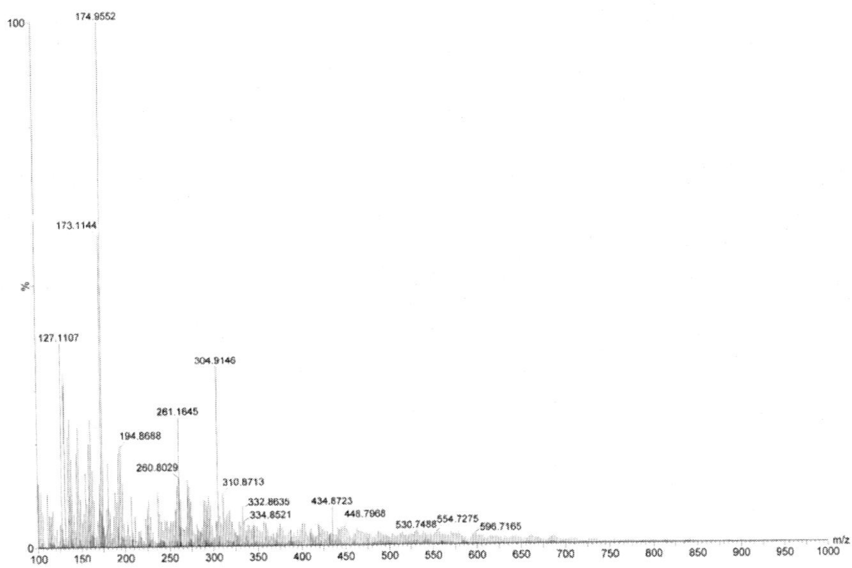

(a)

Figure 2. (Continued).

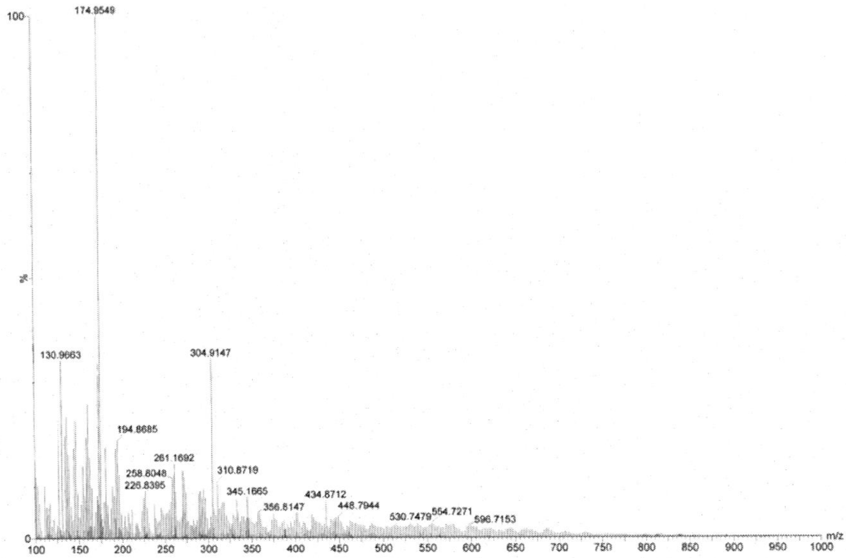

(b)

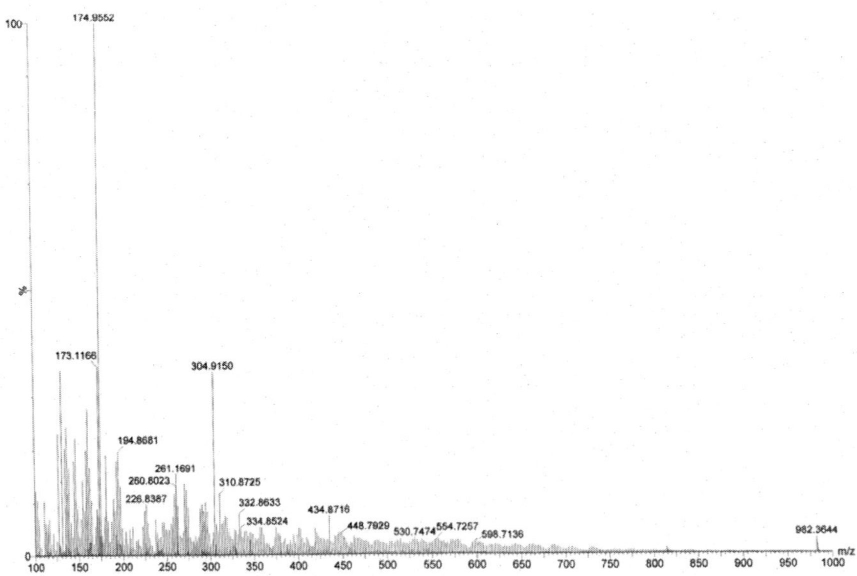

(c)

Figure 2. (Continued).

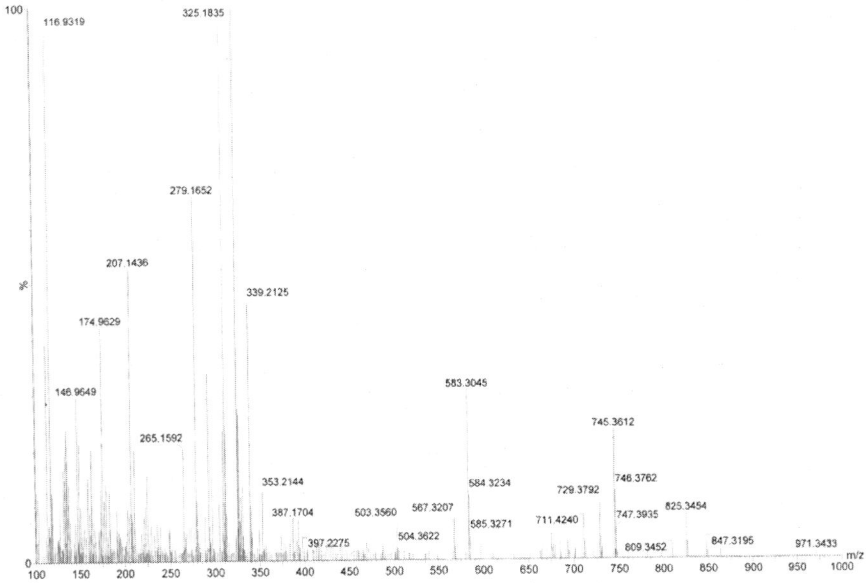

(d)

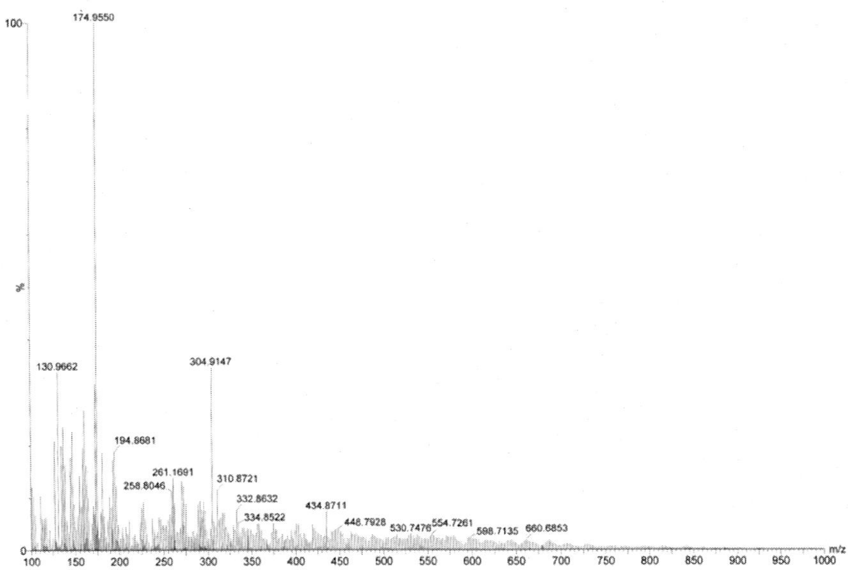

(e)

Figure 2. (Continued).

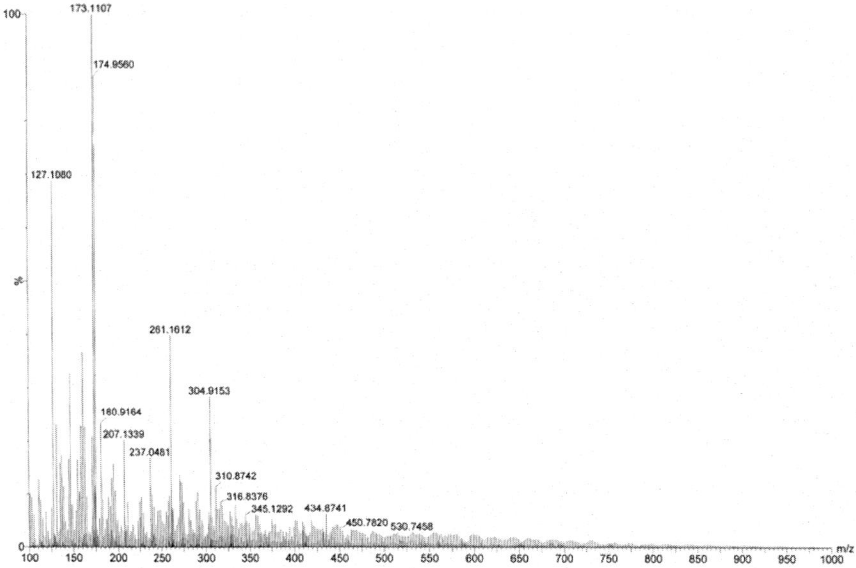

(f)

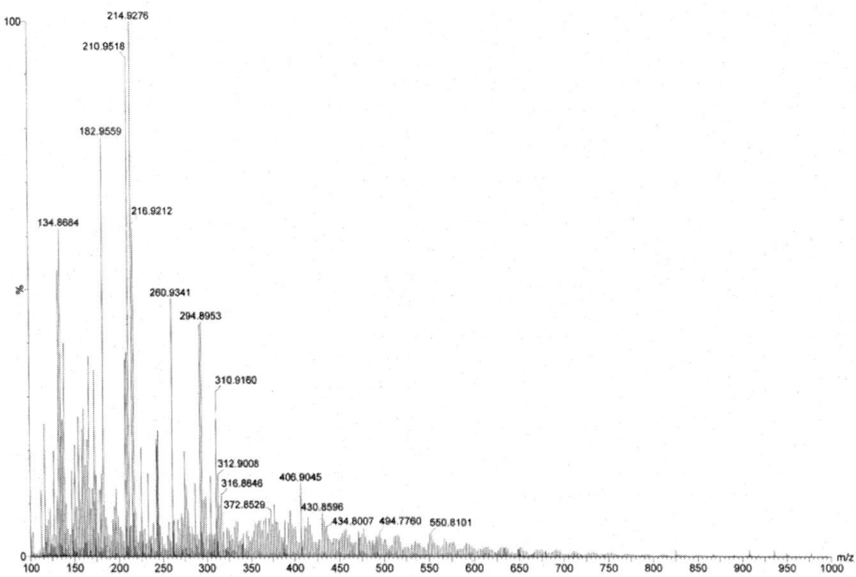

(g)

Figure 2. (Continued).

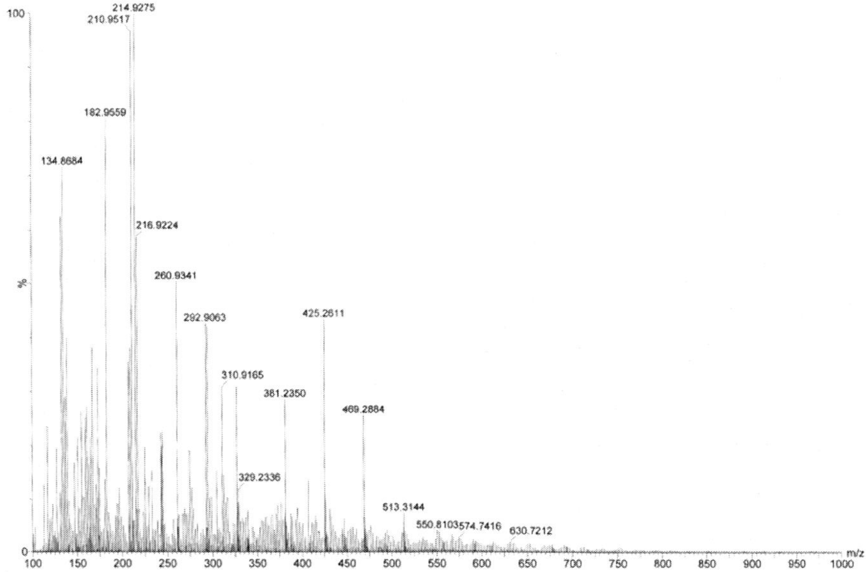

(h)

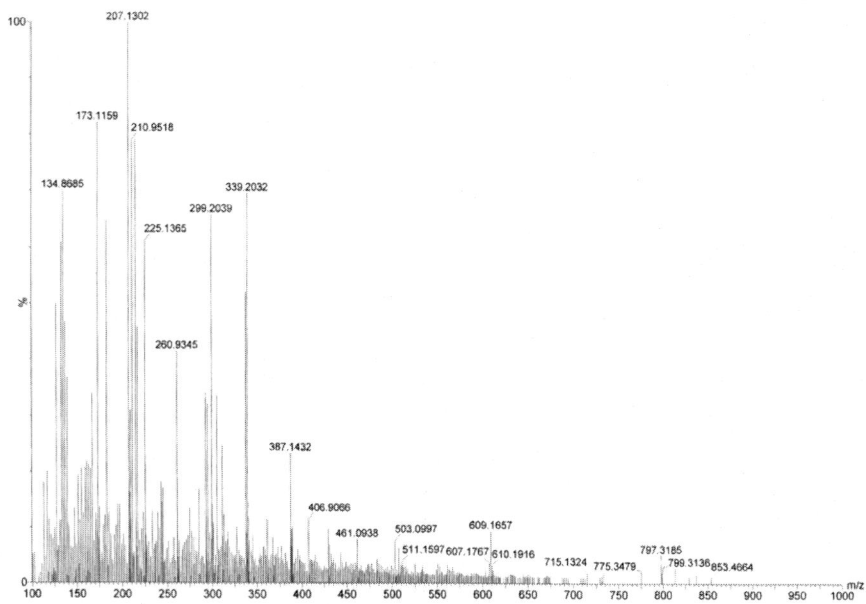

(i)

Figure 2. (Continued).

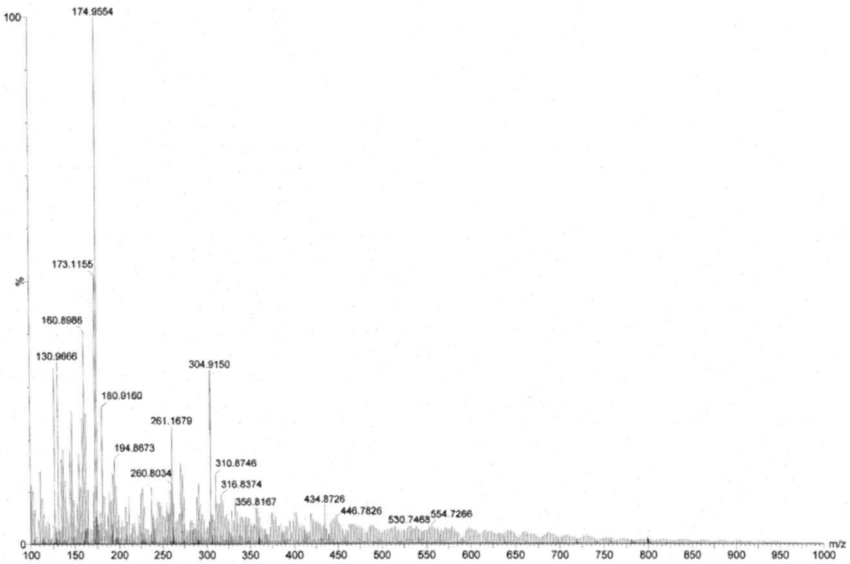

(j)

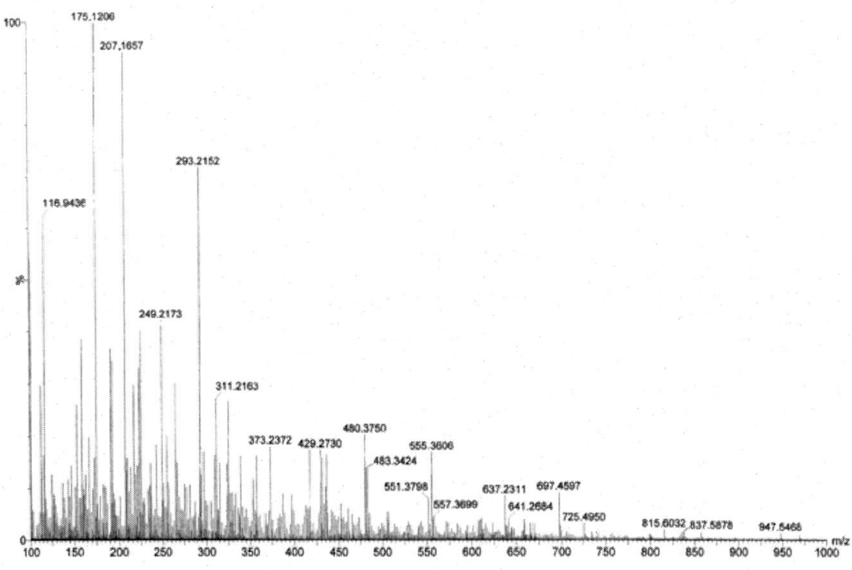

(k)

Figure 2. (Continued).

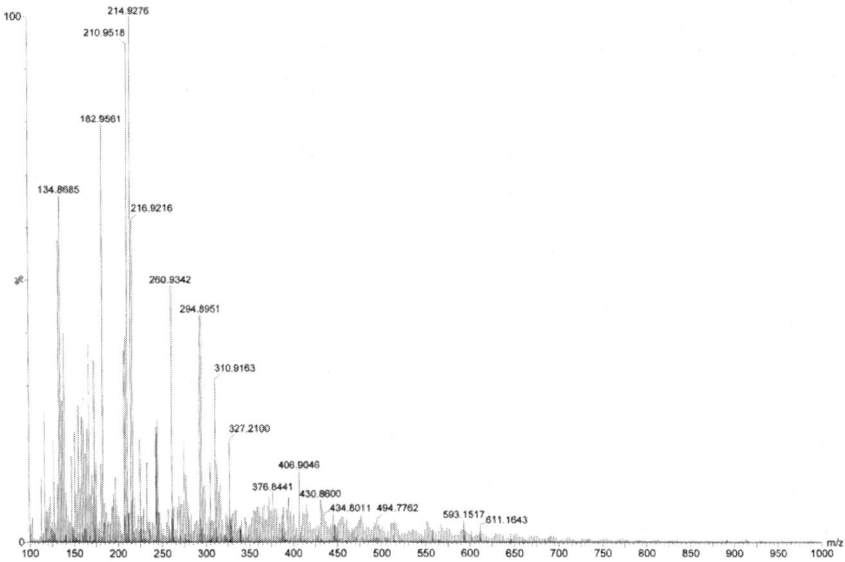

(l)

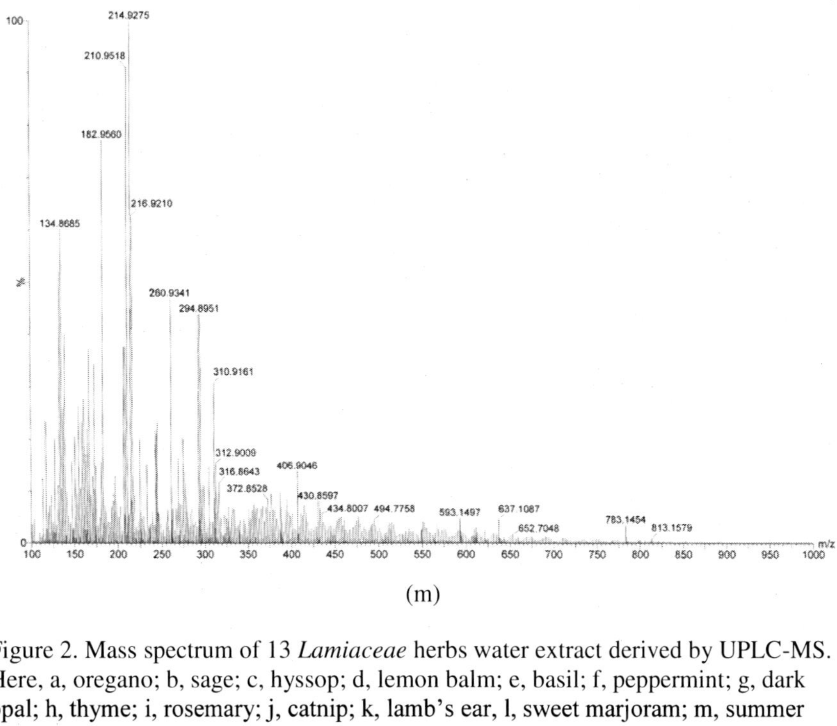

(m)

Figure 2. Mass spectrum of 13 *Lamiaceae* herbs water extract derived by UPLC-MS. Here, a, oregano; b, sage; c, hyssop; d, lemon balm; e, basil; f, peppermint; g, dark opal; h, thyme; i, rosemary; j, catnip; k, lamb's ear, l, sweet marjoram; m, summer savory.

DISCUSSION

A large number of secondary metabolites are produced in plants that serves them as defense compounds against various biotic and abiotic challenges through direct prevention or by acting as a signal compound (Wink, 2015). Variation in secondary metabolites in plants under same family are often found which results from various stimulus (Pavarini et al., 2012). The medicinal plants under *Lamiaceae* family are rich in secondary metabolites and such variation are also present in the family (Shanaida et al., 2018). However, most of the information regarding this is recorded on the basis of organic extraction. Researchers have used different organic solvents for extracting these secondary metabolites from plant materials (Hakkim et al., 2008; Yesiloglu et al., 2013; Karau et al., 2015). Some of these secondary metabolites are partially hydrophilic in nature hence can be extracted by water. Information regarding the major components found in water extract of *Lamiaceae* herbs thus is a less ventured area.

In the present study, the major compounds present in the water extract of 13 *Lamiaceae* herbs were investigated though UPLC-MS and three most important compounds were selected based on the retention time. The retention time peak area is an important indicator of compound concentration in an extract. The major compounds identified in each herb extracts were present at a high concentration as observed from the peak size except for lamb's ear. Although the three compounds present in lamb's ear extract were identified as apigenin, kaempferol and shikimic acid, their concentration was very low as observed from the chromatogram. This could possibly impact the effectivity of the extract in practical use.

The herb extracts were found to contain a range of different secondary compounds, some of which were found individually to a specific herb whereas some were common among several herbs. As a common understanding, plants under same family or clads often contain similar secondary metabolites (Liu et al., 2017), although variation in content was observed in different species (Zgórka and Glowniak, 2001). Among the identified compounds, apigenin was found in maximum number of herbs. As a flavone subclass of flavonoids, apigenin is widely distributed in many plant species which possess various biological activities like anticancer, antibacterial, antioxidant, anti-apoptosis and anti-inflammatory (Wei et al., 1990; Sharma et al., 2014, Wang et al., 2014; Banerjee et al., 2015). Importance of apigenin in human health promotion has gained importance in recent years due to its low intrinsic toxicity and anticancer effect compared to other flavonoids (Gupta et al., 2001). Besides it has a number of therapeutic uses against several diseases. Furthermore, the antioxidant and antigenotoxic effect in human health functionality of apigenin is also well known (Ali et al., 2017). The role of apigenin in plant defense mechanism is also quite important as it acts as antifeedant to pests (Sosa et al., 2004) and plays a role in UV damage reduction to plants by regulation of growth hormones (Potters et al., 2009).

The flavonoid kaempferol was found to be present in 7 out of the 13 herbs that were evaluated in the current study. It is a yellow compound with low molecular weight (286.2 g/mol) and is commonly found in plants used in traditional medicine such as *Lamiaceae* herbs (Calderón-Montaño et al., 2011). For instance, kaempferol shows activity against gram-positive and gram-negative bacteria, as well as against the fungus *Candida glabrata* (Saleem et al., 2010). Besides, consumption of kaempferol rich foods were found to reduce risk of cancer, cardiovascular disease, pancreatic disease etc. in humans (Hertog et al., 1993). Along with human health protection, the potential of kaempferol in plant defense mechanism is also found out. For instance, kaempferol, quercetin and myricetin (flavonols) act as deterrents against *Radopholus similis* and *Meloidogyne incognita* and kaempferol in addition restricts their hatching (Wuyts et al., 2006). It plays an important role in protection of plants from UV radiation (Fini et al., 2011) and acts as antioxidant under stress condition. Kaempferol also plays an important role in formation of mycorrhizal association and it has been reported to have promoting effect on hyphal growth and spore germination of mycorrhizal fungi (Tsai et al., 1991).

Geniposide, identified in 6 out of the 13 herbs is also a very important flavonoid which has multifaceted effect in both human and plant system. In case of human health aspect, geniposide was identified to play an important role in neuroprotection and as antidiabetic (Liu et al., 2015), anti-inflammatory (Deng et al., 2018) and against liver disease (Meng et al., 2013). Furthermore, this compound also plays an important role in plant defense response due to its insecticidal, antibacterial and antifungal potential as well as high radical scavenging activity (Prashith et al., 2017). Geniposide and related iridoid compounds has the ability to create barrier that releases enzymes against fungi and might cause direct modification to structural and functional properties of the fungal cell wall (Roldán-Arjona et al., 1999; Anisimov et al., 2000) which is responsible for the antimicrobial activity.

Shikimic acid, an important intermediate of lignin biosynthesis, is a natural compound exclusively found in plants which is also involved in aromatic amino acid and alkaloid synthesis in both plants and microorganisms (Enrich et al., 2008). It has great importance in medicinal industry as it is used for manufacturing of antiviral and anticancer drugs (Bradley, 2005; Zhang et al., 2006). On the other hand, shikimic acid also garners great interest in agricultural production system due to having herbicidal as well as antibacterial effect (Song et al., 2001)

Luteolin is one of the most common flavones found in plant kingdom. A large amount of vegetables fruits and herbs contains luteolin. *Lamiaceae* herbs also are rich in luteolin content (Gutiérrez-Grijalva et al., 2017, Ordaz et al., 2018). The anticarcinogenic effect of luteolin makes it a valuable nutraceutical. It prevents cancer cell development through neutralizing carcinogenic stimuli, tumor cell proliferation, by induction of cell cycle arrest and by induction of apoptosis via intrinsic and extrinsic signaling pathways (Seelinger et al., 2008). Possible mechanisms for its different pharmacological properties include but not limited to scavenging of reactive oxygen species (Lien et al., 1999) and enhancement of endogenous antioxidants (Leung et al., 2006). In case of plant defense aspect, luteolin was demonstrated to show stronger inhibition of spore germination of *C. sublineolum* than apigenin. Because of their pathogen-inducible and antifungal nature, both these flavone aglycones are considered phytoalexins (Du et al., 2010). Besides luteolin was also found to have antifungal activity against *Aspergillus niger, Trichophyton mentagrophytes* and *Candida albicans* (Abad et al., 2007). The membrane disruption ability of these flavonoids is responsible for their characteristic antimicrobial abilities against the pathogen (Górniak et al., 2019).

Anthocyanins like oenin are known to be strong antioxidants and thus have various health benefits including disease prevention (Olejnik et al., 2016; Björk et al., 2016). The potent antioxidant ability of oenin have been found to be considerably higher than synthetic antioxidants (Huyut et al., 2017) and its uses in human health protection like prevention of oxidative damage to retinal pigment epithelial cell (Yacout and Gaillard, 2017) and prevention of lyme disease (Goc et al., 2015) were recorded. Furthermore, oenin was also found to have food preservative effect (Sathishkumar et al., 2015).

Bicyclic monoterpenoids like camphene are reported to have plasma cholesterol and triglyceride reducing properties that has a potential use in combating cardiovascular disease (Vallianou, 2011). It was found to be one of the most common compounds in leaf extract of certain plants and also contributes to the anti-inflammatory effect as found by Tosun et al. (2014). The antioxidative activity of rosemary extracts are also partially due to the presence of camphene in it (Benyoucef et al., 2018). In case of antimicrobial activity, camphene showed to have a strong complementary effect with limonene against microorganisms such as *Bacillus cereus*, *Bacillus cereus*, *Staphylococcus aureus*, *Pseudomonas aeruginosa* and *Escherichia coli* (Aliberti et al., 2016).

Epigallocatechin and its gallates are commonly found as prime components in herbal teas of different medicinal plants and praised for their antioxidative ability (Yoo et al., 2009) and preventive potential against oral disease (Melok et al., 2018) as well as anti-HIV activity (Singh et al., 2011). Besides, the potential antimicrobial effect of these compounds against fungal and bacterial pathogens are also reported (Das et al., 2014). Furthermore, epigallocatechin is also found to have the ability to suppress phytopathogens which provides its applicability in agricultural systems (Yang and Zhang, 2019).

Rosmarinic acid is one of the most abundantly found phenolics in *Lamiaceae* herbs and common in species of *Boraginaceae* and the subfamily *Nepetoideae* of the L*amiaceae* (Bulgakov et al., 2012).

Rosmarinic acid has a range of biological activities such as anti-inflammatory, anti-oxidant, anti-tumor, anti-viral,and anti-microbial activities, making it an interesting material for the pharmaceutical, food, and cosmetics industries (Al-Dhabi et al., 2014). This phenolic acid act as both an antioxidants as well as defense compounds in plants (Capecka et al. 2005).

Limocitrin is commonly found in citrus fruits belonging to the family rutaceae (Favela-Hernández et al., 2016) although it's presence is also observed in other families including *Lamiaceae*. The most important aspect of this compound is found in human health improvement such as hypotension regulation (Matsubara et al., 1985) as well as anticancer effect (Guthrie et al., 2000).

From the discussion, it becomes evident that the water extracts of *Lamiaceae* herbs contains a wide range of secondary metabolites that plays an important role both in the plant development and defence as well as have implication in human health concern. The process of extract preparation is relatively more simpler than that of organic extraction which shows the superiority of using such extracts in practical condition. Furthermore, the compounds present in the water extracts are not inferior compared to that of organic extracts and so can be a reliable and economic alternative for use.

REFERENCES

Abad, M. J., Ansuategui, M. and Bermejo, P. (2007). Active antifungal substances from natural sources. *Arkivoc*, 2: 116-145.

Al-Dhabi, N. A., Arasu, M. V., Park, C. H., and Park, S. U. (2014). Recent studies on rosmarinic acid and its biological and pharmacological activities. *EXCLI journal* 13: 1192-1195.

Alexa, I. D., Pancu, A. G., Moroşanu, A. I., Ghiciuc, C. M., Lupuşoru, C., Prada, G. I. and Cepoi, V. (2014). The impact of self-medication with NSAIDs/analgesics in a north-eastern region of Romania. *Farmacia*, 62: 1164-1170.

Ali, F., Rahul, Naz, F., Jyoti, S. and Siddique, Y. H. (2017). Health functionality of apigenin: A review. *International Journal of Food Properties*, 20: 1197-1238.

Aliberti, L., Caputo, L., De Feo, V., De Martino, L., Nazzaro, F., and Souza, L. F. (2016). Chemical composition and in vitro antimicrobial, cytotoxic, and central nervous system activities of the essential oils of *Citrus medica* L. cv. 'Liscia' and *C. medica* cv. 'Rugosa' cultivated in southern Italy. *Molecules*, 21: 1244.

Anisimov, M. M., Strigina, L. I., Gorovoi, P. G., Aminin, D. L. M. and Agafonova, I. G. (2000). Chemical composition and medical biological properties of triterpene glycosides of the far eastern plant *Caulophyllum robustum* Maxim. (the family Berberidaceae). *Rastitelnye Resusry*, 36: 107-129.

Bahmani, M., Shirzad, H., Majlesi, M., Shahinfard, N and Rafieian-Kopaei, M. (2014). A review study on analgesic applications of Iranian medicinal plants. *Asian Pacific Journal of Tropical Medicine*, 7: 43-53.

Banerjee, K., Banerjee, S., Das, S. and Mandal, M. (2015). Probing the potential of apigenin liposomes in enhancing bacterial membrane perturbation and integrity loss. *Journal of Colloid and Interface* Science, 1: 48-59.

Benyoucef, F., El Amine, D. M., Arrar, Z., Costa, J. and Muselli, A. (2018). Synergistic antioxidant activity and chemical composition of essential oils from *Thymus fontanesii, Artemisia herba-alba* and *Rosmarinus officinalis. Journal of Applied Biotechnology Reports*, 5: 151-156.

Björk, C., Wilhelm, U., Mandrup, S., Larsen, B. D., Bordoni, A., Heden, P., Ryden, M., Arner, P. and Laurencikiene, J. (2016). Effects of selected bioactive food compounds on human white adipocyte function. *Nutrition & Metabolism*, 13: 1-10.

Bradley, D. (2005). Star role for bacteria in controlling flu pandemic. *Nature Reviews Drug Discovery*, 4: 945–946.

Bulgakov, V. P., Inyushkina, Y. V. and Fedoreyev, S. A. (2012). Rosmarinic acid and its derivatives: biotechnology and applications. *Critical Reviews in Biotechnology*, 32: 203-217.

Calderón-Montaño, J. M., Burgos-Morón, E., Pérez-Guerrero, C. and López-Lázaro, M. (2011). A review on the dietary flavonoid kaempferol. *Mini-Reviews in Medicinal Chemistry*, 11: 298-344.

Capecka, E., Mareczek, A. and Leja, M. (2005). Antioxidant activity of fresh and dry herbs of some *Lamiaceae* species. *Food Chemistry*, 93: 223-226.

Das, S., Tanwar, J., Hameed, S., Fatima, Z. and Manesar, G. (2014). Antimicrobial potential of epigallocatechin-3- gallate (EGCG): a green tea polyphenol. *Journal of Biochemical and Pharmacological Research*, 2: 167-174.

Enrich, L. B., Scheuermann, M. L., Mohadjer, A., Matthias, K. R., Eller, C. F., Newman, M. S., Fujinaka, M. and Poon, T. (2008). Liquidambar styraciflua: a renewable source of shikimic acid. *Tetrahedron Letters*, 49: 2503–2505.

Fini, A. Brunetti, C. di Ferdinando, M. Ferrini, F. and Tattini, M. (2011). Stress-induced flavonoid biosynthesis and the antioxidant machinery of plants. *Plant Signaling & Behavior*, 6: 709-711.

Goc, A., Niedzwiecki, A. and Rath, M. (2015). In vitro evaluation of antibacterial activity of phytochemicals and micronutrients against *Borrelia burgdorferi* and *Borrelia garinii*. *Journal of Applied Microbiology*, 119: 1561–1572.

Górniak, I., Bartoszewski, R. and Króliczewski, J. (2019). Comprehensive review of antimicrobial activities of plant flavonoids. *Journal of Phytochemistry Reviews*, 18: 241.

Gupta, S., Afaq, F. and Mukhtar, H. (2001). Selective growth-inhibitory, cell-cycle deregulatory and apoptotic response of apigenin in normal versus human prostate carcinoma cells. *Biochemical and Biophysical Research Communications*, 87: 914-920.

Guthrie, N., Horowitz, R. M. D. and Manthey, J. A. (2000). *Compositions and methods of inhibiting neoplastic diseases with compounds related to limocitrin and 5-desmethyl sinensetin.* Patent No. WO2000019998 A1.

Gutiérrez-Grijalva, E. P., Picos-Salas, M. A., Leyva-López, N., Criollo-Mendoza, M. S., Vazquez-Olivo, G. and Heredia, J. B. (2018). Flavonoids and Phenolic Acids from Oregano: Occurrence, Biological Activity and Health Benefits. *Plants*, 7: 2.

Hakkim, F. L., Arivazhagan, G. and Boopathy, R. (2008). Antioxidant property of selected *Ocimum* species and their secondary metabolite content. *Journal of. Medicinal Plants Research*, 2: 250-257.

Hertog, M. G., Feskens, E. J., Hollman, P. C., Katan, M. B. andKromhout, D. (1993). Dietary antioxidant flavonoids and risk of coronary heart disease: the Zutphen Elderly Study. *Lancet,* 342: 1007-1011.

Huyut, Z., Beydemir, S. and Gülçin, İ. (2017). Antioxidant and Antiradical Properties of Selected Flavonoids and Phenolic Compounds. *Biochemistry Research International*, 2017: 7616791.

Karau, G. M., Njagi, E. N. M, Machocho, A. K., Wangai, L. N., Nthinga, M. J. (2015). Chemical composition and in vitro antioxidant activities of *Ocimum americanum. Advances in Analytical Chemistry*, 5: 42-49.

Leung, H. W., Kuo, C. L., Yang, W. H., Lin, C. H. and Lee, H. Z. (2006). Antioxidant enzymes activity involvement in luteolin-induced human lung squamous carcinoma CH27 cell apotosis. *European Journal of Pharmacology*, 534: 12-18.

Lien, E. J., Ren, S., Bui, H. H. and Wang, R. (1999). Quantitative structure–activity relationship analysis of phenolic antioxidants. *Free Radical Biology & Medicine*, 26: 285-294.

Liu, K., Abdullah, A. A., Huang, M., Nishioka, T., Altaf-Ul-Amin, M., and Kanaya, S. (2017). Novel Approach to Classify Plants Based on Metabolite-Content Similarity. *BioMed research international*, 2017: 5296729.

Liu, W., Li, G., Hölscher, C. and Li, L. (2015). Neuroprotective effects of geniposide on Alzheimer's disease pathology. *Reviews in the Neurosciences*, 26: 371-383.

Matsubara, Y., Kumamoto, M., Iizuka, Y., Murakami, K., Okamoto, H., Miyake, H. and Yokoi, K. (1985) Structure and hypotensive effect of flavonoid glycosides in *Citrus unshiu* peeling. *Agricultural and Biological Chemistry*, 49: 909-914.

Meng, S., Peng, J., Feng, Q., Cao, J. and Hu, Y. (2013). The role of genipin and geniposide in liver diseases: a review. *Alternative & Integrative Medicine*, 2:117.

Olejnik, A., Kowalska, K., Kidon, M., Czapski, J., Rychlik, J., Olkowicz, M., and Dembczynski, R. (2016). Purple carrot anthocyanins suppress lipopolysaccharide induced inflammation in the co-culture of intestinal Caco-2 and macrophage RAW264.7 cells. *Food & Function*, 7: 557-564.

Özkan, M. (2008). Glandular and eglandular hairs of *Salvia recognita* Fisch. & Mey. (*Lamiaceae*) in Turkey. *Bangladesh Journal of Botany*, 37: 93-95.

Pavarini, D. P., Pavarini, S. P., Niehues, M. and Lopes, N. P. (2012). Exogenous influences on plant secondary metabolite levels. Animal Feed Science and Technology, 176: 5-16.

Potters, G., Pasternak, T. P., Guisez, Y. and Jansen, M. A. (2009). Different stresses, similar morphogenic responses: integrating a plethora of pathways. *Plant Cell Environ*, 32: 158-169.

Prashith, K. T. R, Raghavendra, H. L., Shilpa, M., Pushpavathi, D., Tejaswini P., Ayesha S. (2017). Antimicrobial, antiradical and insecticidal activity of *Gardenia gummifera* L. F. (Rubiaceae). *International Journal of Pharmacy and Pharmaceutical Sciences*, 9: 265-272.

Raja, R. R. (2012). Medicinally potential plants of *Labiatae* (*Lamiaceae*) Family: An Overview. *Research Journal of Medicinal Plants*, 6: 203-213.

Roldán-Arjona, T., Perez-Espinosa, A. and Ruiz-Rubio, M. (1999). Tomatinase from *Fusarium oxysporum* f.sp. *lycopersici* defines a new class of saponinases. *Molecular Plant-Microbe Interactions*, 12: 852-861.

Saleem, M. Nazir, M. Ali, M. S. Hussain, H. Lee, Y. S. Riaz, N. and Jabbar, A. (2010). Antimicrobial natural products: An update on future antibiotic drug candidates. *Natural Product Reports*, 27: 238-254.

Sathishkumar, P., Hemalatha, S., Arulkumar, M., Ravikumar, R., Yusoff, A. R. M., Hadibarata, T. and Palvanna, T. (2015). Curcuminoid extraction from turmeric (*Curcuma longa* L.): efficacy of bromine-modified curcuminoids against food spoilage flora. *Journal of Food Biochemistry*, 39: 325-333.

Shanaida, M., Golembiovska, O., Hudz, N. and Wieczorek, P. P. (2018). Phenolic compounds of herbal infusions obtained from some species of the *Lamiaceae* family. *Current Issues in Pharmacy and Medical Sciences*, 31: 194-199.

Sharma, H., Kanwal, R., Bhaskaran, N. and Gupta, S. (2014). Plant flavone apigenin binds to nucleic acid bases and reduces oxidative DNA damage in prostate epithelial cells. *PLoS One*, 9: e91588.

Singh, J., Kumar, A. and Kumar, V. (2011). Plant derived anti-human immunodeficiency virus (HIV) compounds. *International Journal of Natural Product Science*, 2: 1-9.

Song, C., Jiang, S. and Singh, G. (2001). Syntheses of (6S)-6-fluoro- and (6R)-6-hydoxyshikimic acids. *Tetrahedron: Assymetry*, 42: 9069–9071.

Sosa, T., Chaves, N., Alias, J. C., Escudero, J. C., Henao, F. and Gutierrez-Merino, C. (2004). Inhibition of mouth skeletal muscle relaxation by flavonoids of *Cistus ladanifer* L.: a plant defense mechanism against herbivores. *Journal of Chemical Ecology*, 30: 1087-1101.

Tosun, A., Khan, S., Kim, Y. S., Calín-Sánchez, A. Hysenaj, X. and Carbonell-Barrachina, A. (2014). Essential oil composition and anti-inflammatory activity of *Salvia officinalis* L. (*Lamiaceae*) in murin macrophages. *Tropical Journal of Pharmaceutical Research*, 13: 937-942.

Tsai, S. M. and Phillips, D. A. (1991). Flavonoids released naturally from alfalfa promote development of symbiotic glomus spores *in vitro*. *Applied and Environmental Microbiology*, 57: 1485-1488.

Vaishali, R. M., Ramanath, P. V., Pratapchandra, K. H. and Hegde, S. (2013). Preliminary phytochemical screening of members of *Lamiaceae* family: *Leucas linifolia*, *Coleus aromaticus* and

Pogestemon patchouli. *International Journal of Pharmaceutical Sciences Review and Research*, 21: 131-137.

Vallianou, I., Peroulis, N., Pantazis, P. and Hadzopoulou-Cladaras, M. (2011). Camphene, a plant-derived monoterpene, reduces plasma cholesterol and triglycerides in hyperlipidemic rats independently of HMG-CoA reductase activity. *PLoS ONE* 6: e20516.

Wang, J., Liu, Y. T., Xiao, L., Zhu, L., Wang, Q. and Yan, T. (2014). Anti-inflammatory effects of apigenin in lipopolysaccharide-induced inflammatory in acute lung injury by suppressing COX-2 and NF-kB pathway. *Inflammation*, 37: 2085-2090.

Wei, H., Tye, L., Bresnick, E. and Birt, D. F. (1990). Inhibitory effect of apigenin, a plant flavonoid, on epidermal ornithine decarboxylase and skin tumor promotion in mice. *Cancer Research*, 50: 499-502.

Wink, M. (2015). Modes of action of herbal medicines and plant secondary metabolites. *Medicines*, 2: 251-286.

Wuyts, N., Swennen, R. and de Waele, D. (2006). Effects of plant phenylpropanoid pathway products and selected terpenoids and alkaloids on the behaviour of the plant-parasitic nematodes *Radopholus similis*, *Pratylenchus penetrans* and *Meloidogyne incognita*. *Nematology*, 8: 89-101.

Yacout, S. M. and Gaillard, E. R. (2017). The anthocyanins, oenin and callistephin, protect RPE cells against oxidative stress. *Photochemistry and Photobiology*, 93: 747-751.

Yang, Y., and Zhang, T. (2019). Antimicrobial activities of tea polyphenol on phytopathogens: a review. *Molecules*, 24: 816.

Yesiloglu, Y., Sit, L. and Kilic, I. (2013). *In vitro* antioxidant activity and total phenolic content of various extracts of *Satureja hortensis* L. collected from Turkey. *Asian Journal of Chemistry*. 25: 8311-8316.

Zgórka, G. and Glowniak, K. (2001). Variation of free phenolic acids in medicinal plants belonging to the *Lamiaceae* family. *Journal of Pharmaceutical and Biomedical Analysis*, 26: 79-87.

Zhang, Y., Liu, A., Ye, Z. G., Lin, J., Xu, L. Z. and Yang, S. L. (2006). New approach to the total synthesis of (-)-zeylenone from shikimic acid. *Chemical and Pharmaceutical Bulletin*, 54: 1459-1461.

In: The Lamiaceae Family: An Overview
Editor: Alexander Adler
ISBN: 978-1-53617-078-8

Chapter 6

HOMOPLOID HYBRIDIZATION AND ITS ROLE IN EMERGENCE AND DIVERSITY OF THE GENUS *ORIGANUM* L. (LAMIACEAE)

Tuncay Dirmenci[1,*], PhD, Turan Arabacı[2], PhD, Taner Özcan[1], PhD, Türker Yazıcı[1], MD, Esra Martin[3], PhD, Sevcan Çelenk[4], PhD, Mikail Açar[5], PhD and Devrim Üzel[6] PhD

[1]Department of Biology Education, Necatibey Education Faculty, Balıkesir University, Balıkesir, Turkey
[2]Department of Pharmaceutical Botany, Faculty of Pharmacy, Inonu University, Malatya, Turkey
[3]Department of Biotechnology, Faculty of Science, Necmettin Erbakan University, Konya, Turkey
[4]Department of Biology, Faculty of Arts and Sciences, Uludag University, Görükle Campus, Bursa, Turkey
[5]Department of Plant and Animal Production,

* Corresponding Author's Email: dirmenci@balikesir.edu.tr.

Tunceli Vocational School, Munzur University, Tunceli, Turkey
[6]Department of Mathematic Education Necatibey Education Faculty, Balıkesir University, Balıkesir, Turkey

ABSTRACT

Hybrid specimens can be occurred as same ploidy level (homoploidy) or not (polyploidy) with its parents. Homoploid hybrids are originated without any change in their ploidy level (2n) and polyploid hybrids are produced with the change in ploidy level (3n, 4n etc.). According to recent studies, the frequency of natural hybridization (homoploid and polyploid) has been estimated to 25% of all flowering plants. Recent studies show that the genus *Origanum* has homoploid hybrids. Hybridization is a common phenomenon among the species of *Origanum* in overlapping distribution habitats (natural or cultivated areas). Ietswaart suggested that the hybridization process is a significant factor in the speciation of *Origanum*, and *Origanum* is probably caused by the hybridization of the old tribe Saturejeae (now tribe Mentheae) genera. Based on the hypothesis of Ietswaart, we can discuss the emergence of the genus *Origanum*, its intrageneric speciation and its hybrids having the possibility for speciation via hybridization using morphological and cytological evidence. During the field studies, it was seen that many populations of wide-distributed species were overlapping, and some *Origanum* putative hybrid specimens were also observed in these overlapping areas with morphological differences from their parents. All studied *Origanum* species and hybrids have 2n = 30 chromosome numbers to define homoploidy. Palynological characters were revealed, and intermediate pollen structure (size, P/E ratio etc.) was seen in the hybrids between their parents. In addition, DNA extractions were done from all the hybrids and their parents. nrITS DNA was sequenced. Hybrid specimens have some heterozygous DNA loci in comparison to their parents. This study presents some evidence about homoploid the hybridization of the genus *Origanum* put forward by Ietswaart. If habitat, ecological and geographic isolations occur between the hybrids and their parents, these hybrids can be speciation via homoploid hybridization in the future.

Keywords: homoploid hybridization, homoploid speciation, *Origanum*, Lamiaceae

Introduction

Lamiaceae, one of the largest Angiosperm families, contains more than 240 genera and approximately 8000 species distributed all over the world. Harley et al., [1] recognized seven subfamilies as Ajugoideae, Lamioideae, Nepetoideae, Prostantheroideae, Scutellarioideae, Symphorematoideae and Viticoideae in the Lamiaceae family. After Harley et al., (2004), five new subfamilies have been described as Cymarioideae, Peronematoideae, Premnoideae, Callicarpoideae and Tectonoideae [2-3]. The Mediterranean basin is one of the most important areas of the Lamiaceae family in terms of species richness. Among the countries in the Mediterranean basin, Turkey is one of the countries with the most species and has approximately 10% percent of the Lamiaceae species in the world. Lamiaceae is the third largest family based on the taxon number and fourth-largest family based on the species number in Turkey. The family has 782 taxa (603 species, 179 subspecies and varieties), 346 taxa (271 species, 75 subspecies and varieties) of which are endemic (ca. 44%) in Turkey [4-5]. They are included in five subfamilies, namely Ajugoideae, Lamioideae, Nepetoideae, Scutellarioideae and Viticoideae [4].

Origanum L. is a well-known genus in the Lamiaceae family and classified in subfamily as Nepetoideae, tribe Mentheae, subtribe Menthineae and consists of 43 species (50 taxa) and 20 hybrids [1, 6-12]. in the world (Table 1). They are mainly distributed in the Mediterranean basin and the southwest Asia (except for a few species, see biogeography) (Figure 1). Only *Origanum vulgare* L. *s.l.* is widespread from Europe to China [1, 8, 10-12] (Figure 1). Turkey and East Aegean Islands are important centres of diversity for the whole genus and Turkey has 21 species with 3 subspecies (24 taxa) and 13 hybrids. Among them, 25 taxa (including hybrids) are endemic.

The species are classified into ten sections; Sect. *Amaracus* (Gleditsch) Vogel (8 species), Sect. *Anatolicon* Benth. (8 species), Sect. *Breviflamentum* Ietsw. (7 species), Sect. *Longitubus* Ietsw. (1 species),

Table 1. ***Origanum* hybrids and their parents with distribution areas and references**

Hybrids	Parents	Distribution	References
O. × *adae* Dirmenci & T.Yazici	*O. ayliniae* Dirmenci & T. Yazici × *O. sipyleum* L.	W. of Turkey (Figure g)	Turkish J. Bot. 42(1): 80 (2018).
O. × *adanense* Baser & H.Duman	*O. brevidens* Bornm. × *O. laevigatum* Boiss.	S. of Turkey (Figure b)	Turkish J. Bot. 22: 55 (1998).
O. × *adonidis* Mouterde	*O. libanoticum* Boiss.× *O. syriacum* L. subsp. *bevanii* (Holmes) Greuter & Burdet	Lebanon (Figure d)	Ann. Fac. Franc. Med. Beyrouth 6 (Reimpr.): 4 (1935).
O. × *barbarae* Bornm.	*O. ehrenbergii* Boiss. × *O. syriacum* subsp. *bevanii*	Lebanon (Figure d)	Verh. K. K. Zool.-Bot. Ges. Wien 48: 615 (1898).
O. × *dolichosiphon* P. H.Davis	*O. amanum* Post. × *O. laevigatum*	S. of Turkey, (Figure b)	Kew Bull. 6: 86 (1951).
O. × *haradjanii* Rech.f.	*O. laevigatum* × *O. syriacum* subsp. *bevanii*	S of Turkey (Figure b)	Oesterr. Bot. Z. 99: 64 (1952).
O. × *intercedens* Rech.f.	*O. onites* L. × *O. vulgare* L. subsp. *hirtum* (Link.) A. Terracc.	Greece to W. of Turkey (Figure f, h)	Bot. Jahrb. Syst. 80: 395 (1961).
O. × *intermedium* P. H..Davis,	*O. onites* × *O. sipyleum* L.	W. of Turkey (Figure b)	Kew Bull. 4: 410 (1949).
O. × *lirium* Heldr. ex Halácsy	*O. scabrum* Boiss. & Heldr. × *O. vulgare* subsp. *hirtum*-	Greece (Figure i)	Verh. K. K. Zool.-Bot. Ges. Wien 49: 192 (1899).
O. × *majoricum* Cambess.	*O. majorana* L. × *O. vulgare* subsp. *virens* (Hoffmans. & Link) Ietsw.	Spain (Figure l)	Mém. Mus. Hist. Nat. 14: 296 (1827).
O. × *malyeri* Dirmenci & T.Yazici	*O. boissieri* Ietsw. × *O. vulgare* subsp. *hirtum*	S. of Turkey (Figure b)	Phytotaxa 371: 148 (2018).
O. × *minoanum* P. H. Davis	*O. microphyllum* (Benth.) Vogel × *O. vulgare* subsp. *hirtum*	Kriti (Figure j)	Notes Roy. Bot. Gard. Edinburgh 21: 137 (1953).
O. × *munzurense* Kit Tan & Sorger	*O. acutidens* (Hand.-Mazz.) Ietsw. × *O. vulgare* subsp. *gracile* (K. Koch) Ietsw.	E. of Turkey (Figure a)	Notes Roy. Bot. Gard. Edinburgh 41: 534 (1984).
O. × *nebrodense* Tineo ex Lojac.	*O. majorana* × *O. vulgare* subsp. *viridulum* (Martrin-Donos) Nyman	Sicilia (Figure k)	Fl. Sicul. 2(2): 196 (1907).
O. × *pabotii* Mouterde	*O. bargyli* Mauterde × *O. syriacum* subsp. *bevanii*	Syria (Figure c)	Saussurea 4: 22 (1973).

Hybrids	Parents	Distribution	References
O. × *sevcaniae* Dirmenci, Arabaci & T.Yazici	*O. vogelii* Greuter & Burdet × *O. vulgare* subsp. *hirtum*	S. of Turkey (Figure b)	Phytotaxa 371: 150 (2018).
O. × *aytacii* Dirmenci, T.Yazıcı & Arabacı	*O. sipyleum* × *O. vulgare* subsp. *hirtum*	W. of Turkey (Figure f)	Unpublished data
O. × *bilgili* Dirmenci, T.Yazıcı & Arabacı	*O. saccatum* P. H. Davis × *O. vulgare* subsp. *hirtum*	S. of Turkey (Figure e)	Unpublished data
O. × *dumanii* Dirmenci, Arabacı & T.Yazıcı	*O. saccatum* P. H. Davis × *O. husnucan-baseri* H. Duman, Aytaç, A. Duran	S. of Turkey (Figure e)	Unpublished data
O. × *malatyanum* Yıldız, Arabacı & Dirmenci	*O. hausknechtii* Boiss. × *O. vulgare* subsp. *gracile*	E. of Turkey (Figure a)	Unpunlished data
	Artificial made hybrids with cultivated Parents		
	O. calcaratum Juss. × *O. dictamnus* L.	Edinburgh Botanical Garden, Kriti?	Ietswaart 1980
	O. amanum Post × *O. dictamnus*	Jena Botanical garden	Ietswaart 1980
	O. dictamnus × *O. sipyleum*		Ietswaart 1980
	O. majorana × *O. vulgare* subsp. *vulgare*		Ietswaart 1980
	Hybrids with at least one parent cultivated		
	O. vylgare subsp. *vulgare* × *Thymus* sp. ?	Austria	Ietswaart 1980

Figure 1. Distribution map of the *Origanum* species (*O. vulgare* s.l.-black line and other species-yellow line).

Sect. *Chilocalyx* (Briq.) Ietsw. (4 species), Sect. *Majorana* (Mill.) Benth. (3 species), Sect. *Origanum* (1 species), and Sect. *Prolaticorolla* Ietsw. (3 species), Sect. *Elongatispica* Ietsw. (2 species), Sect. *Campanulaticalyx* Ietsw. (6 species) [1, 8, 10-12].

Hybridization is common in the genus *Origanum* and it has one of the most natural hybrids with 47% percent amongst the genera of the Lamiaceae family [10-12]. Hybridization in the genus *Origanum* is possible, in natural habitats, cultivated areas or botanical gardens, between closely related species or distinct species belonging to different sections. Twelve hybrids were listed in the revision of the genus *Origanum* by Ietswaart [6], where he noticed that many more hybrids are expected to occur in nature. In the opinion of Ietswaart [6] hybridization is an important evolutionary force such as role emergence of the genus and speciation within the genus. In the recent studies, nine new hybrids have been identified in the natural habitats and five of them have been described and have been found to be homoploid with their parents, 2n = 30 [10-14].

Recent investigations claims that the frequency of natural hybridization (polyploid and homoploid) have been estimated as 2%–4% by Otto & Whitton [15], as 15% by Wood et al., (16] and as 25% by Mallet [17] and Folk et al., [18]. Homoploid hybridization originates hybrid species without a change in their ploidy level [19-24] and by polyploid hybridization process, unlike homoploid hybridization, produce hybrid species with the change in their ploidy level [22, 25-26]. These hybridization types, especially polyploid hybridization [25, 27-29] have been considered as examples of sympatric speciation via hybridization [17, 30-31]. Some recent studies show that [17, 31- 35] allopolyploid hybrid speciation appears much more common than homoploid hybrid speciation. On the other hand, homoploid hybridization can be important for plant evolution [21, 24-26, 36-37]. Three important criteria have been put forward for speciation to occur via hybridization by Schumer et al., [24] hybrids should have (1) a reproductive isolation (RI) barrier from their parental species, (2) evidence of genetical difference from their parents in their genome, and (3) evidence about RI causing by hybridization.

When we analyze the recent studies, it can be seen that a large number of studies focus on founding out genetic evidence of hybridization (Schumer et al., Criterion 2). Although, genetical studies are very important to understand hybridization, using a small DNA region may make it confounding to find hybridization evidence on hybrids or on their parents (Schumer et al., 2014), and also, hybridization is common in the absence of hybrid speciation (Mallet 2007). Studies using dense genome-wide markers or even whole-genome sequences to characterize hybridization throughout the genome, allow researchers to distinguish between ILS and hybridization using gene tree methods, model-based approaches, or the length of ancestry tracts associated with discordant gene trees [18, 38, 39, 40, 41].

Helianthus annuus and *H. petiolaris* produce hybrid individuals (homoploid, 2n = 34) in the same growing area and the hybrids show genomic mosaicism [42] which some of the phenotypes are derived via DNA recombination between parent [20]. The same condition was shown in different important studies [43-44]. It is very difficult to show genetical recombination, if hybridization continues and repeats, extensive gene flow can occur and this may result in the extinction of some hybrids via genetic assimilation [37, 45-46], or two different taxa can have same genetic structure and merge into same single evolutionary lineage ([37, 47-48]. As mentioned above, polyploid hybridization was observed much more, and the fertility of polyploid hybrids might actually increase with increasing parental genetic divergence [37, 49, 50].

In this study, we firstly aimed to test Ietswaart's hybridization hypothesis based on homoploid hybridization. For this purpose these followings are intended: 1- to contribute the formation hypothesis of *Origanum* genus, 2- to discuss occurring of *Origanum* genus via homoploidy or polyploidy, 3- to speculate occurring of *Origanum* genus at which chromosomal level, 4- to give morphological, palynological, cytological and molecular evidence to *Origanum* hybrids, 5- to reveal the specimens of possible *Origanum* hybrids in the speciation, 6- to provide some contributions diversification, evolutionary process of *Origanum* genus.

METHODS

Plant Materials

The specimens examined in this study were collected from Turkey in revision study of *Origanum* and the herbarium material deposited in ANK, B, E, EGE, GAZI, HUB, INONU, ISTE, ISTF, ISTO, K, KNY A, L, LE, MA, W, and WU herbaria. Distinguished characters of the some parental species and hybrids were drawn under stereo microscope and photographed in natural habitats (Figures 2 and 3).

Palynological Studies

Palynological studies were done by using the light (LM) and scanning electron microscope (SEM). Wodehouse's (1959) method was used for LM slides. Palynological features of 30 pollen grains (P and E exception of *O.* × *haradjanii*) were measured using a LM. The measurements included the following parameters: polar axis (P), equatorial axis (E), colpus length (Clg) and width (Clt), apocolpium diameter (T), the distance between two nearest colpi edge (t), mesocolpium, exine, and intine thickness. Pollen grains were placed on stubs using the double-sided adhesive tape and they were coated with gold-palladium for two minutes in a BAL–TEC SCD 005 sputter-coater for SEM investigation and XL–30 ESEM-FEG/PHILIPS microscope was used at 20 kV.

Statistical Analyses

For morphometric studies, measurements were taken from *O. laevigatum*, *O. sipyleum*, *O. vulgare* subsp. *hirtum*, *O. ayliniae*, *O. syriacum* subsp. *bevanii*, *O. vogelii*, *O.* × *adae*, *O.* × *haradjanii*, O. × *sevcaniae*. For each taxon and putative natural hybrid, a total of 9 pollen characters mentioned above were measured 30 times for each character[1]. Principal component analysis (PCA) and Cluster analyses were performed for the 9 taxa with 9 selected quantitative pollen characters were. The selected characters were: PCA was applied based on the data matrix [51]. In order to determine the taxonomic status of *Origanum* taxa, Cluster analysis based on the Gower similarity index was used [52]. All analyses were carried out with PAleontoSTatistics (PAST) [53].

Also, ANOVA and Tukey tests were used to determine the significance of the relationships between taxa. The data obtained for each species and hybrid was reduced to a single value with the arithmetic mean of the data obtained. The data was collected by making the same operation for all taxa. These analyses were performed with SPSS 22. 0 program.

Karyological Studies

Chromosome analyses were performed on mitotic metaphases by using Image System Analysis. Root meristems from germinating seeds collected in the wild were used. Root tips were pretreated with α-

[1] Measured 10 times due to lack of material in the Origanum × haradjanii.

monobromonaphthalene at 4°C for 16 h. The tips were fixed with Carnoy solution for 24 h at 4°C. Before staining, the materials were hydrolyzed with 1 N HCl for 12 min at room temperature. The chromosomes were stained with 2% acetic orcein and mounted in 45% acetic acid. Permanent slides were made using the standard liquid nitrogen method. Photographs were taken with an Olympus BX51 microscope.

Molecular Studies

DNeasy® Plant Mini Kit (Qiagen, Germany) was used to be conducted DNA extraction reaction from different hybrid and parent specimens. After DNA isolation, ITS5a (5′-CCT TAT CAT TTA GAG GAA GGA G-3′) [54] and ITS4 (5′-TCC TCC GCT TAT TGA TAT GC-3′) (White et al., 1990) primers were used to amplify of the internal transcribed spacer (ITS) region of the nuclear ribosomal DNA (nrDNA). The PCR solutions were sent to Genoks (Gene Research and Biotechnology Company, Turkey) for sequencing, and DNA sequences were checked and edited using Sequencer version 4.9 [55]. Gene Codes Corporation, Ann Arbor, MI, USA), aligned with MEGA software [56], and some ambiguous nucleotides from the 5′ end of ITS1 and 3′ end of ITS2 (about 15-20 nucleotides) were cut. Phenetic analyses were performed by the NJ method using PAUP* software [57].

RESULTS AND DISCUSSION

Morphology

The genus *Origanum* could be described morphologically as following characters: Subshrubs or perennial herbs, branches usually present in 1-3 orders, glabrous to stalked glandular, with different types of pubescence, strongly aromatic herbs. Leaves rather small, entire or toothed; inflorescence branched, paniculate to corymbose, spicate or spiciform, or compact; flowers single or 2-8 flowered cymes in axils of bracts, erect or nodding; bracts imbricate, distinct from the leaves in shape and size; calyx actinomorphic and 5-lobed to strongly zygomorphic 5- to 1-lobed; either 2- lipped, 3/2 to ½, lobes sometimes reduced or calyx 1-lipped with posterior lip entire, calyx tube when present, cylindrical to campanulate, 10-13 nerved, pilose in throat; corolla 2-lipped, usually 5-lobed (2/3), white, pink or purple, tube straight or curved sometimes saccate; stamens 4, the lower 2 longest, protruding from the corollas or (sub)included, sometimes reduced or absent (gynodioecious); stigma lobes subequal; nutlet ovoid to apiculate, smooth [1, 6-7].

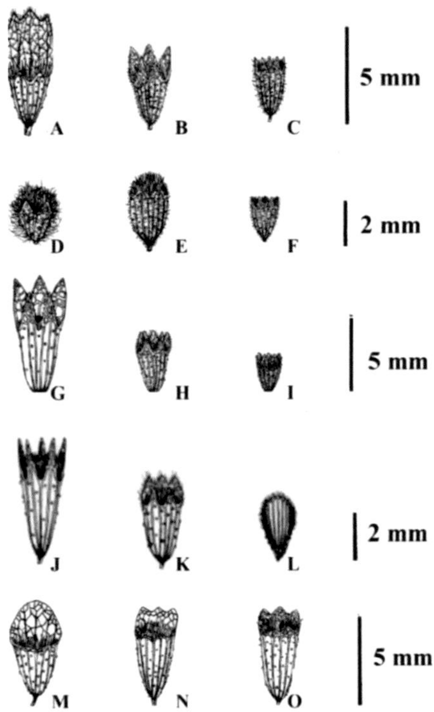

Figure 2. A-*O. boissieri*, B-*O.* × *malyeri*, C-*O. vulgare* subsp. *hirtum*, D-*O. vogelii*, E-*O.* × *sevcaniae*, F-*O. vulgare* subsp. *hirtum*, G-*O. acutidens*, H-*O.* × *munzurense*, I-*O. vulgare* subsp. *gracile*, J-*O. laevigatum*, K-*O.* × *haradjanii*, L-*O. syriacum* subsp. *bevanii*, M-*O. ayliniae*, N-*O.* × *adae*, O-*O. sipyleum*.

Figure 3. A-*O. boissieri*, B-*O.* × *malyeri*, C-*O. vulgare* subsp. *hirtum*, D-*O. vogelii*, E-*O.*× *sevcaniae*, F-*O. vulgare* subsp. *hirtum*, G-*O. acutidens*, H-*O.* × *munzurense*, I-*O. vulgare* subsp. *gracile*, J-*O. laevigatum*, K-*O.* × *haradjanii*, L-*O. syriacum* subsp. *bevanii*, M-*O. ayliniae*, N-*O.* × *adae*, O-*O. sipyleum*.

The genus *Origanum* is divided into ten sections because of showing a large variation in stems, leaves, inflorescence, indumentums, and especially floral characters. The calyx and corolla shape and size are very variable (Figures 2A-O and 3A-O). For example, calyx and corolla characters of sections *Amaracus*, *Majorana* and *Origanum* as follows. Sect. *Amaracus*: calyx tubular 1- or 2-lipped for 2/5-3/5; teeth in upper or lower lips often reduced, corolla large, pinkish to purple and saccate (Figures 2A, M and 3A, M); Sect. *Majorana*: calyx flattened, 1-lipped for 9/10; upper lips entire or denticulate, corolla small size, white and tubular (Figure 2L and 3L); Sect. *Origanum*: Calyx actinomorphic, tubular with 5 equal teeth, corolla small size and white to pinkish (Figures 2C, F, I and 3C, F, I). Morphological wide variations have revealed different opinions in the classification of the genus. The genus *Origanum* has been divided into three genera as *Amaracus*, *Majorana*, and *Origanum* by some authors, and has been united by some authors under the name of *Origanum* in the broader sense [6-7, 58-67].

Hybridization is commonly encountered among species of *Origanum* (Lamiaceae) in the natural habitats where the distribution areas are overlapped or in cultivation habitats (such as botanical gardens, fields, etc.) [6, 9-12]. Species with different morphological characteristics in different sections in the genus *Origanum* can make a hybrid between them. In the natural habitats of the species, a total of 20 hybrids are described among 20 species (23 taxa) in 8 different sections (Table 1 and 2). Some hybrids emerged as a result of the cultivation of species whose natural distribution areas don't overlap in the same botanical garden (two hybrids, Table 1). In the genus *Origanum*, most of the hybrids co-exist with their parents in the same area. However, it is observed that some hybrids do not share the same area or geographic proximity with their parents [12]. Sometimes one parent shares the same habitat with the hybrid, while the other parent related material may have moved away. Therefore, only one of the parents can be identified (Table 1). Generally, morphological characters of hybrids are intermediate between their parents [10, 11, 12] (Figures 2A-O and 3A-O). Calyx drawings and flower photographs of 5 hybrids and their parents are given in Figures 2 and 3. For instance, calyx and corolla shapes of hybrids have an intermediate forms between their parents. *O.* × *haradjanii,* a hybrid is formed between *O. laevigatum* (in Sect. *Prolaticorolla*) and *O. syriacum* subsp. *bevanii* (in Sect. *Majorana*). The distinguishing floral characters of *O. syriacum* subsp. *bevanii* and *O. laevigatum* are; calyx 1-lipped for 9/10 and upper lip entire, corolla white (Figures 2L and 3L) and calyx actinomorphic with 5-subequal teeth, corolla purple (Figure 2J and 3J) respectively. Shapes and sizes of floral characters of *O.* × *haradjanii* are intermediate forms between its parents. Namely, calyx shape 2-lipped and sizes intermediate (Figure 2K), corolla colour pinkish and corolla sizes intermediate (Figure 3K). *O.* × *haradjanii* distinguished from *O. laevigatum* by its calyx c. 4 mm (not 5-6 mm) 2-lipped (not regular and teeth 1-2 mm), upper lip teeth deltoid, c. 0.5 mm, lower teeth triangular, c. 1 mm, corolla pinkish and 9-12 mm (not purple and 10-16 mm). It differs from other parents, *O. syriacum* subsp. *bevanii*, by its calyx c. 4 mm and 2-lipped, upper teeth c. 0.5 mm, lower teeth c. 1 mm (not 2-2.5 mm and 1-lipped for 9/10), corolla 9-12 mm and pinkish (not 4.5-7.5 mm and white). The morphological comparison of the other hybrids and their parents in Figure 2 are given in the protologues of the hybrids [10, 11, 12].

Biogeography and Ecology

The species of the genus *Origanum* are distributed northern hemisphere. 85% of the species are growing in the Mediterranean basin (Figures 1). Only a few species (*O. haussknechtii* and *O. acutidens* in eastern Turkey, and *O. rotundifolium* towards to the Caucasus from the north of Turkey) spreading beyond the Mediterranean region. Also, *O. vulgare* s.l. is a cosmopolitan species and divided five subspecies distributed from Europe to China (Figure 1) [6, 9]. A large part of the species is the element of the Eastern Mediterranean subregion. Turkey has 24 taxa (13 endemic) and most of the species amongst the countries.

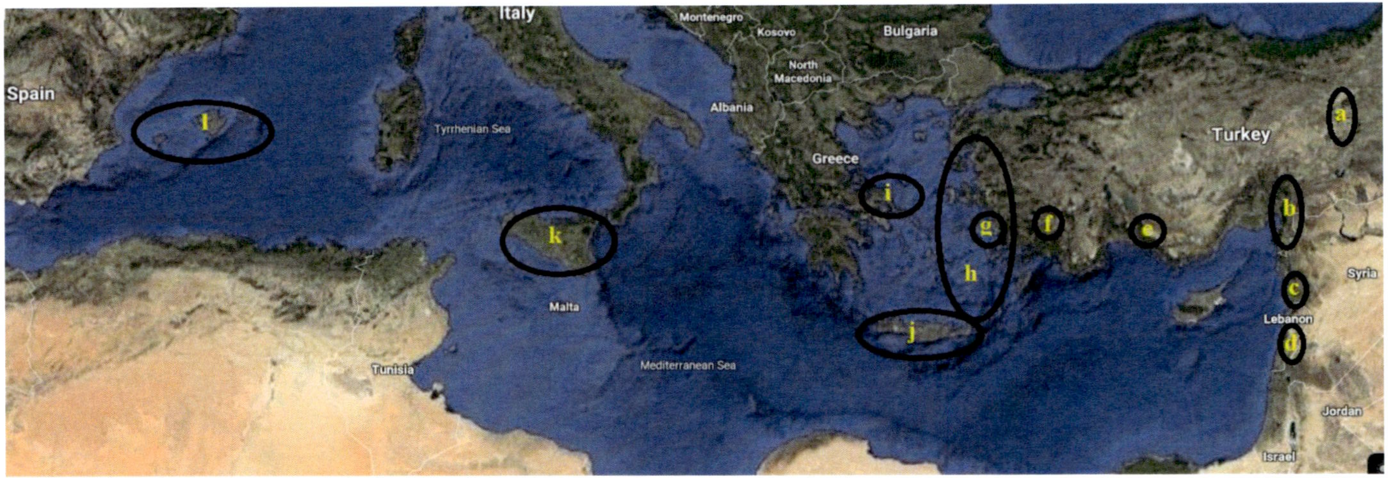

Figure 4. Natural distribution map of the *Origanum* hybrids (a- *O.* × *malatyanum* and *O.* × *munzurense*, b-*O.* × *adanense*, *O.* × *haradjanii*, *O.* × *sevcaniae*, c-*O.* × *pabotii*, d-*O.* × *adonidis* and *O.* × *barbarae*, e-*O.* × *bilgilii* and *O.* × *dumanii*, f-*O.* × *aytacii*, *O.* × *intermedium* and *O.* × *intercedens*, g-*O.* × *adae*, h-*O.* × *intercedens*, i-*O.* × *lirium*, j-*O.* × *intercedens* and *O.* × *minoanum*, k-*O.* × *nebrodense*, l-*O.* × *majoricum*.

Table 2. Hybridization scheme between *Origanum* species

	Section	*Amaracus*			*Anatolicon*			*Breviflamentum*					*Longitubus*	*Chilocalyx*		*Majorana*			*Origanum*				Prolaticorolla	
Section	Taxa \ Taxa	*O. ayliniae*	*O. boissieri*	*O. saccatum*	*O. libanoticum*	*O. scabrum*	*O. sipyleum*	*O. acutidens*	*O. bargyli*	*O. brevidens*	O. *haussknechtii*	*O. husnucan-baserii*	*O. amanum*	*O. microphyllum*	*O. vogelii*	*O. majorana*	*O. onites*	*O. syriacum* subsp. *beva*	*O. vulgare* subsp. *gracile*	*O.vulgare* subsp. *hirtum*	*O. vulgare* subsp. *virens*	*O.vulgare* subsp. *viridulum*	*O.laevigatum*	*O. ehrenbergii*
Amaracus	*O. ayliniae*						X																	
	O. boissieri																			X				
	O. saccatum											X								X				
	O. libanoticum																	X						
	O. sacbrum																			X				
	O. sipyleum	X															X			X				
Breviflamentum	*O. acutidens*																		X					
	O. bargyli																						X	
	O. brevidens																						X	
	O. haussknechtii																		X					
	O. husnucan-baserii			X																				
Longitubus	*O. amanum*																						X	
Chilocalyx	*O. microphyllum*																			X				
	O. vogelii																			X				
Majorana	*O. majorana*																				X	X		
	O. onites						X													X				
	O. syriacum subsp. *bevanii*				X				X	X													X	X
Origanum	*O. vulgare* subsp. *gracile*							X			X													
	O. vulgare subsp. *hirtum*		X	X		X	X							X	X		X							
	O. vulgare subsp. virens															X								
	O. vulgare subsp. *viridulum*															X								
Prolaticorolla	*O. laevigatum*												X					X						
	O. ehrenbergii																	X						

Also, 92% of endemic taxa in Turkey are distributed in the Mediterranean region of Turkey. As seen in Tables 1 and 2, the genus *Origanum* comprises 20 hybrids forming amongst the 23 taxa belonging to 8 sections [6, 9]. All hybrids are distributed in the Mediterranean basin, except two hybrids, namely *O.* × *munzurense* and *O.* × *malatyanum* (unpublished) are growing apart from the Mediterranean basin climate (Table 1 and Figure 4a-l). Most of the hybrids are the element of the Eastern Mediterranean subregion and two hybrids, *O.* × *majoricum* (Figure 4l) and *O.* × *nebrodense* (Figure 4l), are the element of Western Mediterranean subregion. Hybrids are generally narrow spreading areas. Populations are composed of a small number of individuals. The most widespread hybrid is *Origanum* × *intercedens*. *O.* × *intercedens* is distributed from Crete to more north Aegean Islands and from southwestern to northwestern of Turkey (Figures 4f, j, h). *O.* × *majoricum* is indigenous to the west Mediterranean area (Figure 4l). But in many countries, it is cultivated in small areas as botanical and personal gardens, etc.

Species and hybrids are growing between the altitudes from sea level to 2100 m. Most of them, the most abundantly occur between 700-1500 m. They are distributed in open areas of forest (coniferous, *Quercus* sp., or mixed forest), maquis, garigue, stony places, steppe and near the stream banks. They are generally grown on the soil that consists from clay, clay-loam or loam with pH value between 6.62 and 7.49 and have electrical conductivity which is saltless or slightly salty, $CaCO_3$ content among 0.08% to 48.97% and the amount of organic matter are between 0.01% and 11.54%, in Turkey. The soil of taxa have grown on low, medium, adequate or high-level availability of P, K, Cu, Fe, Zn and Mn [68].

Origanum ayliniae, *O. boissieri* and *O. calcaratum* are chasmophytes on limestone cliffs. On the other hand, some species (*O. acutidens*, *O. dayi*, *O. elongatum*, *O. vulgare* s.l., *O. hypericifolium*) occur to serpentine substrates [6, 9]. Most of the hybrids grow stony places in open areas of coniferous forest (*O.* × *adanense, O.* × *bilgilii*, *O.* × *dumanii*, *O.* × *intercedens*, *O.* × *malyeri, O.* × *sevcaniae*) and maquis (*O.* × *adae*, *O.* × *aytacii*, *O.* × *haradjanii*, *O.* × *intercedens*, *O.* × *intermedium*). Also,

O. × *intercedens* occurs in cultivation areas of *Origanum vulgare* subsp. *hirtum* and *Origanum onites* together [9].

Palynology

Investigations of pollen morphology have been essential as an aid to classification within Lamiaceae. Pollen morphology of Lamiaceae has been studied since Erdtman [69] suggested a division of the family into two subfamilies based on some nuclei and aperture number in the pollen grains [69, 70-73]. Palynological studies on *Origanum* taxa were limited to a few papers [74-75] and hybridization within the genus reported by Dirmenci et al. [9-11]. Palynological studies on the *Origanum* hybrids and parental species provided some important and new data concerning their shape of size and surface ornamentation and in comparison with the parental species, enlarged variations in morphological features were observed in hybrids [10-11]. The results of these studies show that pollen grains of the hybrids are intermediate in sizes between its two parents. Pollen sizes of *O.* × *sevcaniae* are smaller than its parents; 73% of the grains measured as suboblate (similarly to parents) and 27% oblate (differentiated from parents, Figure 5A-L, Table 3). Sizes of pollen grains of *O.*× *adae* are between its parents, *O. ayliniae* and *O. sipyleum*. Pollen grains shapes are varied from suboblate (80%) to oblate spheroidal (20%). When the P/E ratios are taken into account, it is seen that *O.* × *adae* is between *O. ayliniae* and *O. sipyleum* (Figure 6A-L, Table 3) The sizes of pollen grains of *O.* × *haradjanii* were measured as smaller than its parental species, similarly (Figure 7A-L, Table 3). The reason for this unexpected situation was the lack of pollen grains for measurement. Heterogenic pollen grains shape was observed in all hybrid taxa (Figures 5 and 7). This is one of the characteristics of hybrid taxa in *Origanum* species. However, there are also two different sizes found in the pollen grains of *O. vogelii*. This suggests that introgression may also occur in the *O. vogeli* (Figure 5 and 7). The obtained data from the studies confirmed that the hybrid characters of the *Origanum* species have a greater similarity of hybrids to one of the parental species. The hybrids have fewer reticule numbers than the ancestral species in case of the primary reticulum, even though the exine surface of hybrid taxa defined as bireticulate. This reduction in the primary reticulum in the pollen grains of *O. vogeli* was evaluated as the similarity to hybrid characters. This feature was also observed in other hybrid members of the Lamiaceae whose ancestors had a biretucalate ornamentation [73]. A possible source of internal instability of pollen grains size and surface structure might be chromosome rearrangement for hybridization.

Statistical Analysis

In the PCA of the morphological characters of pollens (Figure 8 and Table 4), the first component explains a significant part of the variation among the species (84,697%). The next components explain respectively 11,276% of the variation. Principal component analysis (PCA) shows that the first two components contribute to explain 95,973% of the total variation of examined quantitative pollen characters in *Origanum* species. The ‘P’ and the ‘E’ and ‘Clg’ are the major explanatory of total variation quantitative pollen characters in PCA.

The clustering analysis based on the Gower index shows us two main groups were distinguished (Figure 9): *Origanum*× *haradjanii*, *O. syriacum* subsp. *bevanii*, *O. laevigatum*, *O. vulgare* subsp. *hirtum*, *O.* × *sevcaniae* and *O. vogelii* are clustered together, and *O. ayliniae*, *O.* × *adae* and *O. sipyleum* are classified under another group. Especially *Origanum* × *haradjanii*, *O. syriacum* subsp. *bevanii* and *O. laevigatum* are seen to be close to each other in these classifications. *O. ayliniae*, *O.* × *adae* and *O. sipyleum* are also classified as close to each other. In addition, the PCA scatter plot is similar to the results of the cluster analysis (Figure 8 and 9).

Table 3. Polar and equatorial axis and pollen shape measurements of *O.* × *sevcaniae, O.* × *adae, O.* × *haradjanii* and parents (all measurements are given in µm)

Taxa	P			E			P/E	Pollen shape
	Min	Max	Mean/St.	Min	Max	Mean/St.		
Origanum vogelii	20.22	23.55	22.27 ± 0.79	24.20	28.60	25.96 ± 0.98	0.79-0.94	Suboblat to oblate spheroidal
Origanum × *sevcaniae*	18.49	21.03	19.47 ± 0.64	22.20	26.63	25.01 ± 1.12	0.73-0.85	Oblate to suboblate
Origanum vulgare subsp. *hirtum*	19.10	21.30	20.56 ± 0.51	23.34	26.45	25.24 ± 0.58	0.76-0.86	Suboblat
Origanum ayliniae	23.03	26.42	24.53 ± 0.80	27.76	31.43	29.62 ± 0.86	0.78-0.89	Suboblate
Origanum sipyleum	29.26	34.31	30.68 ± 0.82	31.33	36.08	33.14 ± 0.85	0.89-0.97	Oblatspheroidal
Origanum × *adae*	20.60	28.77	23.67 ± 2.00	24.58	32.12	28.11 ± 2.00	0.72-0.94	Suboblate to oblatspheroidal
Origanum laevigatum	19.12	21.14	20.26 ± 0.53	21.05	22.94	22.16 ± 0.44	0.72-0.94	
Origanum × *haradjani*	11.49	14.73	13.41 ± 1.28	11.76	13.36	12.58 ± 0.57	0.72-0.94	
Origanum syriacum subsp. *bevanii*	18.06	20.08	18.91 ± 0.39	18.41	20.50	19.59 ± 0.42	0.72-0.94	

Table 4. Principal component analysis of the *Origanum* taxa showing the eigenvalues of total variance

PC	1	2	3	4	5	6	7	8	9
Eigenvalue	66.1248	8.80318	1.31921	1.02405	0.562831	0.137016	0.0806535	0.0182383	0.00200086
% variance	84.697	11.276	1.6897	1.3117	0.72091	0.1755	0.10331	0.023361	0.0025628

Table 5. Tukey test of hybrids and their parents

	Taxa	Mean of differences	p
O. × sevcaniae	*O. vogelii*	-.081	.152
	O. vulgare subsp. *hirtum*	-1.081	.000
O. × haradjanii	*O. laevigatum*	-3.073	.000
	O. syriacum subsp. *bevanii*	-2.127	.000
O. × adae	*O. ayliniae*	-.381	.000
	O. sipyleum	-2.454	.000

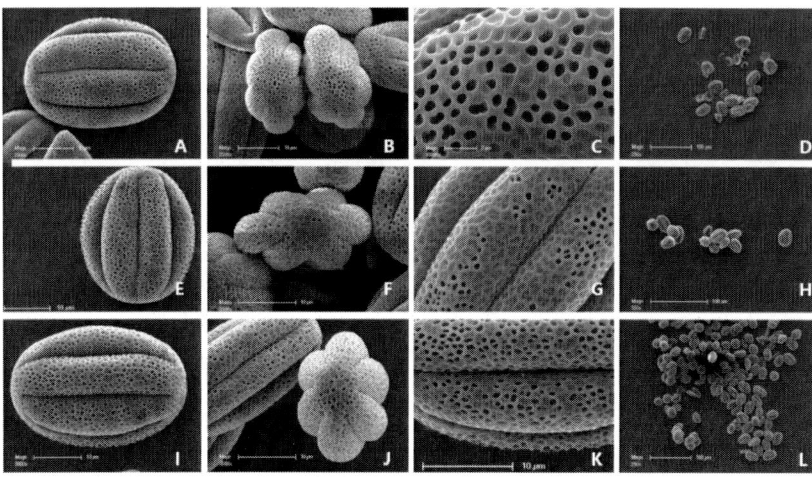

Figure 5. SEM micrographs of investigated taxa. *O. × sevcaniae* (A-D), *O. vogelii* (E-H) and *O. vulgare* subsp. *hirtum* (I-L). Mesocolpium and colpi (A, E, I), apocolpial area (B, F, J), exine surface (C, G, K), pollen grains with different size (D, H, L).

To determine if there is a significant relationship between taxa one-way analysis of variance (ANOVA) was used and the value of p was found .000 < .01. This means that there was a significant relationship between taxa. Tukey test was performed to show which of these taxa were significantly different from each other (Table 5). When we examined the Table 5. It was seen that there was no significant relation between parents, but it was seen that there was a significant relationship between parents and hybrids. The TUKEY test of parents and hybrids showed that there was a significant relation between *O. × sevcaniae* and *O. vogelii* but it was found that there was no significant relationship with its the other parent.

The results showed that *O.* × *haradjanii* was more similar to *O. syriacum* subsp. *bevanii* than *O. laevigatum* and also *O.*× *adae* was more similar to *O. ayliniae* than *O. sipyleum* (Table 5).

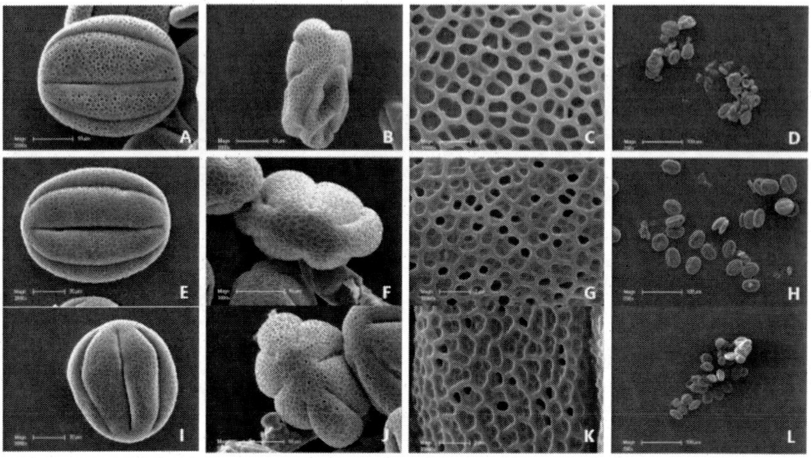

Figure 6. SEM micrographs of investigated taxa. *O.* × *adae* (A-D), *O. ayliniae* (E-H) and *O. sipyleum* (I-L). Mesocolpium and colpi (A, E, I), apocolpial area (B, F, J), exine surface (C, G, K), pollen grains with different size (D, H, L).

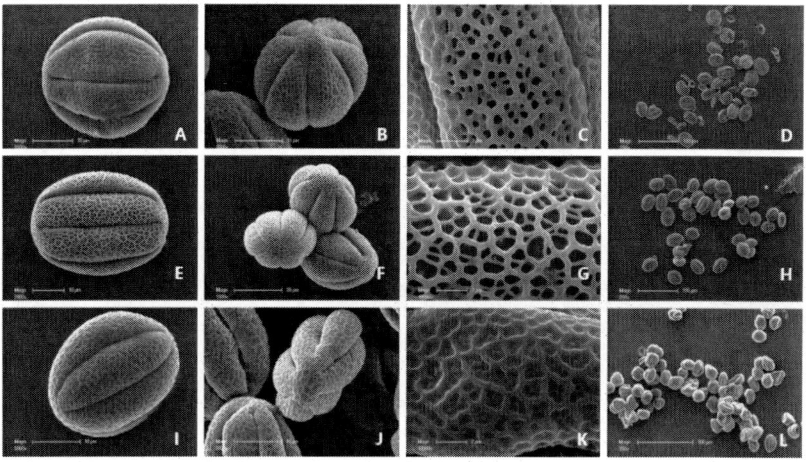

Figure 7. SEM micrographs of investigated taxa. *O.* × *haradjanii* (A-D), *O. laevigatum* (E-H) and *O. syriacum* subsp. *bevanii* (I-L). Mesocolpium and colpi (A, E, I), apocolpial area (B, F, J), exine surface (C, G, K), pollen grains with different size (D, H, L).

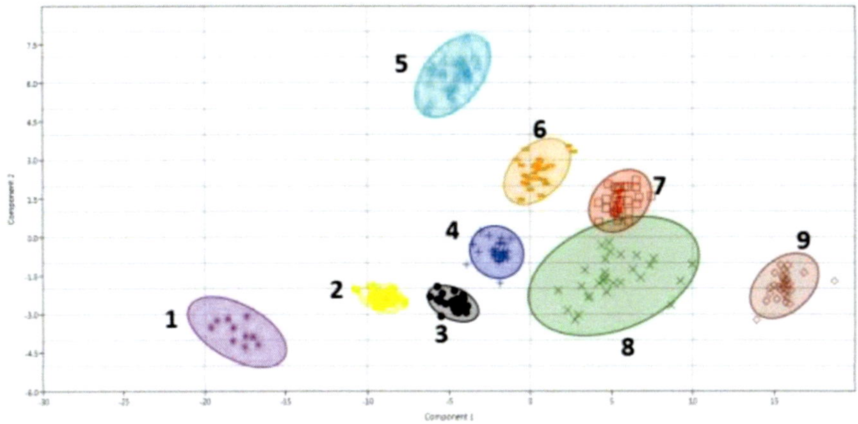

Figure 8. PCA scatter plot (1- *O.* × *haradjanii*, 2- *O. syriacum* subsp. *bevanii*, 3- *O. laevigatum*, 4- *O. vulgare* subsp. *hirtum*, 5-*O.* × *sevcaniae*, 6- *O. vogelii*, 7-*O. ayliniae*, 8- *O.* × *adae*, 9-*O. sipyleum*).

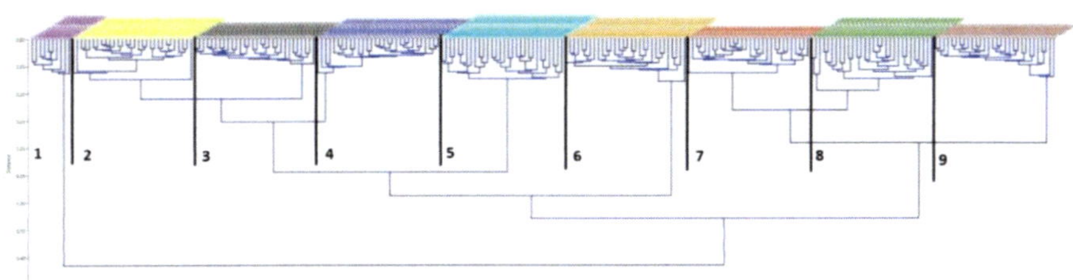

Figure 9. Cluster analysis according to quantitative pollen characters (1- *O.* × *haradjanii*, 2- *O. syriacum* subsp. *bevanii*, 3- *O. laevigatum*, 4- *O. vulgare* subsp. *hirtum*, 5-*O.* × *sevcaniae*, 6- *O. vogelii*, 7-*O. ayliniae*, 8- *O.* × *adae*, 9-*O. sipyleum*).

Cytology

According to the cytological studies on the genus *Origanum* so far, all species (except *O. vulgare* s.l., 2n = 28, 30, 32) have the same somatic chromosome number 2n = 30 [6, 10-12, 14, 76]. Karyologically investigated four *Origanum* hybrids (*O.* × *adae*, *O.* × *haradjanii*, *O.* × *munzurense*, *O. sevcaniae*) have the same chromosome number (2n = 30) with their parents.

Molecular Phylogeny

nrITS DNA region was examined to define the phylogenetic relationship of *Origanum* species. 583 characters were obtained after alignment and cutting redundant nucleotides. 425 characters were constant in all the specimens, and 116 characters were parsimony-informative and very important to separate different species. Tree scores of Figure 10 are the followings: CI: 0.646, RI: 0.826 and HI: 0.354. It is desirable for us to have a low homoplasy index (HI) and a high retention index (RI).

Three different genera members (*Satureja* ssp., *Thymus* ssp. and *Thymbra* ssp.) were used to define the phylogenetic relationship of the hybrids and their parents (Figure 10). *Thymbra* species are the nearest outgroup of examined *Origanum* species. *O. ramonense*, *O. elongatum* and *O. petraeum* species are selected as the closest outgroups, and according to our findings two specimens of *O. ramonense* and *O. petraeum* are placed in Clade 1 (Figure 10). The other *Origanum* species have a monophyletic structure. All Turkish *Origanum* members constitute Clade 2, 3, 4 and 5 including a foreign taxon *O. elongatum* the closest species of *O. vulgare* group.

It can be seen four distinctive clades containing Turkish *Origanum* taxa with *O. elongatum* in Figure 10 Clade 2 contains *O. ayliniae*, *O. sipyleum* and their putative hybrids *O.* × *adae*. *O. vulgare* spp. and *O. boissieri* and three putative hybrid specimens related to these species *O.* × *malyeri*, *O.* × *munzurense* and *O.* × *adananse* constitutes Clade 3, and Clade 4 is composed of *O. laevigatum* and its putative hybrid specimens *O.* × *haradjanii*, and latest Clade (5) has *O. majorana, O. vogelii*, *O. syriacum* subsp. *bevanii*, *O. scabrum*, *O. acutidens* and *O. brevidens* species, and also *O.* × *sevcaniae* related *O. vogelii*, *O.* × *lirium* related *O. scabrum*, *O.* × *munzurense* related *O. acutidens*. *O. acutidens*-*O.* × *munzurense* group is obviously distinguished from other Clade 5 members.

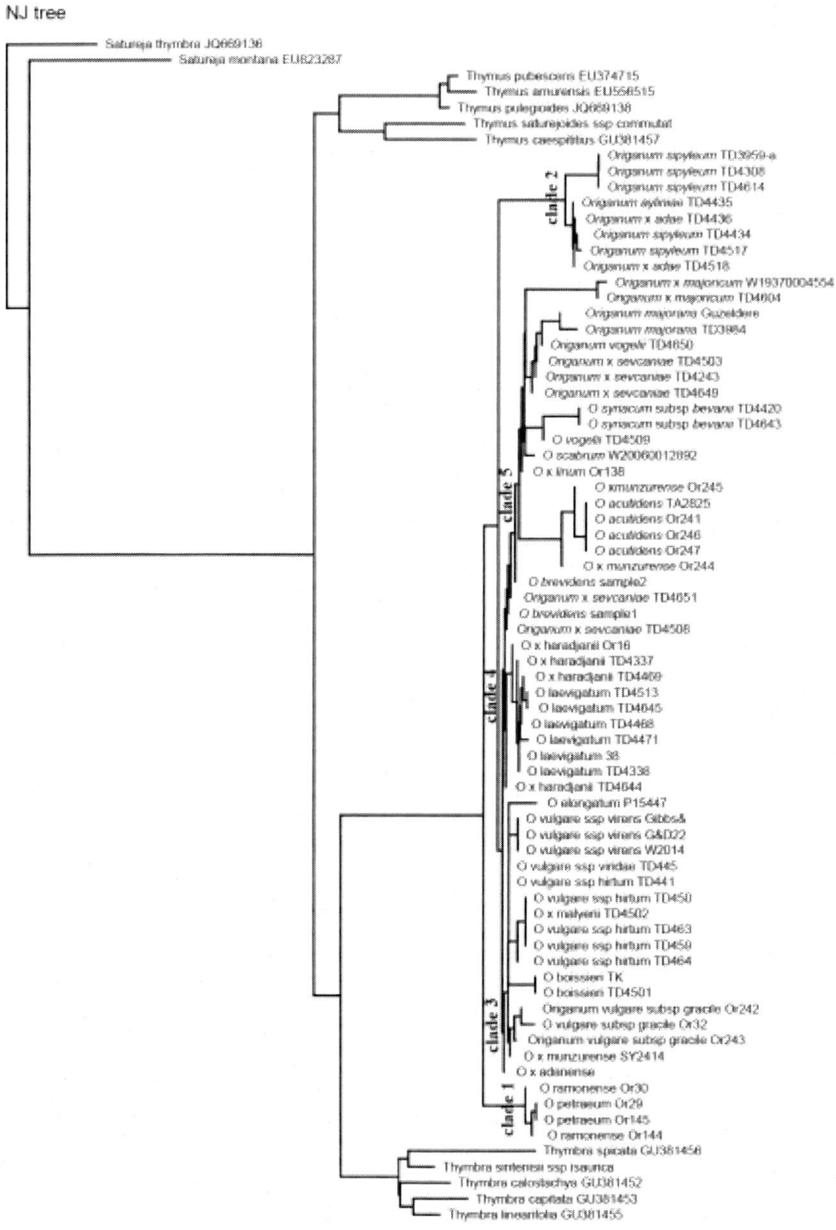

Figure 10. Neighbour-Joining tree showing phylogenetic position of *Origanum* members (based on nrITS data).

Table 6. Separative nrITS DNA loci of the hybrids and their parents

Taxa	Isolation Codes	Nucleotide Loci																													
		11	23	24	28	95	117	118	126	143	149	350	368	370	371	375	384	386	422	424	435	449	456	513	518	523	537	538	549	555	561
O. boissieri	TK s.n.	C	T	C	C	A	G	A	-	C	A	C	C	C	G	A	T	T	T	T	C	G	T	T	T	A	-	A	C	C	G
	TD4501	C	T	C	C	A	G	A	-	C	A	C	C	C	G	A	T	T	T	T	C	G	T	T	T	A	-	A	C	C	G
O. × *malyerii*	TD4502	C	T	C	C	A	G	A	-	C	A	C	C	C	G	G	T	T	T	T	C	T	T	T	T	A	-	A	C	C	G
O. vulgare subsp. *hirtum*	TD4500	C	T	C	C	A	G	A	-	C	A	C	C	C	G	G	T	T	T	T	C	T	T	T	T	A	-	A	C	C	G
	TD4482	C	T	C	C	A	G	A	-	C	A	C	C	C	G	G	T	T	T	T	C	T	T	T	T	A	-	A	C	C	G
	TD4309	C	T	C	C	A	G	A	-	C	A	C	C	C	G	G	T	T	T	T	C	T	T	T	T	A	-	A	C	C	G
	TD4635	C	T	C	C	A	G	A	-	C	A	C	C	C	G	G	T	T	T	T	C	T	T	T	T	A	-	A	C	C	G
	TD4591	C	T	C	C	A	G	A	-	C	A	C	C	C	G	G	T	T	T	T	C	T	T	T	T	A	-	A	C	C	G
	TD4648	C	T	C	C	A	G	A	-	C	A	C	C	C	G	R	T	T	T	T	C	K	T	T	T	A	-	A	C	C	G
	TD4419	C	T	C	C	A	G	A	-	C	A	C	C	C	G	A	T	T	T	T	C	G	T	T	T	A	-	A	C	C	G
	TD4507	C	Y	C	C	A	G	A	-	C	A	C	C	C	G	R	Y	W	T	T	C	K	T	A	Y	R	-	A	C	C	G
O. × *sevcaniae*	TD4651	C	T	C	C	A	G	A	-	C	A	C	C	C	G	R	Y	W	T	T	C	K	T	A	Y	R	-	A	C	C	G
	TD4508	C	Y	C	C	A	G	A	-	C	A	C	C	C	G	R	Y	W	T	T	C	K	T	A	Y	R	-	A	C	C	G
	TD4503	C	Y	C	C	A	G	A	-	C	A	C	C	C	G	A	C	A	T	T	C	G	T	A	C	G	-	A	C	C	G
	TD4243	C	Y	C	C	A	G	A	-	C	A	C	C	C	G	A	C	A	T	T	C	G	T	A	C	G	-	A	C	C	G
	TD4649	C	Y	C	C	A	G	A	-	C	A	C	C	C	G	A	Y	A	T	T	C	G	T	A	C	G	-	A	C	C	G
O. vogelii	TD4650	C	C	C	C	A	G	A	-	C	A	C	C	C	G	A	C	A	T	T	C	G	T	A	C	G	-	A	C	C	G
	TD4509	C	T	C	C	A	G	A	-	C	A	C	C	T	G	A	C	A	T	T	C	G	T	A	C	A	-	A	C	C	G
O. ayliniae	TD4435	C	T	C	T	A	A	G	C	T	A	C	C	T	G	A	T	A	C	T	C	G	T	A	T	A	-	A	T	C	G
O. × *adae*	TD4436	C	T	C	T	A	A	G	C	T	A	C	C	T	G	A	T	A	C	T	C	G	T	A	T	A	-	A	T	C	G
	TD4518	C	T	C	T	A	A	G	C	T	A	C	C	T	G	A	T	A	C	T	C	G	T	A	T	A	-	A	T	C	G

Taxa	Isolation Codes	Nucleotide Loci																													
		11	23	24	28	95	117	118	126	143	149	350	368	370	371	375	384	386	422	424	435	449	456	513	518	523	537	538	549	555	561
O. sipyleum	TD3959-a	C	T	C	T	A	A	G	C	T	A	G	C	T	G	A	T	A	C	T	C	G	T	A	T	A	-	A	T	C	A
	TD4308	C	T	C	T	A	A	G	C	T	A	G	C	T	G	A	T	A	C	T	C	G	T	A	T	A	-	A	T	C	A
	TD4614	C	T	C	T	A	A	G	C	T	A	G	C	T	G	A	T	A	C	T	C	G	T	A	T	A	-	A	T	C	A
	TD4434	C	T	C	T	A	A	G	C	T	A	C	C	T	G	A	T	A	C	T	C	G	T	A	T	A	-	A	T	C	G
	TD4517	C	T	C	T	A	A	G	C	T	A	C	C	T	G	A	T	A	C	T	C	G	T	A	T	A	-	A	T	C	G
O. acutidens	TA2825	C	T	C	C	T	G	A	-	C	A	C	T	C	T	A	C	A	T	C	T	G	T	A	C	A	A	A	C	T	A
	TD4956-b	C	T	C	C	T	G	A	-	C	A	C	T	C	T	A	C	A	T	C	T	G	T	A	C	A	A	A	C	T	A
	TD4956-c	C	T	C	C	T	G	A	-	C	A	C	T	C	T	A	C	A	T	C	T	G	T	A	C	A	A	A	C	T	A
	TD4956-a	C	T	C	C	T	G	A	-	C	A	C	T	C	T	A	C	A	T	C	T	G	T	A	C	A	A	A	C	T	A
O. × *munzurense*	TD4957-a	Y	T	C	C	W	G	A	-	C	A	C	T	C	T	A	Y	A	T	C	Y	G	T	A	C	A	A	A	C	T	A
	TD4957-b	C	T	C	C	T	G	A	-	C	A	C	T	C	T	A	Y	A	T	Y	Y	G	T	A	Y	A	A	A	C	T	R
	SY2414	Y	T	C	C	A	G	A	-	C	A	C	C	C	G	A	T	T	T	T	C	G	T	T	T	A	-	A	C	C	G
O. vulgare subsp. *gracile*	TD4958-b	T	T	C	C	A	G	A	-	C	A	C	C	C	G	A	T	T	T	T	C	G	T	T	T	A	-	A	T	C	G
	TD4958-a	Y	T	C	C	A	G	A	-	C	A	C	C	C	G	A	T	T	T	T	C	G	T	T	T	A	-	A	T	C	G
	TD4117	C	T	C	C	A	G	A	-	C	A	C	C	C	G	A	T	T	T	T	C	G	T	T	T	A	-	A	T	C	G
O. majorana	TD3083	C	Y	C	C	A	G	A	-	C	A	C	C	C	G	A	C	A	T	T	C	G	T	A	C	G	-	A	C	C	G
	TD3984	C	C	C	C	A	G	A	-	C	A	C	C	C	G	A	C	A	T	T	C	G	T	A	C	G	-	A	C	C	G
O. × *majoricum*	W19370004554	C	T	C	C	A	G	A	-	C	A	C	C	C	G	A	C	A	T	T	C	G	T	A	C	A	-	A	C	C	G
	TD4604	C	T	C	C	A	G	A	-	C	A	C	C	C	G	A	C	A	T	T	C	G	T	A	C	A	-	A	C	C	G
O. vulgare subsp. *virens*	Gibbs&Dominguez 224-1	C	T	C	C	A	G	A	-	C	A	C	C	C	G	A	T	T	T	T	C	G	T	T	T	A	-	A	C	C	G
	Gibbs&Dominguez 224-2	C	T	C	C	A	G	A	-	C	A	C	C	C	G	A	T	T	T	T	C	G	T	T	T	A	-	A	C	C	G
	W20140004728	C	T	C	C	A	G	A	-	C	A	C	C	C	G	A	T	T	T	T	C	G	T	T	T	A	-	A	C	C	G

Table 6. (Continued)

Taxa	Isolation Codes	Nucleotide Loci																													
		11	23	24	28	95	117	118	126	143	149	350	368	370	371	375	384	386	422	424	435	449	456	513	518	523	537	538	549	555	561
O. vulgare subsp. *viridulume*	TD4456-a	C	T	C	C	A	G	A	-	C	A	C	C	C	G	A	T	T	T	T	C	G	T	T	T	A	-	A	C	C	G
O. syriacum subsp. *bevanii*	TD4470	C	T	C	C	A	G	A	-	C	A	C	C	T	G	A	C	A	T	T	C	G	A	A	C	A	-	A	C	C	G
	TD4643	C	T	C	C	A	G	A	-	C	A	C	C	T	G	A	C	A	T	T	C	G	A	A	C	A	-	A	C	C	G
O. × *haradjanii*	TD3444	C	T	S	C	A	G	A	-	C	A	C	C	Y	G	A	Y	W	T	T	C	G	W	A	Y	A	-	R	C	C	G
	TD4337	C	T	G	C	A	G	A	-	C	A	C	C	C	G	A	Y	W	T	T	C	G	T	W	T	A	-	G	C	C	G
	TD4469	C	T	G	C	A	G	A	-	C	R	C	C	C	G	A	T	T	T	T	C	G	T	T	T	A	-	G	C	C	G
	TD4644	C	T	S	C	A	G	A	-	C	R	C	C	C	G	A	Y	W	T	T	C	G	T	A	Y	A	-	A	C	C	G
O. laevigatum	TD4471	C	T	G	C	A	G	A	-	C	R	C	C	C	G	A	T	T	T	T	C	G	T	T	T	A	-	G	C	C	G
	TD4468	C	T	G	C	A	G	A	-	C	R	C	C	C	G	A	T	T	T	T	C	G	T	T	T	A	-	G	C	C	G
	TD3445	C	T	G	C	A	G	A	-	C	A	C	C	C	G	A	T	T	T	T	C	G	T	T	T	A	-	G	C	C	G
	TD4338	C	T	G	C	A	G	A	-	C	A	C	C	C	G	A	T	T	T	T	C	G	T	T	T	A	-	G	C	C	G
	TD4513	C	T	G	C	A	G	A	-	C	G	C	C	C	G	A	T	T	T	T	C	G	T	T	T	A	-	G	C	C	G
	TD4645	C	T	G	C	A	G	A	-	C	G	C	C	C	G	A	T	T	T	T	C	G	T	T	T	A	-	G	C	C	G
O. × *adanense*	Or163	C	Y	S	C	A	G	A	-	C	A	C	C	C	G	A	Y	W	T	T	C	G	T	W	Y	R	-	A	C	C	G
O. brevidens	TD3574-1	C	Y	S	C	A	G	A	-	C	A	C	C	C	G	A	Y	W	T	T	C	G	T	A	Y	R	-	A	C	C	G
	TD3574-2	C	Y	S	C	A	G	A	-	C	A	C	C	C	G	A	Y	W	T	T	C	G	T	A	Y	R	-	A	C	C	G
O. × *lirium*	Or138	C	T	C	C	A	G	A	-	C	A	C	C	C	G	A	C	A	T	T	C	G	T	A	Y	A	-	A	C	C	G
O. scabrum	W20060012892	C	T	C	C	A	G	A	-	C	A	C	C	C	G	A	C	A	T	T	C	G	T	A	C	A	-	A	C	C	G

When we analyzed eight hybrids and their parental species using nrITS DNA sequences (Table 6), 30 different loci could be detected occurring single nucleotide polymorphisms (SNPs). *O. ayliniae-O.* × *adae-O. sipyleum* group has the second biggest differentiation at eight loci (at the positions 28, 117, 118, 126, 143, 370, 422, 549) having SNPs. That means *O. ayliniae-O.* × *adae-O. sipyleum* are distinguished as molecularly from the other examined species as molecular in these loci. For example, all the examined specimens have "C" nucleotide at the position 28 but *O. ayliniae*, *O.* × *adae* and *O. sipyleum* have "T" nucleotide. The same situation can be seen at other loci except the nucleotide position 126. A single-nucleotide-insertion (C nucleotide) is observed at the 126th position. These differences are given in Table 6 also distinguish *O. ayliniae*, *O.* × *adae* and *O. sipyleum* from the other examined taxa in the phylogram (Figure 5). In addition to this information, three different *O. sipyleum* individuals (numbered as TD3959-a, TD4308 and TD4614) differ the rest of examined two different *O. sipyleum* (TD4434 and TD4517), one *O. ayliniae*, and two different *O.* × *adae* (TD4436 and TD4518) specimens at the nucleotide positions of 350 (C-G polymorphism) and 561 (A-G polymorphism).

O. × *sevcaniae* is a hybrid generating between *O. vogelii* and *O. vulgare* subsp. *hirtum* species. According to our nrITS data results, all five examined *O.* × *sevcaniae* specimens have a close relationship with *O. vogelii* rather than *O. vulgare* subsp. *hirtum*. Seven heterozygous loci are detected in *O.* ×*sevcaniae* nrITS sequences. Nucleotide positions of 23, 384 and 518 are samples of C-T heterozygosity, 325th and 523rd nucleotides have A-G heterozygosity and also A-T and G-T polymorphisms are seen at the nucleotide positions of 386 and 449, respectively. *O. vogelii* and *O. vulgare* subsp. *hirtum* distinguish at these seven nucleotides and the position of 513, so eight single nucleotide polymorphisms are very important to separate from two parents.

O. acutidens and *O.* × *munzurense* have the most polymorphic loci based on examined nrITS DNA sequences. *O. acutidens* and *O.* × *munzurense* can be distinguished from other examined species at 10 different loci. C-T single nucleotide polymorphisms are observed at the positions of 368, 384, 424, 435, 518 and 555. Other separative polymorphisms are obtained at the positions of 95, 371 and 561. In addition, there is an A-nucleotide-insertion region at the 537th nucleotide position. *O.* × *munzurense* coded Or244 and Or245 specimens have also polymorphisms at these loci. This means that *O.* × *munzurense* individuals have two different nucleotide sets probable stemming from their ancestors. For instance, at position 384 we can see that *O. acutidens* has C nucleotide and the other parent *O. vulgare* subsp. *gracile* has T nucleotide, and *O.* × *munzurense* Or244 and Or245 have an international nucleotide code-named "Y" and that means this hybrid has C and also T nucleotide at the same time in its DNA sequence. The same situation can be seen for different code named as "R" at the position of 561 meaning as A-G polymorphism (Table 6). In addition, one specimen of *O.* × *munzurense* SY2414 has a close relationship with one of its parents, *O. vulgare* subsp. *gracile* rather than acutidens (Figure 10). When we analyzed the sequences of *O.* × *munzurense* SY2414, it differs the other examined *O.* × *munzurense* individuals and *O. acutidens* at the nucleotide positions of 95, 368, 371, 384, 386, 424, 435, 518, 537, 555 and 561 and has same nucleotides with *O. vulgare* subsp. *gracile* (Table 6). So, this information supports that all hybrids are not similar with the same parents.

O. × *haradjanii* is another hybrid having some important morphological characters similar to its parents. According to our nrITS results, four different specimens of *O.* × *haradjanii* (TD4337, 4469, 4644, Or16) are placed with six different individuals of *O. laevigatum*, one of its parents (Figure 10). *O.* × *haradjanii* TD4644 has a heterozygous structure at the nucleotide positions of 24, 149, 384, 386 and 518, and also this specimen has similar nucleotides with another parent *O. syriacum* subsp. *bevanii* at the nucleotide positions of 513 and 538. In addition, *O. syriacum* subsp. *bevanii* and *O. laevigatum* can be distinguished at nine nucleotide positions (24, 149, 370, 384, 386, 456, 513, 518 and 539). Nine different nucleotides are very important to separate two species, but these species interestingly generate a hybrid taxon.

O. × *adananse* is the other hybrid generating between *O. laevigatum* and *O. brevidens*. As mentioned above, *O.* × *haradjanii* is more similar with *O. laevigatum*, on the contrary *O.* × *adananse* has eight totally similar nucleotides with *O. brevidens*, and two specimens of *O. laevigatum, O. brevidens* and *O.* × *adananse* specimens have A nucleotide at the nucleotide position of 149 and this is the only same nucleotide among three taxa according to our nrITS data. On the other hand, although DNA extractions of the same *O. brevidens* specimen were performed twice, we observed some heterozygous nucleotides after sequencing. The nucleotide positions of 23, 24, 384, 386, 518 and 523 are heterozygote and similar to its hybrids (Table 6).

The last two examined hybrids, *O.* × *majoricum* and *O.* ×*lirium* have more similarity with their parents, *O. majorana* and *O. scabrum*. These two hybrids share the same clade, but they absolutely differ from the clade 3 containing of *O. vulgare* subspecies (Figure 10). When we analyzed the Table 6. *O.* ×*majoricum* has totally the same nucleotides with *O. majorana* at the nucleotide positions of 384, 386, 513 and 518, and these nucleotides separate *O. majorana* and *O. vulgare* subsp. *virens*. On the other hand, *O.* × *lirium* has the same nucleotides at six positions (375, 384, 386, 449, 513, 518) with *O. scabrum* and differs *O. vulgare* subsp. *hirtum* specimens.

What Is Role of the Hybridization on the Speciation of the Genus Origanum?

The mechanism of speciation through hybridization occurs via polyploidy and homoploidy. Hybrid speciation is more common in plants than it can be thought (Mullet, 2007). Homoploid hybrid speciation is the origin of a hybrid species without a change in chromosome numbers [18, 21, 24, 31]. Polyploid species are often thought to occur after the fusion of reduced (n) and unreduced (2n) gametes and producing a triploid (3n) offspring, or sometimes in different ploidy levels [77]. Allopolyploids are largely reproductively isolated by their ploidy level (Mullet, 2007) but homoploid hybrid speciation is unlikely unless ecological and geographical isolation occurs between the hybrid and its ancestors [21, 78-80] Differentiation of the hybrids from their parents in terms of chromosome number, habitat isolation, ecological isolation or geographical isolation is required for the speciation of hybrids [21-22, 25,-26, 35, 37, 41, 43-44, 49, 78-81].

If gene flow among ecologically separated hybrids and parents is reduced with different conditions, hybrid speciation is more likely to occur in this area. There is some lacking agreement on the importance of hybrid speciation process, and the relative proportions of each parental genome present in a hybrid species are one of them [80]. As mentioned above, **the** ploidy level and chromosome number have significant importance of the taxonomical and evolutionary studies [21-22, 31-35, 81-82]. Ietswaart's [6] *Origanum* hybridization studies are examples of important evolutionary studies for the genus. Ietswaart (1980) argued that *Origanum* is caused by the hybridization of the old tribe Saturejeae (now tribe Mentheae) genera. Based on the hypothesis of Ietswaart, we can discuss the emergence of the genus *Origanum*, its intrageneric speciation and its hybrids having the possibility for speciation via hybridization using morphological and cytological evidence. The genus *Origanum* is a well-known member of the Mentheae tribe and the Menthinae subtribe [1]. The chromosome numbers of the Menthinae subtribe vary from $2n = 12$ (*Hyssopus*) to $2n = 144$ (*Mentha*). The chromosome numbers were $2n = 30$ in most of the *Thymbra* L., *Satureja* L., and *Micromeria* Benth. s.str., subsections *Vulgares* L. and *Piperella* Willk of the genus *Thymus* species belonging to this subtribe. Ietswarts [6] put forward in his hypothesis that *Origanum* might have emerged in the Pliocene, and that formerly, the tribe Saturejeae genera were probably the main genera from which *Origanum* could be derived [6]. When we analyzed chromosome numbers of the closest genera species suggested by Ietswaart and *Origanum* species, we found that most of them have 2n = 30 chromosomes number. Chromosome number of the identified species and hybrids of *Origanum* (*O.* × *adae*, *O.* × *haradjanii, O.* × *munzurense*, *O.* × *sevcaniae* and other some hybrids-unpublished data) was determined as 2n = 30. So, it seems that,the homoploid hybridization process causes the homoploid hybrids. If the genus is derived from the genera of Menthinae subtribe by hybridization, there is great evidence that it is derived by homoploid speciation. Ietswaart [6] also suggested that the hybridization phenomenon causes the infrageneric divergence of *Origanum* genus, and arising of some sections of the hybridization with other sections within itself or some *Thymus*, *Satureja* and *Micromeria* s.str. species [6]. According to this hypothesis, the genera mentioned above are important regarding chromosome number compatibility, and also, *Micromeria*, *Satureja,* and *Thymus* are morphologically closer to *Origanum*, but *Thymbra* is different. The genus *Origanum* may be derived from one or more of these genera and then continued to speciation via hybridization until today. The compatibility of chromosome numbers between mentioned genera, *Origanum* species, and hybrids supports Ietswaart's hypothesis [6].

In infrageneric divergence of the genus, the sections *Amaracus*, *Majorana*, and *Origanum* (*O. vulgare* s.l.) can be considered as the ancestral sections (see Table 2). Hybridization easily occurs in different sections. Later, species in some sections have intermediate characteristics between mentioned different ancestral sections. For example, Sect. *Anatolicon* and Sect. *Chilocalyx* species have intermediate characters between the sections *Amaracus* and *Origanum*, and *Majorana* and *Origanum*, respectively [6, 9].

These are the main characteristics of the above mentioned sections [6]: Sect. *Majorana* species are characterized by its branches first order always present, second-order often so, third-order sometimes so. Spikes small, subglobose to cylindrical. Bracts differ from the leaves, densely imbricate, as long as calyces, enclosing these marginally, whitish, greyish or green, hairy. Flowers 2 per verticillaster, bisexual or female, very small. Calyces flattened, 1-lipped for 9/10 or more, upper lip entire or denticulate, throat not pilose. Corollas 2-lipped for 2/3, c. × 2 calyces (Figures 2L and 3L). Sect. *Chilocalyx* species are characterized by its branches first order always present, second or third-order often so. Spikes are very small. Bracts often more ore less continuing from the leaves, densely imbricate, greyish or green, hairy. Flowers 2 per verticillaster, bisexual or female, very small. Calyces tubular or slightly campanulate, 1-or 2-lipped for 1/5-2/5, small teeth or lobes in upper and lower lips often present; throats conspicuously pilose. Corollas 2-lipped for 2/3, c. x 2 calyces (Figures 2D and 3D). Sect. *Origanum* species are characterized by its branches of the first order and second order always present, third order usually so. Spikes dense, distinct from the rest of the plants, small or medium-sized. Bracts different from the leaves in texture and color, shape and size, imbricate. Flowers 2 per verticillaster, bisexual or female, small or medium-sized. Calyces more or less tubular with 5 equal teeth for c. 1/3, throat pilose. Corollas 2-lipped for c. 1/3, c. 2.5 × calyces (Figures 2C, F, I and 3C, F, I).

In addition to morphological evidence, biogeographic evidence plays an important role in the formation and distribution of hybrids. The distribution of the species belonging to these three sections coincides in the Eastern Mediterranean. Namely, Sect. *Origanum* (*O. vulgare* with five subspecies) members are distributed from Canary Island, Europe to East Asia (Figure 1 and Figure 11) Sect. *Chilocalyx* species (four species) are distributed Kriti and Southern of Turkey (Figure 11). Sect. *Majorana* species are distributed in southern Spain, Portugal, Algeria, Morocco, Balkans, west and southern Turkey to the Sinai Peninsula (Figure 11). Sect. *Chilocalyx* species are similar to sect. *Majorana* and Sect. *Origanum* species in point of spikes, bracts and small-sized flowers. But they have some intermediate characters as calyx shape and corolla colours (Figures 2D and 3D): Sect. *Majorana* members' calyces flattened and 1-lipped for 9/10 or more, corollas' colour white (Figures 2L and 3L), Sect. *Origanum* members' calyces tubular with 5 equally teeth, corollas' colour white, pinkish-white to purple (Figures 2C, F, I and 3 C, F, I). Sect. *Chilocalyx* members' calyces tubular or slightly campanulate, 1-or 2-lipped for 1/5-2/5, small teeth or lobes in upper and lower lips often present, corollas' color white or pinkish to purple (Figures 2D and 3D). Sect. *Chilocalyx* consists of four species, two species with white flowers (*O. minutiflorum* and *O. bilgeri* are endemic to west Mediterranean part of Turkey) and two species with pinkish to purple flowers (*O. microphyllum*, endemic to Kriti and *O. vogelii*, endemic to east Mediterranean part of Turkey). Both biogeographic and morphological evidence suggesting that *O. majorana*

(Sect *Majorana*) and *O. vulgare* subsp. *hirtum* play a major role in the hybridization of Sect *Chilocalyx* species are much more compared with other taxa. Thus, the biogeographic spreading of *O. vulgare* subsp. hirtum is more compatible compared with the other *O. vulgare* s.l. The spreading with *O. majorana* coincides exactly in the same area where the *Chilocalyx* section species emerge.

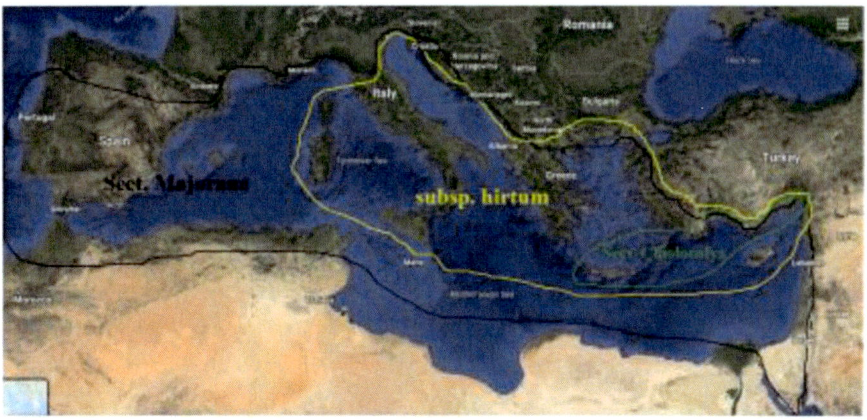

Figure 11. Distribution map of the sections *Majorana* (black line), *Origanum* (only subsp. *hirtum*-yellow line) and *Chilocalyx* (green line)

During the fieldwork we have carried out in Turkey, we obtained data on 11 hybrids, and it was found that most of the studied hybrids share the same area with both parents of them. For instance, *O.* × *sevcaniae* shares the same area with both parents (*O. vogelii* × *O. vulgare* subsp. *hirtum*), in a different region it only shares the same area with *O. vogelii* [11]. An example collected by Sorger was identified by Sorger and Kit Tan as *O. munzurense*. Although no sample belonging to the parents suggests that it acts as an independent species, our studies showed that *O. acutidens* and *O. vulgare* subsp. *gracile* taxa were found in the same area with *O. munzurense*, and *O. munzurense* has intermediate characters between *O. acutidens* and *O. vulgare* subsp. *gracile*, and these properties showed that it was probably a hybrid, and it was rearranged as *O.* × *munzurense* [12]. *Origanum* hybrids can occur in the region where the natural growth areas of two different species coincide, as well as the cultivation of species brought from different geographies in botanical gardens (Iestwaart 1980) (see Table 1). For example, *O. amanum* is endemic to the south of Turkey and *O. dictamnus* is growing Aegean Island and endemic to Greece. There is an unidentified hybrid formed by growing two species in the Jena Botanical Garden [6] (see Table 1). The population of a significant number of hybrids consists of a small number of individuals. Thus, Davis [83] stated that *O.* × *dolicosiphon* consisted of 6 individuals (we did not collect any specimens belonging to this hybrid). In addition, the population of the *O.* × *malyeri* consists of approximately 10 individuals [11]. Based on our observations in the field studies we have conducted for about 5 years, it is the hybrid *O.* × *intercedens* which has the highest number of individuals and is more common in nature than the others.It was determined from north-west to south-west of Turkey, and also discovered in Crete, Mytilene (personal observation) and some Aegean Islands [84]. However, there is another hybrid which is different from other hybrids, named as *O.* × *majoricum*. *O.* × *majoricum* is generated between *O. majorana* and *O. vulgare* subsp. *virens*. Although, *O.* × *majoricum* is a hybrid of natural spread on the Balearic Island (Spain), it is widely cultivated in different countries for spice, herbal tea and ornamental in botanical and personal gardens and rarely fields. *Senecio squalidus* L. (*S. aethenensis* DC × *S. chrsanthemifolius* Poir.), a homoploid hybrid with no post-zygotic barrier, has completed its speciation geographically isolated from its parents [25, 85-86]. Hybrid species adapt to different habitats and are geographically isolated from their parents, causing them to be isolated in terms of reproduction. But, post-zygotic reproductive isolation is relatively weak among homoploid hybrids and their parents. When the expansive feature of *O.* × *majoricum* emerges, it is likely to be separated from their parents by human factor and adapt to new habitats such as *S. squalidus*, and it is the candidate hybrid species closest to speciation by homoploid hybridization among the *Origanum* hybrids.

ACKNOWLEDGMENTS

We would like to thank TÜBİTAK (Project No: 113Z225) for financial support of our project, thank Serap SEMERCİ for checking the chapter in terms of language, and thanks to curators of the following herbaria, which permitted us to examine the specimens: ANK, B, BM, E, EGE, G, GAZI, GOET, HUB, ISTE, ISTF, K, LE, W, and WU.

REFERENCES

[1] Harley, R. M., Atkins, S., Budantsev, A., Cantino, P. D., Conn, B. J., Grayer, R., Harley, M. M., Kok, R., Krestovskaya, T., Morales, R., Paton, A. J., Ryding, O. and Upson, T. (2004). Labiatae. "In *The families and genera of vascular plants*, edited by Kadereit, J. W., Vol. 7:167–275. Berlin: Springer.

[2] Li, B., Cantino, P. D., Olmstead, R. G., Bramley, G. L. C., Xiang, C. L., Ma, Z. H., Tan, Y. H. and Zhang, D. X. (2016). "A large-scale chloroplast phylogeny of the Lamiaceae sheds new light on its subfamilial classification." *Scientific Reports, 6, 34343.*

[3] Li, B. and Olmstead, R. G. (2017). "Two new subfamilies in Lamiaceae." *Phytotaxa, 313, 222-226.*

[4] Celep F. and Dirmenci T. (2017). "Systematic and Biogeographic overview of Lamiaceae in Turkey." *Natural Volatiles & Essential Oils*, 4: 14-27.

[5] Duman, H. and Dirmenci, T. (2017). "A new species of *Micromeria* (Lamiaceae) from Köyceğiz (Muğla, southwest of Turkey)." *Turkish Journal of Botany* 41:383-391.

[6] Ietswaart, J. H. (1980). *A taxonomic revision of the genus Origanum* (Labiatae). Leiden Botanical Series. No. 4. Netherlands: Leiden University Press.

[7] Ietswaart, J. H. (1982). "*Origanum* L." In *Flora of Turkey and the East Aegean Islands,* edited by Davis, P. H., Vol. 7:297–313. Edinburgh: Edinburgh Univ Press.

[8] Goavaerts, R., Paton, A., Harvey, Y., Navarro. T. and Pena, M. R. G. (2011). World Checklist of Lamiaceae, Kew. Facilitated by the Royal Botanic Gardens, Kew. Published on the Internet; http://apps.kew.org/wcsp/ [accessed 30/09/2016].

[9] Dirmenci, T., Kılıç, T., Çelenk, S., Martin, E., Arabacı, T., Yıldız, B. and Yazıcı, T. (2017). "Türkiye'de Yetişen *Origanum* L. (Lamiaceae) Türleri Üzerinde Taksonomik, Moleküler, Karyolojik Ve Kimyasal Araştırmalar" TÜBİTAK-KBAG Project (113 Z 225) (in Turkish).

[10] Dirmenci, T., Yazıcı, T., Özcan, T., Çelenk, S. and Martin, E. (2018a). "A new species and a new natural hybrid of *Origanum* L. (Lamiaceae) from the west of Turkey". *Turkish Journal of Botany* 42:73-90.

[11] Dirmenci, T., Özcan, T., Yazıcı, T., Arabacı, T., and Martin, E. (2018b). "Morphological, cytological, palynological and molecular evidence on two new hybrids: an example of homoploid hybridization in *Origanum* (Lamiaceae)." *Phytotaxa* 371:145-167.

[12] Dirmenci, T., Özcan, T., Açar, M., Arabacı, T., Yazıcı, T., and Martin, E. (2019a). "A rearranged homoploid hybrid species of *Origanum* L. (Lamiaceae): *O.* × *munzurense* Kit Tan & Sorger." *Botany Letters* 166: 153-162. DOI:10.1080/23818107.2019.1585283.

[13] Duman, H., Başer, K. H. C. and Aytaç, Z. (1996). "Two new species and a new hybrid from Anatolia." *Turkish Journal of Botany* 2"2:51-55.

[14] Bakha, M., Al Faiz, C., Daoud, M., El Mtili, N., Aboukhalid, K., Khiraoui, A., Machon, N. and Silijak-Yakovlev, S. (2017). "Genome size and chromosome number for six taxa of *Origanum* genus from Morocco." *Botany Letters* 164 (4): 361-370.

[15] Otto, S. P. and Whitton, J. (2000). "Polyploid incidence and evolution. "*Annual Review of Genetics* 34: 401–437. http://dx.doi.org/10.1146/annurev.genet.34.1.401.

[16] Wood, T. E., Takebayashi, N., Barker, M. S., Mayrose, I., Greenspoon, P. B. and Rieseberg, L. H. (2009). "The frequency of polyploid speciation in vascular plants." *Proceeding of the National Academy of Sciences U.S.A.* U.S.A. 106: 13875–13879. http://dx.doi.org/10.1073/pnas.0811575106.

[17] Mallet, J. 2007. "Hybrid speciation." *Nature* 446: 279–283.

[18] Folk, R. A., Soltis, P. S., Soltis, D. E. and Guralnick R. (2018). "New prospects in the detection and comparative analysis hybridization in the tree of life. " *American Journal of Botany* 105: 364-375.

[19] Rieseberg, L. H. (1997). "Hybrid origins of plant species." *Annual Review of Ecology and Systematics* 28, 359–389. http://dx.doi.org/10.1146/annurev.ecolsys.28.1.359..

[20] Rieseberg, L. H., Raymond, O., Rosenthal, D. M., Lai, Z., Livingstone, K., Nakazato, T., Durphy, J. L., Schwarzbach A. E., Donovan, L. A. and Lexer, C. (2003). "Major ecological transitions in wildsunflowers facilitated by hybridization." *Science* 301: 1211–1216. http://dx.doi.org/10.1126/science.1086949.

[21] Abbott, R. J., Hegarty, M. J., Hiscock S. J. and Brennan A. C. (2010). "Homoploid hybrid seciation in action." *Taxon* 59:1375-1386.

[22] Abbott, R. Albach, D., Ansell, S., Arntzen, J. W., Baird, S. J. E., Bierne, N., Boughman, J., Brelsford, A., Buerkle, C.A., Buggs, R., Butlin, R. K., Dieckmann, U., Eroukhmanoff, F., Grill, A., Cahan, S. H., Hermansen, J. S., Hewitt, G., Hudson, A. G., Jiggins, C., Jones, J., Keller, B., Marczewski, T., Mallet, J., Martinez-Rodriguez, P., Möst, M., Mullen, S., Nichols, R., Nolte, A., Parisod, C., Pfennig, K., Rice, A. M., Ritchie, M. G., Seifert, B., Smadja, C. M., Stelkens, R., Szymura, J. M., Väinölä, R., Wolf, J. B. W. and Zinner, D. (2013). "Hybridization and speciation." *Journal of Evolutionary Biology* 26: 229–246. http://dx.doi.org/10.1111/j.1420-9101.2012.02599.x.

[23] Abbott, R. J. (2017). "Plant speciation across environmental gradients and the occurrence and nature of hybrid zones." *Journal of Systematics and Evolution* 55 (4): 238–258.

[24] Schummer, M., Rosenthal, G. G. and Andolfatto, P. (2014). "How common is homoploid hybrid speciation?" *Evolution* 68: 1553–1560. http://dx.doi.org/10.1111/evo.12399.

[25] Abbott, R. J., and Lowe, A. J. (2004). "Origins, establishment and evolution of new polyploid species: *Senecio cambrensis* and *S. eboracensis* in the British Isles." *Biological Journal Linnean Society* 82:467–474.

[26] Soltis, P. S., and Soltis, D. E. (2009). "The role of hybridization in plant speciation." *Annual Review Plant Biology* 60: 561–588. http://dx.doi.org/10.1146/annurev.arplant.043008.092039.

[27] Soltis, P. S. and Soltis, D. E. (2000). "The role of genetic and genomic attributes in the success of polyploids." *Proceeding of the National Academy of Sciences U.S.A.* 97: 7051–7057. http://dx.doi.org/10.1073/pnas.97.13.7051.

[28] Soltis, D. E., Soltis, P. S. and Tate J. A. (2003). "Advances in the study of polyploidy since Plant speciation." *New Phytologist* 161: 173–191. http://dx.doi.org/10.1046/j.1469-8137.2003.00948.x.

[29] Soltis, D. E., Soltis, P. S., Schemske, D. W., Hancock, J. F., Thompson, J. N., Husband, B.C., and Judd, W. S. (2007). "Autopolyploidy in angiosperms: Have we grossly underestimated the number of species?" *Taxon* 56: 13–30.

[30] Grant, V. (1981). *Plant speciation*, 2nd ed. New York: Columbia Univ. Press.

[31] Kadereit, J. W. (2015). "The geography of hybrid speciation in plants." *Taxon* 64: 673–687.

[32] Soltis, D. E., Buggs, R. J. A., Doyle, J. D. and Soltis, P. E. (2010). "What we still don't know about polyploidy." *Taxon* 59: 1387–1403.

[33] Soltis, D. E., Buggs, R. J. A., Barbazuk, W. B., Chamala, S., Chester, M., Gallaghar, J. P., Schnable, P. S. and Soltis, P. S. (2012). "The early stages of polyploidy: Rapid and repeated evolution in Tragopogon." In *Polyploidy and genome evolution*. Edited by Soltis, P. S., and Soltis, D. E., 271–292 Heidelberg: Springer. http://dx.doi.org/10.1007/978-3-642-31442-1_14.

[34] Soltis, D. E., Visger, C. J. and Soltis, P. S. (2014). "The polyploidy revolution then … and now: Stebbins revisited." *American Journal of Botany* 101: 1057–1078. http://dx.doi.org/10.3732/ajb.1400178.

[35] Yakimowski, S. B. and Rieseberg, L. H. (2014). "The role of homoploid hybridization in evolution: A century of studies synthesizing genetics and ecology." *American Journal of Botany* 101: 1247–1258. http://dx.doi.org/10.3732/ajb.1400201.

[36] Stebbins, G. L. (1959). "The role of hybridization in evolution." P Am Philos Soc 103: 231-251.

[37] Chapman, M. A. and Burke, J. M. (2007). "Genetic divergence and hybrid speciation." *Evolution* 61: 1773–1780. http://dx.doi.org/10.1111/j.1558-5646.2007.00134.x.

[38] Schumer, M., Cui, R., Boussau, B., Walter, R., Rosenthal, G. and Andolfatto, P. (2013). "An evaluation of the hybrid speciation hypothesis for Xiphophorusclemenciae based on whole genome sequences." *Evolution* 67:1155–1168.

[39] Allman, E. S., Kubatko, L. S. and Rhodes, J. A. (2016). "Split scores: A tool to quantify phylogenetic signal in genome-scale data." *Systematic Biology* 66: 620–636.

[40] Folk, R. A., Mandel, J. R. and Freudenstein, J. V. (2017). "Ancestral gene flow and parallel organellar genome capture result in extreme phylogenomic discord in a lineage of angiosperms." *Systematic Biology* 66: 320–337.

[41] Schummer, M., Xu, C., Powell, D. L., Durvasula, A., Skov, L., Holland, C., Blazier, J. C., Sankararman, S., Andolfatto, P., Rosenthal, G. G. and Przeworski, M. (2018). "Natural selections with recombination to shape the evolution of hybrid genomes." *Science* 360: 656-660. 10.1126/science.aar3684.

[42] Rieseberg, L. H., Van Fossen, C. and Desrochers, A. M. (1995). "Hybrid speciation accompanied by genomic reorganization in wild sunflowers." *Nature* 375: 313–316. http://dx.doi.org/10.2307/2445981.

[43] Gross, B. L. and Rieseberg, L. H. (2005). "The ecological genetics of homoploid hybrid speciation." *Heredity* 96: 241–252. http://dx.doi.org/10.1093/jhered/esi026.
[44] James, J. K. and Abbott, R. J. (2005). "Recent, allopatric, homoploid hybrid speciation: The origin of *Senecio squalidus* (Asteraceae), in the British Isles from a hybrid zone on Mount Etna, Sicily." *Evolution* 59: 2533–2547.
[45] Ayres, D. R., Zaremba, K.. and Strong, D. R. (2004). "Extinction of a common native species by hybridization with an invasive congener." *Weed Technology* 18:1288–1291.
[46] Konishi, M. and Takata, K. (2004). "Impact of asymmetrical hybridization followed by sterile F1 hybrids on species replacement in Pseudorasbora." *Conservation Genetics*. 5: 463–474.
[47] Hegde, S. D., Nason, J. D., Clegg, J. M. and Ellstrand, N. C. (2006). "The evolution of California's wild radish has resulted in the extinction of its progenitors." *Evolution* 60:1187–1197.
[48] Taylor, E. B., Boughman, J. W., Groenenboom, M., Sniatynski, M., Schluter, D. and Gow, J. L. (2006). "Speciation in reverse: morphological and genetic evidence of the collapse of a three-spined stickleback (*Gasterosteus aculeatus*) species pair." *Molecular Ecology* 15:343–355.
[49] Stebbins, G. L. (1950). *Variation and evolution in plants*. Columbia Univ. Press, New York.
[50] Mable, B. K. (2004). "Why polyploidy is rarer in animals than in plants': myths and mechanisms." *Biological Journal of the Linnean Society* 82:453–466.
[51] Jolliffe, I. T. (2002). *Principal Component Analysis*. Springer Verlag, NewYork, 487 pp.
[52] Romesburg, H. C. (2004). *Cluster analysis for researchers.* Lulu press, North Carolina, pp. 334
[53] Hammer, Q., Harper, D. A. T. and Ryan, P. D. (2001). Past: Paleontological statistics software package for education and data analysis. *Palaeontologia Electronica* 4: 1-9.

[54] Stanford, A. M., Harden, R. and Parks, C. R. (2000). "Phylogeny and biogeography of *Juglans* (Juglandaceae) based on matK and ITS sequence data." *American Journal of Botany* 87: 872–882.

[55] Sequencher version 4.9. (2009). DNA sequence analysis software, Gene Codes Corporation, Ann Arbor, MI USA.

[56] Tamura, K., Stecher, G., Peterson, D., Filipski, A. and Kumar, S. (2013). MEGA6: Molecular Evolutionary Genetics Analysis version 6.0. *Molecular Biology and Evolution 30*(12), 2725–2729. doi:10.1093/molbev/mst197.

[57] Swofford, D. L. (2002). PAUP*. Phylogenetic Analysis Using Parsimony (*and Other Methods). Sinauer Associates, Sunderland, Massachusetts.

[58] Linneaus, C. (1753). *Species Plantarum*. 2:588. Holmiae.

[59] Miller, P. (1754). *The Gardeners Dictionary*. 829-830. London.

[60] Gleditsch, J. G. (1764). *Systema Plantarum a Staminum Situ*. 189. Haude & Spencer, Berlin.

[61] Bentham, G. (1834). *Labiatarum Genera et Species*. 333-339. Ridgway, London.

[62] Bentham, G. (1848). *Origanum* L. In *Prodramus Systematis Naturalis* edited by de Candolle A. 12:191-197, Masson, Paris.

[63] Bentham, G., and Hooker J. D. (1876). *Genera Plantarum* 2:1185-1186. Williams & Norgate, London.

[64] Boissier, E. (1879). *Flora Orientalis* 4:546-554. George, Geneve, Basel.

[65] Briquet, J. (1895). Labiatae. In *Natürlichen Pflanzenfamilien* edited by Engler & Prantl, 4(3a):183-184. Engelmann, Leipzig.

[66] Tutin, T. G., Heywood, N. A., Burges, N. A., Moore, D. M., Valintine, D. H., Walters, S. M. and Webb, D. A. (1972). *Flora Europaea* 3:126-192. Cambridge University Press.

[67] Rechinger, K. H. (1982). *Flora Iranica*. No: 150:527-532.

[68] Arabacı, T., Yazıcı, T., Özcan T. and Dirmenci, T. (2019). "Physicochemical soil properties of genus Origanum (Lamiaceae) L. grown." *Biological Diversity and Conservation*. In press.

[69] Erdtman, G. (1945). "Pollen morphology and plant taxonomy IV. Labiatae, Verbenaceae, and Aviceniaceae." *Svensk Bot Tidskr* 39:279–285.

[70] Harley, M. M. (1992). "The potential value of pollen morphology as an additional taxonomic character in subtribe Ociminae (Ocimeae: Nepetoideae: Labiatae)." In *Advances in Labiate science*, edited by Harley, R. M., and Reynolds T, 125–138. Royal Botanical Gardens, Kew, London.

[71] Abu-Asab, M. S., and Cantino, P. D. (1992). "Pollen morphology in the subfamily Lamioideae (Labiatae) and its phylogenetic implications." In *Advances in Labiatae Science*, edited by Harley R. M., Reynolds T, 97–112. Royal Botanical Gardens, Kew, London.

[72] Çelenk, S., Dirmenci, T., Malyer, H., and Bicakci, A. (2008a). "A palynological study of the genus Nepeta L. (Lamiaceae)." *Plant Systematics and Evolution* 276:105-123. doi:10.1007/s00606-008-0062-7.

[73] Çelenk, S., Tarimcilar, G., Bicakci, A., Kaynak, G. and Malyer, H. (2008b). "A palynological study of the genus *Mentha* L. (Lamiaceae)." *Botanical Journal of the Linnean Society* 157:141-154.

[74] Husain, S. Z. and Heywood VH. (1982). "Pollen Morphology of the Genus *Origanum* L. and Allied Genera". In *Aromatic Plants Basic and Applied Aspectes*, edited by Margaris, N., Koedam., and A. Vokou, D, 39-56. Martinus Nijhoff Publishers, Nijhoff, Saudi Arabia.

[75] Varghese, T. M. and Verma, D. P. S. (1968). "Pollen morphology of some Indian Labiatae." *Journal of Palynology* 4: 77-83.

[76] Ayyangar, K. R. and Vembu B. (1985). "Karyo-specific and karyo-generic affiliations amongst *Mentha arvensis* Benth., *M. piperita* L. and *Origanum vulgare* L." *Proceedings of the Indian Academy of Science* 72: 127.

[77] Ramsey, J. and Schemske D. W. (1998). "Pathways, mechanism, and rates of polyploid formation in flowering plants." *Annual Review of Ecology and Systematics* 29:467-501.

[78] Buerkle, C. A., Morris, R. J., Asmussen, M. A. and Rieseberg, L. H. (2000). "The likelihood of homoploid hybrid speciation." *Heredity* 84: 441–451.

[79] Burke, J. M. and Arnold, M. L. (2001). "Genetics and the fitness of hybrids." *Annual Review of Genetics* 35:31–52.

[80] Nieto, G. F., Alvarez, I., Fuertes-Aguilar, J., Heuertz, M., Marques, I., Moharrek, F., Piñeiro, R., Riina, R., Rosselló, J. A., Soltis, P. S. and Villa-Machíols, I. 2017. "Is homoploid hybrid specieation that rare? An empricist's view" *Heredity* 118:513-516.

[81] Mallet, J. (2007). Hybrid Speciation. Nature 446:279-283.

[82] Nie, Z. L., Wen, J., Gu, Z. J., Boufford, D. E. and Sun, H. (2005). "Poliploidy in the Flora of Henguan Mountain hotspots, Southwestern China." *Annals of the Missouri Botanical Garden* 92:275-306.

[83] Davis, P. H. (1951). "Additamenta ad floram anatoliae I." *Kew Bulletin* 6:404-441.

[84] Bariotakis, M, Koutroumpa, K, Karousou, R. and Pirintsos, S. A. (2016). "Environmental (in)dependence of a hybrid zone: Insights from molecular markers and ecological niche modeling in a hybrid zone of *Origanum* (Lamiaceae) on the island of Crete." *Ecology and Evolution* 6: 8727–8739. doi.10.1002/ece3.2560.

[85] Abbott, R. J., James, J. K., Irwin, J. A. and Comes, H. P. (2000). "Hybrid origin of the Oxford ragwort, *Senecio squalidus* L." *Watsonia* 23: 123–138.

[86] Abbott, R. J., Brennan, A. C., James, J. K., Forbes, D. F., Hegarty, M. J. and Hiscock, S. J. (2009). "Recent hybrid origin and invasion in the British Isles by a self-incompatible species, Oxford ragwort (*Senecio squalidus* L., Asteraceae)." *Biological Invasions* 11: 1145–1158.

Biographical Sketch

Tuncay Dirmenci, PhD

Affiliation: Balıkesir University, Necatibey Education Faculty, Department of Biology Education, Balıkesir, Turkey

Education: PhD. Balıkesir University, Necatibey Education Faculty, Department of Biology Education

Business Address: Balıkesir University, Necatibey Education Faculty, Department of Biology Education, Balıkesir, Turkey

Publications from the Last 3 Years:

1. Dirmenci T, Özcan, T, Arabaci, T, Çelenk S, İsmailyova G, Yildiz B. (2019). Hybridization among three Cirsium (Asteraceae) species and important evidence for three new hybrids from Turkey. *Turkish Journal of Botany* 43 (3), 366-385.
2. Dirmenci T, Özcan T, Açar M, Arabacı T, Yazıcı T, Martin E. (2019). A rearranged homoploid hybrid species of *Origanum* (Lamiaceae): *O.* × *munzurense* Kit Tan & Sorger. *Botany Letters.*
3. Gezek G., Hasdemir P., Kalaycıoğlu Z., Kaygusuz H., Sarıoğlu G., Döker S., Dirmenci T., Erim B. 2019. Evaluation of some Turkish *Salvia* species by principal component analysis based on their vitamin B2, mineral composition, and antioxidant properties. *LWT* 100:287-293.
4. Dirmenci T, Özcan T, Yazıcı T, Arabacı T, Martin E. (2018). Morphological, cytological, palynological and molecular evidence on two new hybrids from Turkey: an example of homoploid hybridization in Origanum (Lamiaceae). *Phytotaxa* 371 (3), 145-167.
5. Çarıkçı S, Kılıç T, Özer Z, Dirmenci T, Arabacı T, Gören AC. Quantitative determination of some phenolics in *Origanum laevigatum* Boiss. extracts via validated LC-MS/MS method and antioxidant

activity. *J. Chem. Metrol.*, 12:2, 121-127, 2018. http://doi.org/10.25135/jcm.21.18.11.1115

6. Ecevit-Genç G, Özcan T, Dirmenci T. (2018). Leaf indumentum in some Turkish species of *Teucrium* (Lamiaceae). *Istanbul Journal of Pharmacy* 48 (1), 6-11.
7. Ecevit-Genç G, Altınbaşak BB, Özcan T, Dirmenci T. (2018). Comparative anatomical studies of some Teucrium sect. Teucrium species: Teucrium alyssifolium Stapf, Teucrium brevifolium Schreb. and Teucrium pestalozzae Boiss. (*Lamiaceae*). *PhytoKeys*, 63.
8. Dirmenci T, Yazıcı T, Özcan, T, Çelenk S, Martin E. (2018). A new species and a new natural hybrid of Origanum L.(Lamiaceae) from the west of Turkey. *Turkish Journal of Botany* 42, 73-90
9. Ecevit-Genç G, Özcan T, Dirmenci T. (2017). Nutlet and leaf micromorphology in some Turkish species of Teucrium L. (Lamiaceae). *Phytotaxa* 312 (1), 71-82.
10. Yılmaz H, Çarıkçı S, Kılıç T, Dirmenci T, Arabacı T, Gören AC. Screening of Chemical Composition, Antioxidant and Anticholinesterase Activity of Section Brevifilamentum of *Origanum* (L.) *Species. Rec. Nat. Prod.* 11:5, 439-455, 2017.

Turan Arabaci, PhD

Affiliation: İnönü University, Faculty of Pharmacy, Department of Pharmaceutical Botany, Malatya, Turkey

Business Address:

İnönü University, Faculty of Pharmacy, Department of Pharmaceutical Botany, 44280, Malatya, Turkey

Publications from the Last 3 Years:

1. Dirmenci T, Özcan T, Arabacı T, Çelenk S, İsmailova G, Yıldız B. Hybridization among three *Cirsium* (Asteraceae) species and important

evidence for three new hybrids from Turkey. *Turk J Bot*, 43: 366-385, 2019. doi:10.3906/bot-1806-53.

2. Dirmenci T, Özcan T, Açar M, Arabacı T, Yazıcı T, Martin E. A rearranged homoploid hybrid species of Origanum (Lamiaceae): *O. ×munzurense* Kit Tan & Sorger. *Botany Letters*, 166:2, 153-162, 2019. DOI:10.1080/23818107.2019.1585283.
3. Çarıkçı S, Kılıç T, Özer Z, Dirmenci T, Arabacı T, Gören AC. Quantitative determination of some phenolics in *Origanum laevigatum* Boiss. extracts via validated LC-MS/MS method and antioxidant activity. *J. Chem. Metrol.*, 12:2, 121-127, 2018. http://doi.org/10.25135/jcm.21.18.11.1115.
4. Dirmenci T, Özcan T, Yazıcı T, Arabacı T, Martin E. Morphological, cytological, palynological and molecular evidence on two new hybrids from Turkey: an example of homoploid hybridization in *Origanum* (Lamiaceae). *Phytotaxa,* 371 (3): 145–167, 2018. https://doi.org/10.11646/phytotaxa.371.3.1.
5. Polat N, Kiran Y, Şahin A, Yildiz B, Arabaci T. Chromosome counts and karyotype analysis of some representatives of genus *Cirsium* Mill. (Asteraceae) in Turkey. *Caryologia,* 71:2, 133-138, 2018. DOI: 10.1080/00087114.2018.1447630.
6. Taşkın T, Taşkın D, Rayaman E, Dikpınar T, Süzgeç-Selçuk S, Arabacı T. Characterization of the Biological Activity and Phenolics in *Achillea lycaonica*. *Analytical Letters*, 51:1-2, 33-48, 2018. DOI: 10.1080/00032719.2017.1318140.
7. Yılmaz H, Çarıkçı S, Kılıç T, Dirmenci T, Arabacı T, Gören AC. Screening of Chemical Composition, Antioxidant and Anticholinesterase Activity of Section Brevifilamentum of *Origanum* (L.) *Species. Rec. Nat. Prod.* 11:5, 439-455, 2017.
8. Kabaktepe Ş, Arabacı T, Kolaç T. A new host for *Puccinia menthae.* Mycotaxon, 132: 1-3, 2017. http://dx.doi.org/10.5248/132.1.
9. Arabacı T, Uzay G, Keleştemur U, Karaaslan MG, Balcıoğlu S, Ateş B. Cytotoxicity, Radical Scavenging, Antioxidant Properties and Chemical Composition of the Essential Oil of Satureja cilicica P.H.

Davis from Turkey. *Marmara Pharmaceutical Journal* 21/3: 500-505, 2017. DOI: 10.12991/marupj.311814

Taner Özcan, PhD

Affiliation: Balıkesir University, Necatibey Education Faculty, Department of Biology Education, Balıkesir, Turkey

Education:

BcS: Balıkesir University, Necatibey Education Faculty, Department of Biology Education

MS: Balıkesir University, Faculty of Science and Arts, Department of Biology

PhD: Balıkesir University, Faculty of Science and Arts, Department of Biology

Business Address: Balıkesir University, Necatibey Education Faculty, Department of Biology Education, Balıkesir, Turkey

Publications from the Last 3 Years:

1. Dirmenci T, Özcan, T, Arabaci, T, Çelenk S, İsmailyova G, Yildiz B. (2019). Hybridization among three Cirsium (Asteraceae) species and important evidence for three new hybrids from Turkey. *Turkish Journal of Botany* 43 (3), 366-385.
2. Dirmenci T, Özcan T, Açar M, Arabacı T, Yazıcı T, Martin E. (2019). A rearranged homoploid hybrid species of *Origanum* (Lamiaceae): *O.* × *munzurense* Kit Tan & Sorger. *Botany Letters*.
3. Özcan T. (2019). Micromorphological comparison of Nepeta viscida, N. nuda subsp. nuda and their putative hybrids N. x tmolea. *Balıkesir Üniversitesi Fen Bilimleri Enstitüsü Dergisi* 21 (1), 30-42.
4. Dirmenci T, Özcan T, Yazıcı T, Arabacı T, Martin E. (2018). Morphological, cytological, palynological and molecular evidence on

two new hybrids from Turkey: an example of homoploid hybridization in Origanum (Lamiaceae). *Phytotaxa* 371 (3), 145-167.

5. Ecevit-Genç G, Özcan T, Dirmenci T. (2018). Leaf indumentum in some Turkish species of Teucrium (Lamiaceae). *Istanbul Journal of Pharmacy* 48 (1), 6-11.
6. Ecevit-Genç G, Altınbaşak BB, Özcan T, Dirmenci T. (2018). Comparative anatomical studies of some Teucrium sect. Teucrium species: Teucrium alyssifolium Stapf, Teucrium brevifolium Schreb. and Teucrium pestalozzae Boiss.(Lamiaceae). *PhytoKeys*, 63.
7. Dirmenci T, Yazıcı T, Özcan, T, Çelenk S, Martin E. (2018). A new species and a new natural hybrid of Origanum L.(Lamiaceae) from the west of Turkey. Turkish Journal of Botany 42, 73-90
8. Ecevit-Genç G, Özcan T, Dirmenci T. (2017). Nutlet and leaf micromorphology in some Turkish species of Teucrium L. (Lamiaceae). *Phytotaxa* 312 (1), 71-82.

Türker Yazici

Affiliation: Balıkesir University, Necatibey Education Faculty, Department of Biology Education, Balıkesir, Turkey

Education: MS: Balıkesir University, Faculty of Science and Arts, Department of Biology

Business Address: Balıkesir University, Necatibey Education Faculty, Department of Biology Education, Balıkesir, Turkey

Publications from the Last 3 Years:

1. Dirmenci T, Özcan T, Açar M, Arabacı T, Yazıcı T, Martin E. (2019). A rearranged homoploid hybrid species of *Origanum* (Lamiaceae): *O. × munzurense* Kit Tan & Sorger. *Botany Letters.*
2. Dirmenci T, Özcan T, Yazıcı T, Arabacı T, Martin E. (2018). Morphological, cytological, palynological and molecular evidence on

two new hybrids from Turkey: an example of homoploid hybridization in Origanum (Lamiaceae). *Phytotaxa* 371 (3), 145-167.
3. Dirmenci T, Yazıcı T, Özcan, T, Çelenk S, Martin E. (2018). A new species and a new natural hybrid of Origanum L.(Lamiaceae) from the west of Turkey. *Turkish Journal of Botany* 42, 73-90.

Esra Martin, PhD

Affiliation: Necmettin Erbakan University, Science Faculty, Department of Biotechology, Konya, Turkey

Education:

BcS: Gazi University, Education Faculty, Department of Biology Education

MS: Gazi University, Education Faculty, Department of Biology Education

PhD: Gazi University, Faculty of Science and Arts, Department of Biology

Business Address: Necmettin Erbakan University, Science Faculty, Department of Biotechnology, Konya, Turkey

Research and Professional Experience: Cytogenetics, Mentheae, Menthinae, Lamiaceae

Publications from the last 3 Years:

1. Martin, E., G. İçyer, Doğan, S. Karaman Erkul, H. E. Eroğlu. 2019. Karyotype analyses of 25 Turkish taxa of *Astragalus* from section *Macrophyllium*, *Hymenostegis*, *Hymenocoleus* and *Anthylloidei* (Fabaceae). *Turkish Journal of Botany*, 43: (1), 232-242.
2. Dirmenci T., T. Özcan, M. Açar, T. Arabacı, T. Yazıcı, E. Martin. 2019. A rearranged homoploid hybrid species of *Origanum*

(Lamiaceae): *O. ×munzurense* Kit Tan & Sorger. *Botany Letters*, 166: (2), 153-162.

3. Kaya A., M. Ünal, A. Sefalı, E. Martin. 2019. Fruit and seed morphology of the genus *Matthiola* (Brassicaceae) in Turkey and their taxonomic value. *Turkish Journal of Botany*, 43: 516-528.
4. Eroğlu, H. E., A. Sefalı, M. Ünal, A. Kelkitoğlu, E. Martin. 2019. Cytogenetical Analyses of some species of the genus *Sisymbrium* (Brassicaceae) in Turkey. *Anatolian Journal of Botany*, 3 (1): 21-24.
5. Martin, E., M. Ünal, M. İleri. 2019. Karyotype analyses in two taxa of *Anchonium* (Brassicaceae) from Turkey. *Karaelmas Science and Engineering Journal*, 9(1):97-100.
6. Martin, E., M. Tekin, G. M. Dirlik, M. Türk. 2019. Karyotype analysis of *Silene muradica* Schischk. (Caryophyllaceae), an endemic species in Turkey. *Karaelmas Science and Engineering Journal,* 9(1):114-117.
7. Dirmenci T., T. Yazıcı, T. Özcan, S. Çelenk, E. Martin. 2018. A new species and a new natural hybrid of *Origanum* L. (Lamiaceae) from the west of Turkey, *Turkish Journal of Botany*, 42 (1), 73-90.
8. Martin E., H. K. Yıldız, A. Kahraman, O. K. Binzat, H. E. Eroğlu. 2018. Detailed chromosome measurements and karyotype asymmetry of some *Vicia* (Fabaceae) taxa from Turkey. *Caryologia,* 71 (3), 224-232.
9. Dirmenci T., T. Özcan, T. Yazıcı, T. Arabacı, E. Martin. 2018. Morphological, cytological, palynological and molecular evidence on two new hybrids: an example of homoploid hybridization in *Origanum* (Lamiaceae). *Phytotaxa,* 371 (3), 145-167.
10. Atasagun, B., A. Aksoy, E. Martin. 2018. Morphological, anatomical, palynological, karyological and ecological remarks of *Astragalus argaeus* (Fabaceae) endemic to Turkey. *Phytotaxa*, 379 (1), 118-130.
11. Tekin, M., G. Özek, E. Martin, T. Özek, G. Yılmaz, K. H. C. Başer. 2018. Essential oil characterization of *Cousinia sivasica* Hub.-Mor. (Asteraceae). *Biological Diversity and Conservation*, 11 (2): 16-21 (2018).
12. Kaya, A., M. Ünal, F. Özgökçe, B. Doğan, E. Martin. 2017. Pollen morphology of six species previously placed in *Malcolmia*

(Brassicaceae) In Turkey. *Bangladesh Journal of Botany*, 46 (2): 623-629.

13. Tekin, M., E. Martin. 2017. Morphology, anatomy and karyology of endangered Turkish endemic *Physoptychis haussknechtii* Bornm. (Brassicaceae) from Central Anatolia. *Acta Botanica Croatica*, 76 (1), 32-40.

Sevcan Çelenk, PhD

Affiliation: Bursa Uludag University, Arts and Science Faculty, Department of Biology, Bursa, Turkey

Education:

BSc: Degree in Biology, Uludag University, Faculty of Arts and Sciences.

MSc: Degree in Botany, Uludag University, Faculty of Arts and Sciences.

PhD: Degree in Botany, Uludag University, Faculty of Arts and Sciences.

Business Address: Bursa Uludag University, Arts and Science Faculty, Department of Biology, Bursa, Turkey

Research and Professional Experience: Pollen morphology, aerobiology, aeroallergens

Publications from the Last 3 Years:

1. Celenk S (2019). Detection of reactive allergens from long-distance transported pollen grains: evidence from *Ambrosia*. *Atmospheric Environment*, 209: 212-219.
2. Jesús Rojo, Jose Oteros, Rosa Pérez-Badia, Patricia Cervigón, Zuzana Ferencova, A. Monserrat Gutiérrez-Bustillo, Karl-Christian Bergmann, Gilles Oliver, Michel Thibaudon, Roberto Albertini, David Rodríguez-

De la Cruz, Estefanía Sánchez-Reyes, José Sánchez-Sánchez, Anna-Mari Pessi, Jukka Reiniharju, Annika Saarto, M. Carmen Calderón, César Guerrero, Daniele Berra, Maira Bonini, Elena Chiodini, Delia Fernández-González, José García, M. Mar Trigo, Dorota Myszkowska, Santiago Fernández-Rodríguez, Rafael Tormo-Molina, Athanasios Damialis, Franziska Haering, Claudia Traidl-Hoffmann, Elena Severova, Elsa Caeiro, Helena Ribeiro, Donat Magyar, László Makra, Orsolya Udvardy, Purificación Alcázar, Carmen Galán, Katarzyna Borycka, Idalia Kasprzyk, Ed Newbigin, Beverley Adams-Groom, Godfrey P. Apangu, Carl A. Frisk, Carsten Skjøth, Predrag Radišic, Branko Šikoparija, Sevcan Celenk, Carsten Schmidt-Weber, Jeroen Buters, (2019). Near-ground effect of height on pollen exposure, *Environmental Research*, 17: 160-169.

3. Dirmenci T, Ozcan T, Arabacı T, Çelenk S, Ismailova G, Yıldız B (2019). Hybridization and introgression between three *Cirsium* species and important evidence for three new hybrids from Turkey, *Turk J Bot.* 43: 366-385.
4. Dirmenci T, Yazıcı T, Ozcan T, Çelenk S, Martin E (2018). A new species and a new natural hybrid of *Origanum* L. (Lamiaceae) from the west of Turkey. *Turk J Bot* 42: 73-90.
5. Ismailova G, Celenk S (2018). Airborne Pollen Spectrum of Domaniç (Turkey). Annals of West University of Timişoara, ser. *Biology,* 21 (1): 47 – 56.
6. Çelenk S, Malyer H (2017). The occurrence of *Ambrosia* pollen in the atmosphere of Northwest Turkey: Investigation of possible source regions. *International Journal of Biometeorology* 61(9): 1499-1510.
7. Sofiev M, Ritenberga O, Albertini R, Arteta J, Belmonte J, Bonini M, Çelenk S et al., (2017). Multi-model ensemble simulations of olive pollen distribution in Europe in 2014. *Atmos. Chem. Phys.,* 17: 12341-12360.
8. Sikoparija B, Skjøth CA, Çelenk S et al., (2017). Spatial and temporal variations in airborne Ambrosia pollen in Europe. *Aerobiologia* 33: 181.

Mikail Açar

Affiliation: Munzur University, Tunceli Vocational School, Department of Plant and Animal Production, Tunceli, Turkey

Education:

BcS: Balıkesir University, Necatibey Education Faculty, Department of Biology Education

MS: Balıkesir University, Institute of Science, Department of Biology

PhD: Balıkesir University, Institute of Science, Department of Biology

Business Address: Munzur University, Tunceli Vocational School, Department of Plant and Animal Production, Tunceli, Turkey

Research and Professional Experience: Lamiaceae, Stachys, Morphology and Anatomy, Statistical Analyses

Publications from the Last 3 Years:

1. Dirmenci T, Özcan T, Açar M, Arabacı T, Yazıcı T, Martin E. (2019). A rearranged homoploid hybrid species of *Origanum* (Lamiaceae): *O.* × *munzurense* Kit Tan & Sorger. *Botany Letters.*

Devrim Üzel, PhD

Affiliation: Balıkesir University, *Necatibey Education Faculty, Department of Mathematics Education, Balıkesir, Turkey*

Education: PhD. Balıkesir University, Necatibey Education Faculty, Department of Mathematics Education

Business Address: Balıkesir University, *Necatibey Education Faculty, Department of Maths Education, Balıkesir, Turkey*

Research and Professional Experience: Mathematics education, Theacher education, Statistics

Publications from the Last 3 Years:

1. Üzel D., Uyangör N., Hasar B., Çakır Ö. 2018. Devoloping the motvation scale fort he Math class. *Journal of Social and Humanities Sciences Research* 5:378-386.
2. Köysüren, M., & Üzel, D. (2018). Matematik Öğretiminde Teknoloji Kullanımının 6. Sınıf Öğrencilerinin Matematik Okuryazarlığına Etkisi. Necatibey Eğitim Fakültesi Elektronik Fen ve Matematik Eğitimi Dergisi, 12 (2), 81-101.
3. Üzel D., Uyangör S. M. 2018. Eğitim Fakültesi Matematik Öğretmen Adayları ve Pedogojik Formasyon Eğitimi Sertifika Programı Adaylarının Teknolojik Pedagojik Alan Bilgilerinin Karşılaştırılması. *Turkish Studies* 13:1593-1607.
4. Üzel, D. (2018). Investigation of Misconceptions and Errors about Division Operation in Fractions. *Universal Journal of Educational Research*, *6*(11), 2656-2662.
5. Özdemir, E., Üzel, D., & Özsoy, N. (2017). An investigation of teachers' views on applicability of modeling in mathematics courses. *Journal of Education and Training Studies*, *5*(5), 145-155.

In: The Lamiaceae Family: An Overview ISBN: 978-1-53617-078-8
Editor: Alexander Adler

Chapter 7

AROMATIC SPECIES OF THE LAMIACEAE FAMILY FROM ECUADOR

Eduardo Valarezo[1,*], Jandry Rosales[2], Nixon Cumbicus[3] and Miguel Angel Meneses[1]

[1]Departamento de Química y Ciencias Exactas, Universidad Técnica Particular de Loja, Loja, Ecuador
[2]Unidad de Fitoquimica, Universidad Nacional de Loja, Loja Ecuador
[3]Departamento de Ciencias Biológicas, Universidad Técnica Particular de Loja, Loja, Ecuador

ABSTRACT

Family Lamiaceae is amongst the largest families of flowering plants with about 250 genera and over 7,000 species distributed around the world. It is considered as the important source of essential oils. In Ecuador 219 species belonging to 27 genera of this family have been registered, of which about 29 species are endemic. In this chapter an over view of aromatic species belonging to the family Lamiaceae is presented,

* Corresponding Author's Email: bevalarezo@utpl.edu.ec.

the chemical composition of the volatile fraction of these existing species in Ecuador is also reported. The plant material was collected in southern Ecuador and botanical specimens was identified by Herbario UTPL. The essential oils were obtained by hydrodistillation and then analysed by gas chromatography/flame ionization detector (GC/FID) and gas chromatography/mass spectrometry (GC/MS). Nine species were studied, including two species of the genus Clinopodium, two of the genus Hyptis, two of the genus Lepechinia, one of the genus Minthostachys, one of the genus Origanum and one of the genus Scutellaria. These species grow in the Andean forests, moors and dry inter-Andean valleys, over 1000 m a.s.l., less frequently in the dry forests of the Coast, Galapagos and Amazon. The principal constituents in essential oil of *Clinopodium nubigenum* was carvacrol acetate (38.1%), in *Clinopodium taxifolium* was Z-caryophyllene (24.1%) and in *Hyptis suaveolens* was eucalyptol (39.0%). Aromadendrene (24.6%) and δ-3-carene (19.9%) were the principal constituents in essential oils of *Lepechinia paniculata* and *Lepechinia radula*, respectively. *Minthostachys mollis* essential oil reported (Z)-dihidro carvone (13.2%). In the essential oil of *Origanum vulgare* was identified γ-terpinene with a relative abundance of 20.8%. Germacrene D was principal constituents in essential oils of *Scutellaria volubilis* with 20.4% and in *Hyptis pectinata* with 34.1%.

Keywords: aromatic species, lamiaceae family, essential oil, genus *clinopodium*, genus *hyptis*, genus *lepechinia*, genus *minthostachys*, genus *origanum* genus *scutellaria*

1. INTRODUCTION

1.1. Family Lamiaceae

The Lamiaceae or Labiatae are a family of flowering plants commonly known as the mint or deadnettle or sage family. It had traditionally been considered closely related to Verbenaceae [1], but in the 1990s, phylogenetic studies showed that many genera classified in Verbenaceae belong instead in Lamiaceae [2]. The currently accepted version of Verbenaceae may not be closer to Lamiaceae than some of the other families in the order Lamiales. It is not known which of the families in Lamiales is closest to Lamiaceae (Table 1).

Table 1. Scientific classification of Lamiaceae Family

Scientific classification	
Kingdom	Plantae – Plants
Subkingdom	Tracheobionta – Vascular plants
Superdivision	Spermatophyta – Seed plants
Division	Magnoliophyta – Flowering plants
Class	Magnoliopsida – Dicotyledons
Subclass	Asteridae
Order	Lamiales
Family	Lamiaceae, Labiatae

The family has a cosmopolitan distribution [3]. The enlarged Lamiaceae contain about 236 genera and have been stated to contain 6,900 to 7,200 species [4]. The largest genera are Salvia (900), Scutellaria (360 spp.), Stachys (300 spp.), Plectranthus (300 spp.), Hyptis (280spp.), Teucrium (250spp.), Vitex (250spp.), Thymus (220spp.), and Nepeta (200spp.) Clerodendrum (150) [4, 5]. Lamiaceae is distributed nearly worldwide, and many species are cultivated_for their fragrant leaves and attractive flowers (Table 2). The family is particularly important to humans for herb plants useful for flavour, fragrance, or medicinal properties.

Table 2. Vegetative characters of Lamiaceae Family

	Vegetative characters
Habit	Mostly aromatic herbs and shrubs, few tree
Root	Tap-root
Stem	Herbaceous, usually erect, squarish and hairy
Leaf.	Simple, opposite and decussate, edstipulate, whorled in few species, due presence of gland secreting volatile oils

The plants are frequently aromatic in all parts and include many widely used culinary herbs, such as basil, mint, rosemary, sage, savory, marjoram, oregano, thyme, lavender and perilla. Some are shrubs, trees, such as teak, or rarely vines. Many members of the family are widely cultivated, owing not only to their aromatic qualities but also their ease of cultivation; these plants are among the easiest plants to propagate by stem cuttings.

Besides those grown for their edible leaves, some are grown for decorative foliage, such as coleus. Others are grown for seed, such as *Salvia hispanica*, or for their edible tubers, such as *Plectranthus edulis*, *Plectranthus esculentus*, *Plectranthus rotundifolius*, and *Stachys affinis*.

The original family name is Labiatae, so given because the flowers typically have petals fused into an upper lip and a lower lip (*labia* in Latin). The flowers are bilaterally symmetrical with five united petals and five united sepals. They are usually bisexual (Table 3) and verticillastrate (a flower cluster that looks like a whorl of flowers, but actually consists of two crowded clusters). Although this is still considered an acceptable alternative name, most botanists now use the name Lamiaceae in referring to this family. The leaves emerge oppositely, each pair at right angles to the previous one (decussate) or whorled. The stems are frequently square in cross section, but this is not found in all members of the family, and is sometimes found in other plant families [6].

Table 3. Floral characters of Lamiaceae Family

Floral characters	
Inflorescence	Verticillaster
Flower	Bisexual, irregular, zygomorphic some time regular (Mentha), hypogynous
Calyx	Five, gamosepalous, bilabiate ringent, inferior
Androecium	Four, posterior one suppressed polyandrous, epipetalous, usually in two sets, inferior
Gynoecium	Bicarpellary suncarpous; ovary superior, bilocular; Placentation: axile, each loculus with one ovules: style: gynobasic: stigma: two-lobed
Fruit	Schizocarpic carcerulus
Seeds	Non-endospermic

The Lamiaceae, is of considerable economic importance, containing several timber trees, such as Tectona, many species of horticultural value, many used as culinary herbs, or in perfumery and many are used medicinally. Also the trade in honey, which is an important part of the rural economy, is reliant on the many nectar-producing species of Lamiaceae, which produce a honey of high quality [1].

In 2004, the Lamiaceae were divided into seven subfamilies with 10 genera not placed in any of the subfamilies. The unplaced genera are: *Tectona*, *Callicarpa*, *Hymenopyramis*, *Petraeovitex*, *Peronema*, *Garrettia*, *Cymaria*, *Acrymia*, *Holocheila*, and *Ombrocharis*. The subfamilies are the Symphorematoideae, Viticoideae, Ajugoideae, Prostantheroideae, Nepetoideae, Scutellarioideae, and Lamioideae. The subfamily Viticoideae is probably not monophyletic. The Prostantheroideae and Nepetoideae are divided into tribes. These are shown in the phylogenetic tree below. Of the seven subfamilies currently recognized, five occur in South America Viticoideae, Ajugoideae, Scutellarioideae, Lamioideae and Nepetoideae [7]

Ecuador is one of the countries with the major number of vegetal species by surface area, while it represents only 0.2% of Earth's territory, it contains about 10% of all plant species existing on the planet. That locates Ecuador in the sixth position in the world [8]. In the Ecuador the family Lamiaceae is represented by 21 genus, 135 species and 33 endemic species [9]

1.1.1. Genus Clinopodium

Clinopodium is a genus of flowering plants in the family Lamiaceae. It is in the tribe Mentheae of the subfamily Nepetoideae. These were names for Clinopodium vulgareIt comprises about 100 specvies, mostly in temperate and tropical areas of América, and temperate Eurasia, but a few in Africa, tropical Asia and Indomalaysia [1]. Various *Clinopodium* species are used as medicinal herbs.

1.1.1.1. Clinopodium Nubigenum

Clinopodium nubigenum (Kunth) Kuntze, an aromatic plant belonging to the family Lamiaceae, is also known with the synonymous *Thymus nubigenus* Kunth, *Micromeria nubigena* (Kunth) Benth, and *Satureja nubigena* (Kunth) Briq. It is widely growing in Latin America between 3000 and 4000 m a.s.l. The plant is popularly known by indigenous people as "tipo de cerro" and, in Ecuador, it is used as a traditional remedy by different communities. The Saraguro people use an aqueous infusion of the plant to treat colds [10]; in the region of Azuay, the plant is used as a remedy for flu [11]; Quechua peoples in the High Sierra apply a decoction of *C. nubigenum* to cure stomach ache; Cañar communities use a plant infusion to avoid urinary incontinence in children; in Tungurahua, Chimborazo, Cañar, and Azuay provinces of Ecuador the plant finds applications also as a digestive, stomachic, a tonic remedy, and against dysentery and menstrual syndromes [8].

1.1.1.2. Clinopodium taxifolium

Clinopodium taxifolium (Kunth) Govaerts is a native shrub of the Andes that grows between 1500 and 3000 m a.s.l, in the ecuadorian provinces of Azuay, Loja and El Oro. It is known as Culantrillo de Cerro or Polea de Castilla. It is used as infusion in folk medicine to treat internal inflammations, flatulence, stomach pain, malaria and cough [8].

1.1.2. Genus Hyptis

Hyptis is a genus of flowering plant in the family Lamiaceae. These plants, known commonly as bushmints, are widespread in tropical North and South America, as well as parts of West Africa. The Hyptis genus includes about 775 species, which are found mainly in tropical America. The plants in this genus have been widely used in traditional medicine for the treatment of various illnesses and have been found to possess tumorigenic, antifertility, antimicrobial, mycotoxic and phytotoxic activities. Most Hyptis plants are characterized by their strong aromatic constituents, regarded as having medicinal properties, and hence, the interest to investigate their essential oils [12].

1.1.2.1. Hyptis Pectinata

The specie *Hyptis pectinata* is an annual to perennial herb that can grow up to 4 metres tall. It is often gathered from the wild for local use as a medicinal plant. in South America it is distributed in Brazil, Bolivia and Peru. In Ecuador this spice is distributed in the provinces of El Oro, Guayas, Loja, Los Ríos, Morona-Santiago, Napo, Tungurahua, Zamora-Chinchipe and Galapagos [9]. Common weed of cultivated land, pastures, plantations, road and trailsides, waste places. Moist or dry, often rocky, brushy hillsides, open fields, pine-oak forest, sandbars along streams, waste or cultivated ground, below 1,900 m a.s.l.

1.1.2.2. Hyptis suaveolens

Hyptis suaveolens is a coarse erect annual or often a perennial with stems that become more or less woody and persist. Usually much branched, it is sometimes as much as 2.5 meters high but is generally much lower. The plant has a range of food and medicinal uses, being often harvested from the wild for local use. This spice is original of American tropics, the plant has spread to tropical and subtropical areas worldwide. In Ecuador this spice is distributed in the provinces of El Oro, Guayas and Loja, principally [9].

1.1.3. Genus Lepechinia

Lepechinia Willd. is a diverse genus in the mint family (Lamiaceae), consisting of approximately 43 species that range from Northern California in the western USA to Central Argentina in South America. Of the 43 *Lepechinia* spp., *c.* 30 occur in South America. Most species of *Lepechinia* in South America are found at high elevations, from 1500 to 4000 m, but within this altitudinal range the genus occupies a fairly broad range of habitats. The majority of South American *Lepechinia* spp. occur in relatively dry open habitats of the Andean highlands, but some species are found in relatively moist sites at the edges of forests and some are found in paramo or puna environments. *Lepechinia* spp. are valued in the horticultural trade, and North and South American indigenous groups commonly use *Lepechinia* for medicinal and antiseptic purposes. The chemical compounds that give most *Lepechinia* a distinct odour may also be responsible for their medicinal and cleansing properties [13].

1.1.3.1. Lepechinia Paniculata

Lepechinia paniculata (Kunth) is an endemic species in Ecuador locally called "Shallshón". *L. paniculata* is used to treat headache by direct application of its buds on head, its flowers are used for the treatment of nerve system affection, and for evil air “mal aire” rituals in body cleaning by Saraguro ethnic group [9, 10].

1.1.3.2. Lepechinia radula

Lepechinia radula is a native shrub found in the Andean region of Ecuador. It is located in growing wild in both Azuay and Loja Provinces at 2000 - 2500 m a.s.l. In Ecuador it has been traditionally used for the inhabitants for the treatment of “espanto” [9].

1.1.4. Genus Minthostachys

The genus *Minthostachys* (Benth.) Spach (Lamiaceae) comprises 17 species, comprising aromatic scandent shrubs. It occurs along the Andes from Northern Venezuela through Colombia, Ecuador, Peru and Bolivia to Central Argentina. The plants are valued by the local population for their content in essential oils, mostly pulegone and menthone, but also limonene, carvone, carvacrol, thymol and similar substances. They are used as condiments, medicinally against illnesses of the respiratory and digestive systems, and traditionally for the protection of stored tubers against pests [14, 15].

1.1.4.1. Minthostachys Mollis

Minthostachys mollis has been reported to be useful as an anthelmintic, digestive, aphrodisiac, anti-diarrheic, in the treatment of respiratory and urinary infections, as a menstruation inducer, and condimen. M. mollis (Kunth) Griseb Vaught is found in Venezuela, Colombia, Ecuador, Peru and Bolivia. This herbaceous plant, 0.5 m high, with aromatic pubescent leaves and white flower spikes, has a distinctive peppermint-like odor. The use of Minthostachys species has been recorded since the 16th century [14].

1.1.5. Genus Origanum

The genus origanum comprises 43 species and 18 hybrids more than two dozens of different plants with leaves and flowers which presents a characteristic spicy aroma. Based on morphological criteria, the genus

Origanum has been classified into 3 groups, 10 sections, 38 species, 6 subspecies and 17 hybrids. Origanum species are usually subshrubs or woody perennial herbs that reach heights of up to 80 cm and have ovate leaves and white or purple flowers. The origanum is commercially recognized around the world by two aromatic qualities, marjoram (*O. majorana* L.) and oregano (usually from *O. vulgare* and *O. onites* L.) widely used all over the world as popular spices [16].

1.1.5.1. Origanum Vulgare

Origanum Vulgare L. is a perenne plant. Originally from the Mediterranean region also cultivated in Europe, Asia and South America. The essential oil presents more than 34 active components, of these carvacrol, thymol, α-terpinene and p-cymene could reach between 80.2% to 98% of the oil composition [17]. The whole plant is used for medical purposes such as gastrointestinal and neurological diseases, and to relief different types of pains.

1.1.6. Genus Scutellaria

The genus *Scutellaria* (Lamiaceae), which includes about 350 species commonly known as skullcaps, is widespread in Europe, the United States and East Asia. Some species are taken to clear away the heat-evil and expel superficial evils in traditional Chinese medicine [18]. The genus has a subcosmopolitan distribution, with species occurring nearly worldwide, mainly in temperate regions, Including the American region.

1.1.6.1. Scutellaria Voluvilis

Scutellaria volubilis is native from the Andes. In Ecuador it has been found in Loja and Azuay at 2250 m a.s.l.. S. volubilis belongs to the species that are traditionally used for nervous system treatment, as well as cure of heart and kidney affections, by ethnos from Azuay, Cañar and Napo regions [8, 9].

2. METHODS

2.1. Plant Material

The plant material was collected according to technical note titled Plant Materials Collection Guide [19]. The preservation of botanical specimens was done according to chapter 5 "Materials to be used for preserving plant samples in the field" of the manual "Laboratory manual for collecting and processing plant species" [20].

2.2. Essential Oil Isolation

The aerial parts of species were divided into at least three equal amounts and steam distilled in four stainless steel Clevenger-type apparatus, for 4 hours. After distillation, each organic layer was separated from the aqueous phase, dried over anhydrous sodium sulfate and stored at 4°C until being used for analysis.

2.3. Physical Properties of Essential Oil

The density and refraction index of essential oils were determined according to the standard AFNOR NF T 75-111 and AFNOR NF T 75-112, respectively. For density, a pycnometer (1 mL) and an analytical balance (model Mettler AC 100, ± 0.0001) were used, whereas for refraction index a refractometer (model ABBE) was used. The optical rotation was determined according to the standard ISO 592:1998 using an automatic polarimeter model Mrc-P810. The procedures were repeated three times and all measurements were performed at 20°C.

2.4. Gas Chromatography/Flame Ionization Detector (GC/FID)

The analyses of the chemical composition of the essential oils were performed on an Agilent gas chromatograph (model 6890N series) equipped with a flame ionization detector (FID). A nonpolar DB-5Ms (5%-phenyl-methylpolyxilosane) 30 m x 0.25 mm, thickness 0.25 μm were used. An automatic injector (series 7683) in split mode was used. The samples, 1 μL of solution (1/100, v/v, essential oil/dicloromethane), were injected with a split ratio of 1:50. Helium was used as a carrier gas at 0.9 mL/min in constant flow mode. The initial oven temperature was held at 50ºC for 3 min. and then it was heated to 210ºC with a ramp of 2.5ºC/min, and the temperature was maintained for 3 min until the end. The injector and detector temperatures were of 210ºC and 250ºC; respectively. Retention index of the compounds was determined based on the homologous of the standard aliphatic hydrocarbons TPH-6RPM of CHEM SERVICE C10-C25, which were injected after the oils at the same conditions.

2.5. Gas Chromatography/Mass Spectrometry (GC/MS)

The GC/MS analyses were performed using an Agilent chromatograph coupled to a mass spectrometer detector (model Agilent series 5973 inert). The spectrometer was operated at 70 eV, electron multiplier 1600 eV, scan rate: 2 scan/s and mass range: 40-350 m/z. It was provided with a computerized system MSD-Chemstation D.01.00 SP1. The same columns described in GC/FID section were used. The ion source temperature was set at 250ºC. The identification of the oil components was based on a comparison of both MS data and relative retention indices (RI) with the published literature [21-23]. The relative amounts of individual components were calculated based on the GC peak area (FID response) without using a correction factor.

3. RESULTS

3.1. Genus Clinopodium

Of the genus Clinopodium two species were studied *Clinopodium nubigenum and Clinopodium taxifolium.* The collection of *C. nubigenum* was performed in village *"Piedra Bola"* (3°47'11.0" S, 79°21'31.5" W), canton and province of Loja, Ecuador. *C. taxifolium* was collected at 2700 m a.s.l. in the mountain "Acacana" canton Saraguro (3°41'36.7" S 79°14'56.4" W), Loja province. The botanical specimens were identified by Dr. Bolivar Merino at the herbarium of the Universidad Nacional de Loja. A voucher specimens are preserved in the Herbarium of the Universidad Técnica Particular de Loja (HUTPL) under codes PPN-la-018 for *Clinopodium nubigenum* and HUTPL-1105 for *Clinopodium taxifolium.*

The essential oil of *C. nubigenum* was a viscous liquid of subjective color amber. The EO yield was 1.41 ± 0.66% (w/w), referred to dried plant material. The essential oil relative density was d^{20} = 0,9556 ± 0.0033 and refraction index was n^{20} = 1,4948 ± 0.0003. The aerial parts of *C. taxifolium* provided an essential oil in yield of 0.04 ± 0.01% (w/w) referred to fresh plant material. The relative density was d^{20} = 0.9245 ± 0.0010 and the refractive index was n^{20} = 1.4905 ± 0.0036.

The oil samples were analyzed by means of Gas Chromatography (GC) and Gas Chromatography/Mass Spectrometry (GC/MS) using the nonpolar capillary columns DB-5Ms. Their components were identified based on the retention index and mass spectra. The percentages of oil components are shown in Table 4, were obtained by averaging the GC-FID peak area data of each sample. One should take into account that they are listed by elution order from the DB-5Ms column.

Forty components were determined in essential oil of *Clinopodium nubigenum*, which represents 98,9% of the total sample. Oxygenated monoterpenes (76.9%) was found to be the principal group of constituents.

Table 4. Chemical composition of the essential oil from *Clinopodium nubigenun* and *Clinopodium Taxifolium*

N	RI	RI[ref]	Compounds[a]	Clinopodium nubigenun[b]		Clinopodium Taxifolium[b]		Type
				%	σ	%	σ	
1	889	876	Isoamyl acetate	0.1	0.0	-	-	OT
2	930	930	α-Thujene	1.8	0.1	-	-	MH
3	937	939	α-Pinene	0.3	0.0	0.3	0.1	MH
4	952	954	Camphene	0.1	0.0	-	-	MH
5	973	975	Sabinene	0.3	0.0	1.1	0.4	MH
6	978	979	β-Pinene	0.2	0.0	0.5	0.2	MH
7	995	990	Myrcene	0.5	0.0	-	-	MH
8	1005	1002	α-Phellandrene	0.2	0.0	-	-	MH
9	1006	1011	δ-3-Carene	0.1	0.0	-	-	MH
10	1016	1017	α-Terpinene	0.9	0.0	-	-	MH
11	1022	1024	p-Cymene	6.6	1.6	-	-	MH
12	1028	1029	Limonene	0.5	0.0	2.6	1.6	MH
13	1029	1031	1,8-Cineole (Eucaliptol)	-	-	1.0	0.3	MO
14	1058	1059	γ-Terpinene	5.0	1.7	-	-	MH
15	1084	1088	α-Terpinolene	0.1	0.0	-	-	MH
16	1092	1088	Undecene	0.4	0.1	-	-	OT
17	1097	1096	Linalool	-	-	1.0	0.3	MO
18	1098	1098	(E)-Sabinene hydrate	0.6	0.1	-	-	MO
19	1108	1112	1-Octen-3-yl acetate	1.3	0.4	-	-	OT
20	1125	1123	3-Octanol acetate	0.3	0.0	-	-	OT
21	1154	1153	Citronellal	0.6	0.1	-	-	MO
22	1159	1162	Iso-menthone	0.1	0.0	-	-	MO
23	1180	1177	Terpinen-4-ol	0.4	0.0	-	-	MO
24	1168	1171	Menthol	-	-	0.2	0.1	MO
25	1188	1188	α-Terpineol	0.1	0.0	0.2	0.1	MO
26	1227	1225	Citronellol	0.4	0.0	-	-	MO

Table 4. (Continued)

N	RI	RI^{ref}	Compounds[a]	Clinopodium nubigenun[b]		Clinopodium Taxifolium[b]		Type
				%	σ	%	σ	
27	1229	1235	Thymol methyl ether	-	-	-	-	MO
28	1234	1237	Pulegone	1.9	0.7	4.5	1.8	MO
29	1251	1256	(E)-Piperitone epoxide	-	-	0.9	0.7	MO
30	1259	1258	2-Phenyl ethyl acetate	0.1	0.0	-	-	OT
31	1273	1267	Geranial	0.1	0.0	-	-	MO
32	1270	1273	ρ-Mentha-1-en-7-al	-	-	3.2	1.2	MO
33	1277	1277	Isopulegyl acetate	-	-	0.4	0.4	MO
34	1304	1285	Isopulegyl acetate	-	-	7.3	1.1	MO
35	1289	1289	Bornyl acetate	0.1	0.0	-	-	MO
36	1290	1290	Thymol	3.1	0.3	-	-	MO
37	1301	1295	Menthyl acetate	-	-	0.9	0.1	MO
38	1298	1299	Carvacrol	29.0	3.8	-	-	MO
39	1323	1313	(E)-2-terp-buty-Cyclohexanol acetate	-	-	0.3	0.1	MO
40	1330	1326	Myrtenyl acetate	0.1	0.0	0.3	0.0	MO
41	1325	1331	Bicycloelemene	-	-	0.2	0.0	SH
42	1331	1334	Carvyl acetate	-	-	-	-	MO
43	1343	1341	3-hydroxy-Benzenemethanol	-	-	0.9	0.3	OT
44	1347	1343	Piperitenone	0.1	0.0	-	-	MO
45	1350	1352	Citronellyl acetate	1.8	0.0	-	-	MO
46	1360	1359	Eugenol	1.1	0.0	-	-	OT
47	1357	1368	Piperitenone oxide	-	-	0.9	0.5	MO
48	1381	1372	Carvacrol acetate	38.1	4.7	-	-	MO
49	1370	1376	α-Copaene	-	-	5.0	0.7	SH
50	1379	1381	Geranyl acetate	0.4	0.0	-	-	MO
51	1381	1388	β-Cubebene	0.1	0.0	0.3	0.1	SH
52	1377	1388	β-Bourbonene	-	-	1.3	0.5	SH

N	RI	RI^{ref}	Compounds[a]	Clinopodium nubigenun[b]		Clinopodium Taxifolium[b]		Type
				%	σ	%	σ	
53	1384	1390	β-Elemene	-	-	0.5	0.1	SH
54	1411	1408	(Z)-Caryophyllene	-	-	24.1	2.6	SH
55	1423	1433	β-Gurjunene	-	-	1.3	0.4	SH
56	1429	1434	(E)-α-Bergamotene	-	-	2.3	0.2	SH
57	1451	1442	(Z)-β-Farnesene	-	-	1.0	0.3	SH
58	1454	1454	α-Humulene	-	-	1.3	0.2	SH
59	1480	1484	Germacrene D	0.1	0.0	12.8	2.1	SH
60	1494	1500	Bicyclogermacrene	0.3	0.0	8.8	0.7	SH
61	1495	1501	Aciphyllene	-	-	0.8	0.3	SH
62	1503	1505	(E,E)-α-Farnesene	0.1	0.0	-	-	SH
63	1513	1523	δ-Cadinene	1.5	0.1	5.9	0.8	SH
64	1547	1546	Selina-3,7(11)-diene	-	-	0.5	0.2	SH
65	1559	1563	E-Nerolidol	-	-	0.2	0.1	SO
66	1569	1575	Germacrene D-4-ol	-	-	-	-	SO
67	1573	1578	Spathulenol	-	-	1.2	0.2	SO
68	1577	1583	Caryophyllene oxide	-	-	1.0	0.3	SO
69	1625	1624	Isospathulenol	-	-	0.2	0.1	SO
70	1649	1654	α-Cadinol	-	-	0.3	0.0	SO
Monoterpene hydrocarbons (MH)				16.6		4.5		
Oxygenated monoterpenes (MO)				76.9		17.8		
Sesquiterpene hydrocarbons (SH)				2.1		66.1		
Oxygenated sesquiterpenes (SO)				0.0		2.9		
Other compounds (OT)				3.3		0.9		
Total identified				98.9		92.2		

RI. Retention Index.

[a]All compounds identified by MS and RI: MS. by comparison of the mass spectrum with those of the computer mass libraries Wiley 7. Adams (21) and NIST 05 (22, 23);
RI. by comparison of RI with those reported in literature; [21, 22] and [23].

[b]Percentage values are means of three determinations.

RI^{ref}. references: [21, 22] and [23].

C. nubigenum essential oil also contains monoterpene hydrocarbons (16,6%) and a low amount of sesquiterpene hydrocarbons (2.1%), other compounds represent the 3,3%. The main constituents (compounds that have a percentage greater than 5.0%) of the oil were carvacrol acetate (38,1 ± 4,7%), carvacrol (29,0 ± 3,8%), p-cymene (6,6 ± 1,6%) and γ-terpinene (5,0 ± 1,7%).

The chemical composition of essential oil of *Clinopodium taxifolium* has been determined and is shown in the Table 4. The principal groups of constituents were sesquiterpene hydrocarbons (66.1%) and oxygenated monoterpenes (17.8%). It was also determined that groups monoterpene hydrocarbons and oxygenated sesquiterpenes are present with 4.5% and 2.9% respectively. Thirty-eight components were identified representing 92.2% of the oil, the major ones being those of (Z)-caryophyllene (24.1 ± 2.6%), Germacrene D (12.8 ± %), Bicyclogermacrene (8.8 ± 0.7%), isopulegyl acetate (7.3 ± 1.1%), δ-Cadinene (5,9 ± 0.8%) and α-Copaene (5.0 ± 0.7%). A low percentage of 3-hydroxy-benzenemethanol (0.9 ± 0.3%) that belongs to the other compounds was also discovered in the sample.

3.2. Genus Hyptis

From the genus Hyptis the essential oil of species *Hyptis pectinata and Hyptis suaveolens* was isolated and studied. The aerial part of the *Hyptis pectinata* was collected in rural district "San Jose" (4°06'44.2" S, 80°07'24.9" W), canton Pindal, province of Loja, Ecuador. On the other hand, Hyptis suaveolens were collected at 249 m a.s.l. in rural district "La Ceiba" (4°20'21.8" S, 80°18'00.9" W) canton Zapotillo, Loja province. The plant material identification was done by Fani Tinitana (Herbarium of The Universidad Técnica Particular de Loja) and Bolivar Merino. Voucher specimens are preserved in the Herbarium of Universidad Técnica Particular de Loja, under codes PPN-la-013 for *H. pectinata* and PPN-LA-033 for *H. suaveolens.*

Table 5. Chemical composition of the essential oil from *Hyptis Pectinata* and *Hyptis suaveolens*

Nº	RI	RI[ref]	Compounds[a]	Hyptis Pectinata[b]		Hyptissuaveolens[b]		Type
				%	σ	%	σ	
1	937	939	α-Pinene	-	-	6.7	0.1	MH
2	952	954	Camphene	-	-	0.3	0.0	MH
3	973	975	Sabinene	-	-	16.6	0.7	MH
4	978	979	β-Pinene	0.4	0.0	-	-	MH
5	995	990	Myrcene	0.1	0.0	1.1	0.1	MH
6	1005	1002	α-Phellandrene	-	-	0.2	0.0	MH
7	1016	1017	α-Terpinene	-	-	0.6	0.1	MH
8	1029	1031	1,8-Cineole (Eucaliptol)	0.7	0.1	39.0	0.5	MO
9	1046	1050	(E)-β-Ocimene	-	-	0.2	0.0	MH
10	1058	1059	γ-Terpinene	0.9	0.0	0.9	0.1	MH
11	1082	1086	Fenchone	-	-	9.7	0.7	MO
12	1118	1116	Fenchyl alcohol	-	-	4.7	0.3	MO
13	1141	1146	Camphor	-	-	0.2	0.0	MO
14	1166	1169	Borneol	-	-	0.5	0.0	MO
15	1180	1177	Terpinen-4-ol	-	-	0.7	0.0	MO
16	1188	1188	α-Terpineol	-	-	0.5	0.0	MO
17	1342	1351	α-cubebene	1.0	0.0	-	-	SH
18	1381	1388	β-Cubebene	1.1	0.1	-	-	SH
19	1377	1388	β-Bourbonene	9.3	0.1	0.6	0.0	SH
20	1384	1390	β-Elemene	6.0	0.5	0.6	0.0	SH
21	1406	1409	α-Gurjunene	0.1	0.0	0.2	0.0	SH
22	1418	1419	(E)-β-Caryophyllene	21.1	0.2	3.0	0.1	SH
23	1454	1454	α-Humulene	3.0	0.1	0.2	0.0	SH
24	1453	1460	allo-Aromadendrene	-	-	1	0.0	SH
25	1471	1479	γ-Muurolene	0.4	0.0	-	-	SH
26	1480	1484	Germacrene D	34.1	0.6	3.0	0.1	SH
27	1492	1492	δ-Selinene	1.2	0.1	-	-	SH
28	1500	1496	Valencene	-	-	0.2	0.0	SH
29	1497	1496	Viridiflorene	8.3	0.1	-	-	SH
30	1507	1513	γ-Cadinene	-	-	0.3	0.0	SH
31	1513	1523	δ-Cadinene	2.1	0.2	1.7	0.2	SH
32	1551	1561	Germacrene B	1.4	0.0	-	-	SH
33	1577	1583	Caryophyllene oxide	1.6	0.1	-	-	SO
34	1650	1640	τ-Cadinol	-	-	0.5	0.0	SO
Monoterpene hydrocarbons (MH)				1.4		26.6		
Oxygenated monoterpenes (MO)				0.7		55.2		
Sesquiterpene hydrocarbons (SH)				89.2		11.0		
Oxygenated sesquiterpenes (SO)				1.6		0.5		
Other compounds (OT)				0.0		0.0		
Total identified				93.0		93.2		

The essential oil of *H. pectinata* yield was 0,04 ± 0.01% (w/w), referred to fresh plant material. The essential oil relative density was d^{20} = 0,8983 ± 0.0084, refraction index was n^{20} =1,5065 ± 0.0001 and specific rotation was $[\alpha]_D^{20}$ = 42.2 ± 3.1. The aerial parts of *H. suaveolens.* provided an essential oil in yield of 0.11 ± 0.06% (w/w) referred to fresh plant material. The relative density was d^{20} = 0.8950 ± 0.0261, the refractive index was n^{20} = 1.4717 ± 0.0012 and the specific rotation was $[\alpha]_D^{20}$ = 5.8 ± 0.7.

The essential oil of *Hyptis Pectinata* was analyzed by GC/MS and GC/FID. The results obtained are summarized in Table 5. Eighteen individual compounds were found in the EO of *H. Pectinata*, which represent a 93.0% of the total essential oil. Sesquiterpene hydrocarbons was the most representative group with 89.2%. Furthermore, a low amount of oxygenated sesquiterpenes (1.6%), monoterpene hydrocarbons (1.4%) and oxygenated monoterpenes were identified. In the essential oil was not possible to determine the presence of compounds belonging to the group other compounds. The main constituents of the essential oil were germacrene D (34.1 ± 0.6%), (E)-β-caryophyllene (21.1 ± 0.2%), β-bourbonene (9.3 ± 0.1%), viridiflorene (8,3 ± 0.1%) and β-elemene (6.0 ± 0.5%).

The chemical components of the essential oils of *Hyptis suaveolens,* their retention indices and percentages are listed in Table 5. Twenty-six individual compounds were identified, which represent a 93.2% of the total essential oil, of these compounds seven belong to the group oxygenated monoterpenes representing a 55.2%, eight compounds belong to the group monoterpene hydrocarbons which represent the 26.6 ± %, ten compounds belong to the group sesquiterpene hydrocarbons (11.0 ± %). Furthermore, was identified a low amount of oxygenated sesquiterpenes which represent only the 0.5%, in the essential oil was not possible to determine the presence of compounds belonging to the groups other compounds. The main constituents of the essential oil were 1,8-cineole (39.0 ± 0.5%), sabinene (16,6 ± 0.7%), fenchone (9,7 ± 0.7%) and α-pinene (6.7 ± 0.1%).

3.3. Genus Lepechinia

The essential oil of species *Lepechinia paniculata* and *Lepechinia radula* belonging to genus Lepechinia was studied. The collection of *L. paniculata* was performed in the mountain "Acacana" canton Saraguro (3°41'36.7" S 79°14'56.4" W), Loja province. *Lepechinia radula* was collected at 2350 m a.s.l. in the hill "Guachanamá" canton Celica (4°03'19.7" S, 79°51'24.5" W), Loja province. The botanical specimens identification was carried out by Bolívar Merino, as well as by Fani Tinitana. Voucher specimens are preserved in the Herbarium of Universidad Técnica Particular de Loja, under codes PPN-la-020 for *L. paniculata* and PPN-LA-034 for *L. radula*.

The essential oil of *L. paniculata* yield was 0,08 ± 0.01% (w/w), referred to fresh plant material. The essential oil relative density was d^{20} = 0,8973 ± 0.0021, refraction index was n^{20} = 1,4957 ± 0.0023 and specific rotation was $[\alpha]_D^{20}$ = -91.3 ± 1.4. The aerial parts of *L. radula.* essential oil yield was 0.37 ± 0.02% (w/w) referred to fresh plant material. The relative density was d^{20} = 0.8713 ± 0.0040 and the refractive index was n^{20} = 1.47133 ± 0.0006.

Essential oils were obtained by steam distilled from the aerial parts of *Lepechinia radula.* Essential oil was subsequently analyzed using both GC/FID and GC/MS systems in which the individual components were identified according to their MS data and retention indices. The constituents of the essential oil of *L. radula* are presented in Table 6. Essential oil was dominated by monoterpene hydrocarbons (52.2%), sesquiterpene hydrocarbons (37.1%) and oxygenated sesquiterpenes (4.4%), containing a low amount of oxygenated monoterpenes (2.3%). Thirty-four individual compounds were characterized in *L. radula* oil, which represents 96.0% of the total essential oil. The main constituents of the oil were δ-3-Carene (19.9 ± 0.9%), β-pinene (17.0 ± 1.3%), (E)-β-caryophyllene (9.7 ± 0.7%), (E,E)-α-farnesene (9.4 ± 1.9%) and β-phellandrene (8.6 ± 0.2%).

The chemical composition of *Lepechinia paniculata* essential oil has been determined. Thirty-one components were identified representing

90.6% of the oil (Table 6). The largest number of compounds belong to the group sesquiterpene hydrocarbons, with seventeen compounds which represent 69.9% of the total percentage, of which the aromadendrene with 24.6% ± 0.6% is the most representative compound, followed by viridiflorene (12.4 ± 0.3%), β-selinene (7.4 ± 0.2%) and valencene (6.7 ± 0.3%). A low percentage of β-phellandrene (7.7%) that belongs to the monoterpenes hydrocarbons (15.6%) was also discovered in the sample. Constituents as camphor (3.9% ± 0.1) and 1-Octen-3-ol (0.4 ± 0.1%) were noticed as minor components, which belong to oxygenated monoterpenes (4.0%) and other compounds (0.4%), respectively.

Table 6. Chemical composition of the essential oil from *Lepechinia radula* and *Lepechinia paniculata*

Nº	RI	RI[ref]	Compounds[a]	Lepechinia radula[b]		Lepechinia paniculata[b]		Type
				%	σ	%	σ	
1	930	930	α-Thujene	0.3	0.0	0.1	0.0	MH
2	937	939	α-Pinene	1.2	0.1	1.4	0.3	MH
3	952	954	Camphene	-	-	2.4	0.4	MH
4	973	975	Sabinene	2.6	0.7	0.1	0.0	MH
5	978	979	β-Pinene	17.0	1.3	0.2	0.1	MH
6	981	979	1-Octen-3-ol	-	-	0.4	0.1	OT
7	995	990	Myrcene	-	-	0.4	0.0	MH
8	1002	1004	ρ-Mentha-1(7), 8-diene	0.2	0.0	0.8	0.1	MH
9	1006	1011	δ-3-Carene	19.9	0.9	-	-	MH
10	1016	1017	α-Terpinene	0.2	0.0	-	-	MH
11	1022	1024	p-Cymene	0.2	0.0	1.2	0.1	MH
12	1028	1029	Limonene	0.4	0.0	-	-	MH
13	1029	1029	β-Phellandrene	8.6	0.2	7.7	0.1	MH
14	1058	1059	γ-terpinene	0.2	0.0	1.2	0.3	MH
15	1084	1088	α-Terpinolene	1.4	0.1	0.1	0.0	MH
16	1097	1096	Linalool	1.4	0.5	-	-	MO
17	1103	1103	2-Methyl butyl 2-methyl butanoate	0.9	0.1	-	-	MO
18	1141	1146	Camphor	-	-	3.9	0.1	MO
19	1166	1169	Borneol	-	-	0.1	0.0	MO
20	1370	1376	α-Copaene	0.4	0.4	0.2	0.0	SH
21	1406	1409	α-Gurjunene	0.1	0.1	1.4	0.3	SH
22	1418	1419	(E)-β-Caryophyllene	9.7	0.7	4.3	0.1	SH
23	1423	1433	β-Gurjunene	0.2	0.1	0.3	0.0	SH

N°	RI	RI^ref	Compounds[a]	Lepechinia radula[b]		Lepechinia paniculata[b]		Type
				%	σ	%	σ	
24	1429	1434	(E)-α-Bergamotene	0.1	0.0	-	-	SH
25	1441	1441	Aromadendrene	-	-	24.6	0.6	SH
26	1454	1454	α-Humulene	1.9	0.1	0.9	0.0	SH
27	1456	1456	α-Patchoulene	-	-	0.9	0.0	SH
28	1453	1460	allo-Aromadendrene	0.2	0.1	2.8	0.0	SH
29	1468	1479	γ-Selinene	-	-	0.6	0.1	SH
30	1454	1471	(+) Epi-Bicyclosesquiphellandrene	0.1	0.0	-	-	SH
31	1480	1484	Germacrene D	7.2	0.7	-	-	SH
32	1485	1490	β-Selinene	0.7	0.1	7.4	0.2	SH
33	1500	1496	Valencene	-	-	6.7	0.3	SH
34	1497	1496	Viridiflorene	-	-	12.4	0.3	SH
35	1490	1498	α-Selinene	-	-	1.6	0.1	SH
36	1494	1500	Bicyclogermacrene	4.0	0.8	-	-	SH
37	1503	1505	(E,E)-α-Farnesene	9.4	1.9	4.3	0.4	SH
38	1498	1509	Germacrene A	0.4	0.0	-	-	SH
39	1507	1513	γ-Cadinene	-	-	0.4	0.1	SH
40	1513	1523	δ-Cadinene	0.7	0.5	-	-	SH
41	1534	1538	α-Cadinene	0.1	0.0	0.5	0.1	SH
42	1547	1546	Selina-3,7(11)-diene	-	-	0.6	0.1	SH
43	1551	1561	Germacrene B	1.9	0.3	-	-	SH
44	1569	1575	Germacrene D-4-ol	1.4	0.1	-	-	SO
45	1577	1583	Caryophyllene oxide	0.7	0.1	-	-	SO
46	1590	1600	Guaiol	0.8	0.2	-	-	SO
47	1649	1654	α-Cadinol	1.5	0.4	-	-	SO
Monoterpene hydrocarbons (MH)				52.2		15.6		
Oxygenated monoterpenes (MO)				2.3		4.0		
Sesquiterpene hydrocarbons (SH)				37.1		69.9		
Oxygenated sesquiterpenes (SO)				4.4		0.0		
Other compounds (OT)				0		0.4		
Total identified				96.0		89.9		

3.4. Genus Minthostachys

The collection of *Minthostachys mollis* was performed in the parish Chuquiribamba (3°51'08.3" S, 79°20'38.8" W) Canton and Province of Loja. The plant material identification was done by Bolivar Merino. Voucher specimens are preserved in the HUTPL under codes PPN-la-009. In regards to physical properties the essential oil relative density of *M.*

molle was d^{20} = 0.9042 ± 0.0093, refraction index was n20 = 1.4740 ± 0.0020 and specific rotation was $[\alpha]D^{20}$ = 8.9510 ± 1.7849. Yield was 0.11 ± 0.04%.

Table 7. Chemical composition of the essential oil from *Minthostachys Mollis*

Nº	LRI	LRI[ref]	Compounds[a]	Minthostachys Mollis[b]		Type
				%	σ	
1	937	939	α-Pinene	1.4	0.4	MH
2	973	975	Sabinene	2.5	1.5	MH
3	978	979	β-Pinene	1.0	0.2	MH
4	995	990	Myrcene	0.6	0.1	MH
5	1001	991	3-Octanol	0.8	0.2	OT
6	1028	1029	Limonene	6.8	2.9	MH
7	1029	1031	1,8-Cineole (Eucaliptol)	0.9	0.9	MO
8	1036	1037	(Z)-β-Ocimene	2.3	0.9	MH
9	1046	1050	(E)-β-Ocimene	2.6	0.6	MH
10	1076	1083	Kewda ether	0.2	0.04	OT
11	1097	1096	Linalool	1.3	0.7	MO
12	1115	1115	2 Octyl acetate	0.9	0.4	OT
13	1150	1152	Menthone	1.8	1.1	MO
14	1154	1153	Citronellal	1.4	0.7	MO
15	1159	1162	Iso-menthone	3.1	1.5	MO
16	1165	1165	Neo-menthol	0.5	0.1	MO
17	1195	1192	(Z)-Dihydro carvone	13.2	0.6	MO
18	1198	1200	(E)-Dihydro carvone	1.2	1.2	MO
19	1222	1229	Nerol	0.6	0.8	MO
20	1227	1225	Citronellol	1.7	0.6	MO
21	1234	1237	Pulegone	2.7	0.7	MO
22	1242	1243	Carvone	5.1	1.1	MO
23	1250	1252	Geraniol	0.9	0.8	MO
24	1273	1267	Geranial	1.7	0.2	MO
25	1270	1273	Neomenthyl acetate	0.6	0.6	MO
26	1321	1307	Dihydrocarvyl acetate	0.9	0.7	MO
27	1331	1334	Carvyl acetate	0.2	0.1	MO
28	1350	1352	Citronellyl acetate	1.1	0.7	MO
29	1359	1361	Neryl Acetate	3.1	5	MO
30	1370	1376	α-Copaene	1.2	0.2	SH
32	1379	1381	Geranyl acetate	6.3	1.4	MO
32	1384	1390	β-Elemene	0.1	0.1	SH
33	1418	1419	(E)-β-Caryophyllene	8.8	0.2	SH

N°	LRI	LRI[ref]	Compounds[a]	Minthostachys Mollis[b]		Type
				%	σ	
34	1454	1454	α-Humulene	1.1	0.1	SH
35	1480	1484	Germacrene D	3.8	1.2	SH
36	1513	1523	δ-Cadinene	1.5	0.3	SH
37	1551	1561	Germacrene B	1.6	0.7	SH
38	1573	1578	Spathulenol	0.1	0.1	SO
39	1577	1583	Caryophyllene oxide	0.4	0.1	SO
Monoterpene hydrocarbons (MH)				17.2		
Oxygenated monoterpenes (MO)				48.3		
Sesquiterpene hydrocarbons (SH)				18.1		
Oxygenated sesquiterpenes (SO)				0.5		
Other compounds (OT)				1.9		
Total identified				86.0		

RI. Retention Index.

[a]All compounds identified by MS and RI: MS. by comparison of the mass spectrum with those of the computer mass libraries Wiley 7. Adams (21) and NIST 05 (22, 23);

RI. by comparison of RI with those reported in literature; [21, 22] and [23].

[b]Percentage values are means of three determinations

RI[ref]. references: [21, 22] and [23].

The chemical composition of essential oil *Minthostachys Mollis* is shown in the Table 7. The principal groups of constituents were oxygenated monoterpenes (48.3%), sesquiterpene hydrocarbons (18.1%) and monoterpene hydrocarbons (17.2%). It was also determined a low amount of other compounds and oxygenated sesquiterpenes, which are present with 1.9% and 0.5%, respectively. Thirty-nine components were identified representing 86.0% of the oil, the major ones are (Z)-dihydro carvone (13.2 ± 0.6%), (E)-β-caryophyllene (8.8 ± 0.2%), geranyl acetate (6.3 ± 1.4%), limonene (6.8 ± 2.9) and carvone (5.1 ± 2.1).

3.5 Genus Origanum

The collection of Origanum vulgare was performed at 2568 m a.s.l. in "Las Lagunas" (3°38'40.2" S, 79°13'02.9" W) Canton Saraguro, Province of Loja. The plant material identification was done by Bolivar Merino. Voucher specimens are preserved in the HUTPL under codes PPN-la-023.

Table 8. Chemical composition of the essential oil from *Origanum Vulgare*

N°	LRI	LRI[ref]	Compounds[a]	Origanum Vulgare[b]		Type
				%	σ	
1	930	930	α-Thujene	0.5	0.1	MH
2	937	939	α-Pinene	0.4	0.1	MH
3	952	954	Camphene	0.2	0.1	MH
4	973	975	Sabinene	3.4	0.3	MH
5	978	979	β-Pinene	0.2	0.1	MH
6	981	979	1-Octen-3-ol	0.2	0.1	OT
7	985	983	3-Octanone	0.3	0.2	OT
8	995	990	Myrcene	1.6	0.2	MH
9	1005	1002	α-Phellandrene	0.1	0.0	MH
10	1016	1017	α-Terpinene	4.1	0.6	MH
11	1024	1026	o-Cymene	5.6	1.3	MH
12	1028	1029	Limonene	1.0	0.6	MH
13	1029	1029	β-Phellandrene	1.3	0.9	MH
14	1036	1037	(Z)-β-Ocimene	5.4	1.4	MH
15	1046	1050	(E)-β-Ocimene	0.8	0.6	MH
16	1058	1059	γ-Terpinene	20.8	1.9	MH
17	1084	1088	α-Terpinolene	1.5	0.7	MH
18	1097	1096	Linalool	19.1	2.5	MO
19	1108	1112	1-Octen-3-yl acetate	0.7	0.4	OT
20	1124	1120	Dehydro-Sabina ketone	0.8	0.4	OT
21	1142	1140	(E)-ρ-Menth-2-en-1-ol	0.3	0.2	MO
22	1180	1177	Terpinen-4-ol	9.5	1.4	MO
23	1188	1188	α-Terpineol	1.8	1.2	MO
24	1229	1235	Thymol methyl ether	3.2	1.2	MO
25	1238	1244	Carvacrol methyl ether	1.9	0.3	MO
26	1250	1256	(E)-Sabinene hydrate acetate	1.6	0.8	MO
27	1290	1290	Thymol	3.4	0.3	MO
28	1298	1299	Carvacrol	0.3	0.3	MO
29	1418	1419	(E)-β-Caryophyllene	4.2	0.9	SH
30	1454	1454	α-Humulene	0.6	0.1	SH
31	1480	1484	Germacrene D	1.7	0.1	SH
32	1494	1500	Bicyclogermacrene	3.1	0.7	SH
Monoterpene hydrocarbons (MH)				50.3		
Oxygenated monoterpenes (MO)				43.2		
Sesquiterpene hydrocarbons (SH)				9.6		
Oxygenated sesquiterpenes (SO)				0		
Other compounds (OT)				2.0		
Total identified				99.5		

RI, Retention Index. RI. Retention Index. [a]All compounds identified by MS and RI: MS. by comparison of the mass spectrum with those of the computer mass libraries Wiley 7. Adams (21) and NIST 05 (22, 23); RI. by comparison of RI with those reported in literature; [21, 22] and [23]. [b]Percentage values are means of three determinations.. RI[ref]. references: [21, 22] and [23].

The essential oil of O. vulgare yield was 0,14 ± 0.03% (w/w), referred to fresh plant material. The essential oil relative density was d20 = 0,9253 ± 0.0134, refraction index was n20 = 1,4792 ± 0.0058 and specific rotation was [α]D20 = 14.7667 ± 1.3405.

The chemical components of the essential oils of *Origanum Vulgare,* their retention indices and percentages are listed in Table 8. Thirty-two individual compounds were identified, which represent a 99,5% of the total essential oil, of these compounds eleven belong to the group monoterpene hydrocarbons representing a 50.3%, nine compounds belong to the group oxygenated monoterpenes which represent the 43.2% and four compounds belong to the group sesquiterpene hydrocarbons (9.6%). Furthermore, was identified a low amount of other compounds which represent only the 2.0%, in this essential oil was not possible to determine the presence of compounds belonging to the groups oxygenated sesquiterpenes. The main constituents of the essential oil were γ-Terpinene (20.8 ± 1.9%), Linalool (19.1 ± 2.5%), Terpinen-4-ol (9.5 ± 1.4%) and (Z)-β-Ocimene (5.6 ± 1.3%).

3.6. Genus Scutellaria

The aerial part of the *Scutellaria volubilis* was recollected at ambient conditions in Cerro el Villonaco (4°00'13.0" S, 79°15'25.4" W) (Loja, Ecuador). The plant material identification was done by Fani Tinitana and Bolivar Merino .Voucher specimen are preserved in the HUTPL under codes PPN-LA-027. The essential oil of *Scutellaria volubilis* yield was 0,03 ± 0.01% (w/w), referred to fresh plant material. The essential oil relative density was d^{20} = 0,8930 ± 0.0008, refraction index was n^{20} = 1,5040 ± 0.0022 and specific rotation was $[\alpha]_{D20}$ = -102.4 ± 2.1.

The components of the essential oil of *Scutellaria volubilis* were identified based on the retention index and mass spectra. The percentages of oil components and their IR are shown in the Table 9. In the essential oil

Table 9. Chemical composition of the essential oil from *Scutellaria volubilis*

Nº	LRI	LRI[ref]	Compounds[a]	Scutellaria volubilis[b]		Type
				%	σ	
1	930	930	α-Thujene	0.2	0.0	MH
2	937	939	α-Pinene	0.8	0.2	MH
3	973	975	Sabinene	1.6	0.3	MH
4	978	979	β-Pinene	1.9	0.0	MH
5	981	979	1-Octen-3-ol	0.2	0.0	OT
6	995	990	Myrcene	0.3	0.0	MH
7	1005	1002	α-Phellandrene	0.5	0.1	MH
8	1016	1017	α-Terpinene	0.1	0.0	MH
9	1022	1024	p-Cymene	0.3	0.1	MH
10	1029	1029	β-Phellandrene	6.9	0.8	MH
11	1046	1050	(E)-β-Ocimene	0.1	0.0	MH
12	1058	1059	γ-terpinene	0.1	0.0	MH
13	1342	1351	α-cubebene	0.2	0.0	SH
14	1370	1376	α-Copaene	0.7	0.0	SH
15	1381	1388	β-Cubebene	0.6	0.1	SH
16	1377	1388	β-Bourbonene	2.3	0.1	SH
17	1384	1390	β-Elemene	1.5	0.2	SH
18	1406	1409	α-Gurjunene	1.4	0.2	SH
19	1418	1419	(E)-β-Caryophyllene	17.5	1.5	SH
20	1429	1434	(E)-α-Bergamotene	1.4	0.2	SH
21	1441	1441	Aromadendrene	3.1	0.3	SH
22	1454	1454	α-Humulene	14.7	1.1	SH
23	1453	1460	allo-Aromadendrene	0.8	0.1	SH
24	1471	1479	γ-Muurolene	0.3	0.1	SH
25	1480	1484	Germacrene D	20.4	0.7	SH
26	1494	1500	α-Muurolene	0.4	0.1	SH
27	1503	1505	(E.E)-α-Farnesene	0.4	0.1	SH
28	1504	1505	β-Bisabolene	5.8	1.0	SH
29	1520	1522	β-Sesquiphellandrene	0.5	0.1	SH
30	1513	1523	δ-Cadinene	2.6	0.4	SH
31	1577	1583	Caryophyllene oxide	2.8	0.4	SO
32	1598	1592	Viridiflorol	0.3	0.0	SO
33	1650	1646	Torreyol (α-Muurolol)	0.8	0.1	SO
34	1649	1654	α-Cadinol	0.6	0.0	SO
Monoterpene hydrocarbons (MH)				12.8		
Oxygenated monoterpenes (MO)				0.0		
Sesquiterpene hydrocarbons (SH)				74.6		
Oxygenated sesquiterpenes (SO)				4.5		
Other compounds (OT)				0.2		
Total identified				92.1		

RI. Retention Index.

[a]All compounds identified by MS and RI: MS. by comparison of the mass spectrum with those of the computer mass libraries Wiley 7. Adams (21) and NIST 05 (22, 23); RI. by comparison of RI with those reported in literature; [21, 22] and [23]. [b]Percentage values are means of three determinations. RI[ref]. references: [21, 22] and [23].

of *S. volubilis* thirty-four components were determined, which represents 92.1% of the total sample. Sesquiterpene hydrocarbons with 74.8% was found to be the principal group of constituents, containing germacrene D (20.4 ± 0.7%), (E)-β-Caryophyllene (17.5 ± 1.5%), α-humulene (14.7 ± 1.1%) and β-bisabolene (5.8 ± 1.0%). A small amount of β-phellandrene (6.9 ± 0.8%) that belongs to the monoterpenes hydrocarbons (12.8%) was also determined to be present. The minor components encountered are caryophyllene oxide (2.8 ± 0.4%) and Torreyol (0.8 ± 0.1%), which belong to oxygenated sesquiterpenes (4.5%), in the essential oil was not possible to determine the presence of compounds belonging to the group oxygenated monoterpenes.

REFERENCES

[1] Harley, R. M. 2012. Checklist and key of genera and species of the Lamiaceae of the Brazilian Amazon. *Rodriguésia,* 63: 129-144.

[2] Wagstaff, S. J., Hickerson, L., Spangler, R., Reeves, P. A., & Olmstead, R. G. 1998. Phylogeny in Labiatae s.l., inferred from cpDNA sequences. *Plant Systematics and Evolution,* 209: 265–274.

[3] Heywood, V. H., Brummitt, R. K., Seberg, O., & Culham, A. 2007. *Flowering plant families of the world.* Ontario, Canada: Firefly Books.

[4] Raymond, M. H., Sandy, A., Andrey, L. B., Philip, D. C., Barry, J. C., Renée, J. G., Madeline, M. H., Rogier, P. J. d. K., Tatyana, V. K., Ramón, M., Alan, J. P., & Olof Ryding, P. 2004. Labiatae. In J. W. K. v. E. Klaus Kubitzki (Editor) (Ed.), *The Families and Genera of Vascular Plants volume VII* (pp. 167-275). Berlin, Heidelberg, Germany: Springer-Verlag.

[5] Yuan, Y. W., Mabberley, D. J., Steane, D. A., & Olmstead, R. G. 2010. Further disintegration and redefinition of Clerodendrum (Lamiaceae): Implications for the understanding of the evolution of an intriguing breeding strategy. *TAXON,* 59: 125-133.

[6] Peters, R. 2011. *Introduction to the Lamiaceae family*. Kandern, Germany: Narayana Publishers.

[7] Bramley, G. L. C., Forest, F., & de Kok, R. P. J. 2009. Troublesome Tropical Mints: Re-Examining Generic Limits of Vitex and Relations (Lamiaceae) in South East Asia. *TAXON,* 58: 500-510.

[8] de la Torre, L., Navarrete, H., Muriel, P., Macía, M. J., & Balslev, H. 2008. *Enciclopedia de las plantas útiles del Ecuador* [*Encyclopedia of useful plants of Ecuador*]. Quito & Aarhus: Herbario QCA de la Escuela de Ciencias Biológicas de la Pontificia Universidad Católica del Ecuador & Herbario AAU del Departamento de Ciencias Biológicas de la Universidad de Aarhus.

[9] Jørgesen, P., & León-Yáñez, S. 1999. *Catalogue of the Vascular Plants of Ecuador*. St. Louis, MO: Missouri Botanical Garden Press.

[10] Andrade, J. M., Armijos, C. P., Malagón, O., & Lucero, H. 2009. *Plantas Medicinales Silvestres Empleadas por la Etnia Saraguro en la Parroquia San Lucas, Provincia De Loja, Ecuador* [*Wild Medicinal Plants Employed by the Saraguro Ethnicity in San Lucas Parish, Loja Province, Ecuador*]. Loja, Ecuador: Editorial UTPL.

[11] Rios, M., Koziol, J., Brorgtoft, P. H., & Granda, G. 2007. *Plantas útiles del Ecuador: aplicaciones, retos y perspectivas* [*Useful plants of Ecuador: applications, challenges and perspectives*]. Quito, Ecuador: Abya-Yala.

[12] McNeil, M., Facey, P., & Porter, R. 2011. Essential Oils from the Hyptis genus- A Review (1909-2009). *Natural Product Communications,* 6: 1934578X1100601149.

[13] Drew, B. T., & Sytsma, K. J. 2012. The South American radiation of Lepechinia (Lamiaceae): phylogenetics, divergence times and evolution of dioecy. *Botanical Journal of the Linnean Society,* 171: 171-190.

[14] Mora, F. D., Araque, M., Rojas, L. B., Ramírez, R., Silva, B., & Usubillaga, A. 2009. Chemical Composition and in Vitro Antibacterial Activity of the Essential Oil of Minthostachys Mollis (Kunth) Griseb Vaught from the Venezuelan Andes. *Natural Product Communications,* 4: 1934578X0900400726.

[15] Schmidt-Lebuhn, A. N. 2008. Ethnobotany, biochemistry and pharmacology of Minthostachys (Lamiaceae). *Journal of Ethnopharmacology,* 118: 343-353.

[16] Skoula, M., Gotsiou, P., Naxakis, G., & Johnson, C. B. 1999. A chemosystematic investigation on the mono- and sesquiterpenoids in the genus Origanum (Labiatae). *Phytochemistry,* 52: 649-657.

[17] Oyarzabal, M. E. B., Schuch, L. F. D., Prestes, L. S., Schiavon, D. B. A., Rodrigues, M. R. A., & de Mello, J. R. B. 2011. Antimicrobial action of Origanum vulgare L. essential oil against bacteria isolated from bovine milk. [Article]. *Revista Cubana de Plantas Medicinales,* 16: 260-266.

[18] Shang, X., He, X., He, X., Li, M., Zhang, R., Fan, P., Zhang, Q., & Jia, Z. 2010. The genus Scutellaria an ethnopharmacological and phytochemical review. *Journal of Ethnopharmacology,* 128: 279-313.

[19] Duvauchelle, D. 2011. *Plant Materials Collection Guide*. Hoolehua, Hawaii: United States Department of Agriculture, Natural Resources Conservation Service, Pacific Islands Area, Hoolehua Plant Materials Center.

[20] Damas, K., & Kiapranis, R. 2017. *Laboratory manual for collecting and processing plant species*. Papua New Guinea: Papua New Guinea Forest Research Institute.

[21] Adams, R. P. 2007. *Identification of Essential Oil Components by Gas Chromatography/Mass Spectrometry* (4th ed.). Carol Stream, IL: Allured Publishing Corporation.

[22] NIST 05. 2005. *Mass Spectral Library (NIST/EPA/NIH).* Gaithersburg, MD National Institute of Standards and Technology.

[23] National Institute of Standards and Technology (NIST). August 2019. *Libro del Web de Química del NIST, SRD 69. in Base de Datos de Referencia Estándar del NIST Número 69* (May 24, 2016 ed.). USA: U.S. Secretary of Commerce.

INDEX

A

acetylcholinesterase, 65
acid, xiii, 19, 27, 46, 47, 49, 67, 70, 83, 88, 97, 99, 101, 102, 111, 112, 113, 116, 118, 119, 128, 130, 166, 186, 187, 212, 215, 216, 230, 232, 233, 234, 236, 240
acidity, x, 142, 145
agar, xi, 142, 150, 157, 158, 188, 191
agrobacterium, 186
Ajuga, vii, ix, 2, 5, 7, 8, 39, 58, 60, 61, 62, 63, 64, 68, 76, 80, 83
alfalfa, 239
alkaloids, viii, 1, 3, 7, 17, 34, 54, 183, 240
angiogenesis, 47, 52, 121
anthocyanin, 126
anti-asthma, 3, 17, 49
antibacterial activity, xi, 40, 45, 53, 71, 74, 78, 86, 90, 93, 142, 143, 149, 155, 157, 208, 236, 324
antibiotic, 182, 238
anticancer drug, 133, 232
anticonvulsant, 52
antidepressant, 48, 50
anti-inflammatory agents, 39, 213
antioxidant, vii, ix, x, 3, 17, 40, 42, 43, 44, 47, 48, 51, 52, 53, 61, 64, 69, 70, 72, 80, 81, 83, 96, 97, 98, 108, 109, 110, 113, 114, 115, 116, 117, 118, 120, 121, 122, 124, 125, 126, 127, 128, 130, 131, 132, 137, 138, 139, 140, 142, 144, 145, 166, 167, 168, 169, 175, 177, 178, 179, 182, 206, 213, 231, 233, 235, 236, 237, 240, 286, 288
antioxidative activity, 233
antipyretic, 3, 39, 43, 47, 53
antitumor, 40, 52, 144, 147
apoptosis, 121, 127, 231, 232
aromatic species, vi, viii, xiv, 297, 298
asthma, ix, 42, 48, 50, 53, 54, 96, 98
astringent, 39, 41, 43, 50, 51
atherosclerosis, 97, 111

B

Bacillus subtilis, 147
bacteria, xi, 45, 81, 142, 145, 147, 150, 151, 155, 157, 158, 165, 168, 170, 185, 231, 235, 325
bacterial cells, 150, 185

bacterial pathogens, 233
bacterial strains, xi, 40, 142, 149, 155, 156, 157, 165
bacterium, 45, 208
bioactivity, v, viii, 1, 2, 5, 39, 81, 129
biodegradability, 146
biodiversity, viii, 2, 3
biogeography, 67, 243, 283
biological activity(ies)ies, viii, 2, 36, 39, 44, 47, 51, 59, 64, 65, 66, 67, 72, 83, 122, 125, 134, 138, 139, 143, 147, 168, 170, 173, 177, 231, 234
biological systems, 183
biomass, vii, xii, 70, 182, 183, 184, 185, 187, 188, 191, 201, 202, 208
biomedical applications, 205
biosynthesis, 71, 187, 232, 236
biotechnological applications, 183
biotechnology, xii, 122, 181, 183, 184, 188, 236
bronchial asthma, 111
bronchitis, ix, 42, 43, 48, 54, 96, 98
bronchodilator, 49
buccal mucosa, 66

C

cancer, 52, 97, 111, 120, 121, 127, 133, 147, 168, 231, 232
candida, 41, 45
capillary, 15, 92, 308
carbohydrates, 106, 182, 187
carbon dioxide, 137, 140, 176, 178
cardiac glycoside, 182
cardiovascular diseases (CVD), 50, 97, 120, 132, 231, 233
cell membranes, 147, 185
Chinese medicine, 305
chitin, 186, 187
chitosan, 137, 175, 186, 187
cholesterol, 44, 213, 233, 240
cholinesterase, 40
chromatograms, 148, 163, 192
chromatography, xiv, 148, 298
chromatography analysis, 148
chromosome, xiii, 242, 259, 263, 271, 272, 273, 278, 292
chronic venous insufficiency, 135, 173
cognitive function, 120, 132
cognitive performance, 51
colitis, x, 142, 145
compounds, ix, x, xii, 3, 17, 27, 42, 46, 68, 74, 82, 96, 97, 98, 105, 108, 110, 113, 115, 116, 120, 121, 125, 126, 128, 129, 130, 131, 133, 134, 146, 147, 149, 153, 155, 168, 173, 182, 183, 204, 211, 213, 215, 216, 230, 232, 233, 234, 235, 236, 239, 303, 307, 311,312, 313, 314, 315, 316, 317, 319, 320, 321, 322, 323
constituents, viii, xiv, 2, 3, 9, 10, 12, 15, 17, 21, 25, 27, 29, 31, 33, 34, 36, 37, 38, 43, 44, 47, 49, 52, 53, 54, 60, 65, 68, 69, 70, 72, 76, 79, 82, 83, 85, 90, 92, 105, 123, 129, 130, 143, 146, 148, 149, 158, 159, 298, 302, 308, 312, 314, 315, 319, 321, 323
cultivars, 14, 24, 26, 55, 59, 71, 73, 74, 77, 86, 131
cultivation, 14, 23, 55, 81, 253, 258, 276, 299
culture, 182, 183, 184, 187, 188, 189, 191, 204, 205, 206, 207, 209, 238
culture conditions, 183, 184, 188
culture media, 189

D

derivatives, 3, 82, 110, 236
detection, 103, 150, 207, 279
diabetes, ix, 39, 53, 96, 98
diarrhea, ix, 41, 43, 96, 98

diseases, x, xii, 29, 43, 48, 51, 97, 120, 122, 142, 145, 165, 211, 212, 231, 236
dispersion, xi, 142, 148, 151, 152, 158, 159, 160, 161, 162, 163, 164, 188
distillation, 64, 76, 306
distilled water, 101, 185, 188, 191
distribution, xi, xiii, 9, 36, 60, 78, 98, 142, 144, 153, 161, 242, 244, 253, 255, 274, 294, 299, 305
diterpenoids, 7, 19, 29, 36, 39, 60, 62, 64, 66, 74, 126
diuretic, 3, 17, 39, 44, 47, 48, 50, 53, 54
divergence, 248, 273, 281, 324
diversity, vii, viii, 2, 4, 7, 17, 21, 24, 27, 29, 30, 55, 63, 67, 74, 87, 92, 98, 243
DNA, xiii, 71, 79, 121, 169, 185, 239, 242, 247, 250, 264, 266, 269, 271, 283
DNA damage, 239
DNA repair, 121
dominance, viii, 2, 9, 10, 21, 26
drug carriers, 146, 168
drug delivery, 185
drug discovery, 5
drugs, vii, xi, 3, 70, 76, 142, 146, 157, 171, 212

E

E. coli, xi, 40, 45, 142, 147, 149, 150, 156, 157, 209
Elsholtzia, vii, ix, 2, 5, 10, 12, 26, 41, 60, 61, 63, 66, 72, 73, 74, 78, 85, 88
encapsulation, vii, x, xi, 142, 146, 152, 160, 163, 164, 171
energy, 127, 138, 176, 185
engineering, 139, 177
environment, xii, 97, 136, 146, 174, 181, 207
environmental conditions, ix, 2, 5
enzymes, 120, 145, 183, 232, 237
essential fatty acids, 159
ethanol, x, 96, 99, 100, 101, 103, 104, 105, 106, 107, 109, 110, 112, 115, 123, 137, 176
extraction, ix, 64, 74, 76, 96, 97, 98, 100, 101, 105, 106, 107, 109, 110, 112, 113, 115, 121, 124, 125, 126, 134, 138, 139, 140, 143, 153, 167, 173, 176, 177, 178, 213, 230, 234, 239, 250
extracts, vii, viii, ix, x, xii, 37, 42, 64, 69, 70, 81, 96, 97, 98, 100, 101, 102, 103, 104, 105, 106, 107, 108, 109, 110, 111, 112, 113, 114, 116, 117, 118, 119, 120, 121, 122, 123, 124, 125, 127, 129, 130, 131, 134, 136, 137, 138, 139, 142, 143, 145, 166, 168, 173, 175, 176, 177, 178, 185, 186, 188, 189, 208, 211, 213, 215, 230, 233, 234, 240, 286, 288

F

fatty acids, 26, 59, 159
flatulence, 43, 302
flavonoids, viii, xii, 1, 3, 7, 26, 29, 34, 52, 54, 70, 106, 113, 114, 120, 121, 129, 130, 132, 133, 183, 187, 212, 230, 233, 236, 237, 239
flowers, 28, 31, 131, 185, 188, 251, 274, 299, 300, 303, 304
fluid, 69, 111, 167
fluid extract, 69, 167
food, viii, ix, 1, 2, 5, 10, 12, 14, 21, 26, 42, 43, 47, 48, 49, 50, 51, 54, 96, 108, 121, 132, 134, 173, 182, 185, 212, 233, 234, 235, 239, 303
food additive, 183
food additives, 183
food industry, 185
food products, ix, 42, 43, 54, 96
food spoilage, 239
forest resources, 4
free radicals, 97, 110, 111

fruits, 128, 130, 138, 177, 232, 234
functional food, 84, 123
fungal infection, 160, 186
fungi, 45, 71, 90, 165, 231, 232

G

GC-FID, 148, 149, 308
genome, 247, 272, 280, 281
genus, viii, ix, xiii, xiv, 5, 7, 9, 10, 12, 14, 20, 24, 26, 28, 30, 32, 34, 35, 36, 37, 47, 48, 58, 60, 68, 80, 82, 83, 84, 88, 96, 98, 122, 123, 138, 143, 177, 242, 243, 246, 248, 250, 253, 254, 258, 263, 272, 273, 277, 278, 283, 284, 288, 292, 298, 301, 302,303, 304, 305, 308, 312, 315, 324, 325
genus *clinopodium*, 298, 301, 308
genus *hyptis*, 298, 302, 312
genus *lepechinia*, 298, 303, 315
genus *minthostachys*, 298, 304, 317
genus *origanum*, vi, 68, 241, 271, 284, 298, 304, 319
genus *scutellaria*, 298, 305, 321
germination, 41, 45, 125, 210, 231, 232
glucoside, 100, 103, 112, 113, 116, 118, 119, 186
glucosinolates, 183
glutathione, 108, 186
glycosylation, 114, 131
gold nanoparticles, 206
green synthesis, vi, 181, 182, 185, 205, 209
growth dynamics, 210
growth hormone, 231
growth temperature, 88

H

harvesting, 70, 77, 94, 97
Helicobacter pylori, 205
hepatitis a, 52, 53
hepatoma, 147, 169
herbal medicine, 41, 125, 240
herbal teas, 233
heterogeneity, 161
homogcncity, 161
homoploid hybridization, vi, viii, xiv, 241, 242, 247, 248, 272, 276, 278, 281, 286, 288, 290, 291, 292
homoploid speciation, 242, 273
human health, ix, 96, 98, 126, 167, 207, 231, 233, 234
human immunodeficiency virus, 66, 239
hybrid, xiii, 23, 242, 246, 247, 249, 250, 253, 257, 258, 264, 269, 270, 271, 272, 275, 278, 279, 280, 281, 282, 285, 286, 287, 288, 289, 290, 291, 292, 294, 295
hybridization, viii, xiii, 242, 246, 247, 248, 258, 271, 272, 273, 275, 276, 278, 279, 280, 281, 282, 286, 288, 290, 291, 292
hydrazine, 108
hydrocarbons, 22, 29, 36, 37, 61, 153, 307, 311, 312, 313, 314, 315, 316, 317, 319, 320, 321, 322, 323
hydrogen atoms, 110
hydrogen peroxide, 88
hydrolysis, 159
hydrophobic properties, 98, 144
hydrophobicity, 158
hydroxyl, 113, 114, 117
hydroxyl groups, 113, 114, 117
hypertension, 39, 50, 53, 120, 132
hypotension, 234
hypotensive, 237

I

immune system, 182, 213
impregnation, 137, 138, 139, 175, 176, 177
in vitro, vii, xii, 41, 45, 114, 121, 127, 131, 136, 138, 159, 166, 169, 171, 175, 176,

181, 182, 184, 186, 188, 192, 197, 201, 204, 205, 207, 208, 213, 235, 237, 239
in vitro biomass production, 182
in vivo, 127, 136, 137, 166, 175
inflammation, ix, 96, 97, 98, 124, 238
inflammatory disease, 52, 53
infusion, 101, 150
inhibition, xi, 40, 45, 58, 71, 104, 121, 130, 142, 145, 150, 156, 157, 158, 185, 232
irritable bowel syndrome, 50

K

kaempferol, xiii, 100, 103, 112, 113, 119, 212, 216, 230, 231, 236
karyotype, 288, 292
kidney, 50, 51, 305

L

Lactobacillus, 206
lamiaceae family, vi, vii, 297, 298, 299, 300
Lavandula officinalis, vi, vii, xii, 181, 182, 187, 188
LC-MS, 215, 286, 288
LC-MS/MS, 286, 288
lignans, viii, 1, 3, 54
lignin, 85, 232
lipid peroxidation, 41, 46, 128, 130
lipids, 146, 151, 182, 186, 187
liposomes, vii, xi, 142, 143, 146, 158, 159, 160, 161, 162, 165, 168, 170, 171, 235
Listeria monocytogenes, 147
loci, xiv, 242, 266, 269
low-density lipoprotein, 130
lung cancer, 147, 169
lyme disease, 233

M

maceration, x, 96, 97, 99, 100, 105, 106, 107, 109, 110, 112, 115
medicinal and aromatic plants, 2, 3, 4, 82
medicine, viii, ix, 1, 2, 4, 7, 9, 26, 29, 32, 34, 36, 39, 40, 42, 43, 44, 47, 48, 49, 50, 51, 52, 53, 54, 96, 97, 98, 105, 111, 212, 213, 231, 302
Mediterranean, 12, 24, 36, 48, 143, 243, 254, 274, 305
Mediterranean countries, 48
melanoma, 147, 169
metabolism, 130, 182, 187
metabolites, viii, xii, 1, 3, 5, 7, 26, 29, 32, 34, 40, 49, 54, 81, 97, 106, 115, 122, 123, 130, 131, 138, 145, 177, 182, 183, 184, 185, 206, 207, 209, 211, 213, 230, 234, 240
metal ions, 186
metal nanoparticles, xii, 181, 209
microorganisms, 184, 232, 233
microspheres, vii, xi, 142, 146
molecules, viii, 2, 5, 33, 51, 55, 185, 187
monoterpenoids, 9, 11, 15, 18, 21, 29, 33, 36, 233
myocardial ischemia, 42

N

nanoparticles, vii, xii, 181, 182, 184, 185, 186, 187, 188, 189, 191, 192, 193, 197, 199, 201, 204, 205, 206, 207, 208, 209, 210
nanotechnology, xii, 181, 184
natural compound, 206, 212, 232
natural habitats, 170, 246, 248, 253
Nepeta, vii, ix, 2, 5, 16, 18, 19, 20, 44, 60, 61, 62, 64, 65, 68, 69, 71, 73, 81, 82, 90, 213, 214, 284, 289, 299
nervous system, 51, 305

Netherlands, 68, 122, 204, 277
neurodegenerative diseases, 120
neurodegenerative disorders, 97
neurological disease, 305
nitrogen, viii, xii, 108, 120, 183, 212, 214, 250
node culture, xii, 182, 199, 204
nucleotides, 250, 264, 269, 270, 271
nutraceutical, 130, 232
nutrient(s), 67, 153, 187
nutrition, viii, 1, 2, 54

O

O. heracleoticum, v, ix, x, 95, 96, 97, 100, 106, 108, 109, 110, 111, 112, 113, 114, 115, 116, 117, 118, 119
Origanum, v, vii, viii, ix, x, xiii, xiv, 2, 5, 24, 25, 48, 76, 78, 86, 87, 95, 96, 97, 98, 107, 109, 110, 121, 123, 124, 125, 126, 128, 129, 131, 141, 142, 143, 167, 168, 169, 170, 171, 214, 242, 243, 244, 246, 248, 249, 250, 253, 254, 255, 256, 257, 258, 259, 260, 263, 264, 265, 272, 273, 274, 275, 277, 278, 283, 284, 285, 286, 287, 288, 289, 290, 291, 292, 294, 295, 298, 305, 319, 320, 321, 325
Origanum compactum, v, x, 141, 142, 143, 167, 168, 169, 170
Origanum vulgare, v, ix, xiv, 78, 86, 87, 95, 96, 97, 98, 123, 124, 125, 128, 129, 131, 214, 243, 258, 260, 284, 298, 305, 319, 320, 321, 325
oxidative damage, 97, 233
oxidative stress, 97, 121, 122, 240
oxide nanoparticles, 206

P

pH, 98, 104, 184, 186, 189, 190, 191, 192, 257
pharmaceutical(s), viii, xi, 1, 2, 12, 14, 24, 37, 42, 43, 47, 51, 54, 108, 134, 142, 159, 173, 182, 213, 234
pharmacokinetics, xi, 142
pharmacological research, 80
pharmacology, 72, 123, 129, 325
phase inversion, 171
phenol, 144, 158
phenolic compounds, ix, xiii, 38, 96, 98, 102, 105, 106, 107, 108, 109, 110, 112, 118, 120, 121, 125, 128, 129, 131, 147, 153, 155, 167, 183, 212
phenotypes, 168, 247
phenyl propanoids, 21
phosphatidylcholine, 159, 161, 162
phospholipids, vii, xi, 142
plant diseases, 45
plants, viii, xii, xiii, xiv, 2, 4, 8, 9, 19, 20, 24, 30, 34, 48, 49, 54, 55, 58, 62, 63, 65, 68, 72, 74, 75, 76, 80, 81, 82, 84, 85, 89, 91, 97, 98, 105, 108, 111, 117, 122, 123, 125, 129, 133, 134, 135, 136, 143, 144, 145, 153, 168, 172, 173, 175, 182, 183, 186, 187, 204, 211, 212, 213, 214, 230, 231, 232, 233, 234, 235, 236, 238, 240, 242, 271, 274, 277, 279, 280, 282, 284, 297, 298, 299, 301, 302, 304
Pliocene, 272
ploidy, xiii, 242, 246, 271, 272
pollen, xiii, 242, 248, 249, 258, 259, 260, 261, 262, 263, 284, 293, 294
polydispersity, 152, 160
polymorphism, 170, 269, 270
polymorphisms, 269, 270
polyphenols, 12, 102, 113, 123, 133
polyploid, xiii, 242, 246, 247, 279, 280, 284
polyploidy, xiii, 242, 248, 271, 280, 281, 282
polypropylene, 137, 176
polysaccharide(s), 66, 186
polyunsaturated fatty acids, 27
proteins, 106, 145, 146, 182, 186, 215

prototype, 3, 127, 138, 176
Pseudomonas aeruginosa, 147, 233

Q

quercetin, 100, 103, 112, 113, 119, 128, 231
quinones, 29

R

radula, xiv, 298, 304, 315, 316
refraction index, 306, 308, 314, 315, 318, 321
regeneration, 120, 205
regions of the world, 14, 32
regression analysis, 104
repellent, 3, 17, 40, 43, 44, 71, 90, 137, 176
reticulum, 259
rheumatic diseases, 135, 174
rheumatic fever, 39
rheumatoid arthritis, 111
room temperature, 100, 101, 104, 159, 186, 189, 191, 250
root(s), vii, xii, 8, 19, 29, 31, 36, 46, 50, 70, 74, 79, 87, 182, 183, 185, 187, 190, 191, 196, 197, 199, 200, 201, 203, 206, 210, 299
Rosmarinic acid, xiii, 47, 97, 111, 116, 118, 212, 216, 233, 236

S

secondary metabolism, 183, 187
sedative, 3, 43, 44, 48, 53, 54
seed, 27, 49, 72, 88, 166, 187, 210, 292, 300
seedlings, 206, 210
SEM micrographs, 261, 262
sequencing, 250, 271
shoot(s), vii, xii, 182, 187, 190, 191, 193, 194, 199, 201, 204, 207, 214
shrubs, 12, 20, 34, 299, 304
silver nanoparticles, vi, vii, xii, 181, 182, 184, 185, 186, 187, 188, 189, 191, 192, 193, 197, 199, 201, 204, 205, 206, 207, 208, 209
solvents, vii, ix, x, 96, 97, 98, 107, 121, 125, 143
Soxhlet extraction, x, 96, 97, 99, 100, 101, 105, 106, 109, 110, 112, 113, 115, 139, 177
stability, vii, x, 138, 142, 143, 146, 159, 161, 162, 165, 171, 177
stress, 47, 51, 52, 68, 81, 97, 108, 122, 124, 126, 182, 186, 231
synthesis, viii, xiii, 2, 3, 52, 75, 145, 182, 185, 186, 189, 205, 206, 207, 209, 212, 232, 240
Syzygium aromaticum, 182, 188

T

tannins, viii, 1, 3, 12, 54, 99, 102, 106, 120
taxa, 14, 24, 30, 63, 77, 78, 87, 243, 247, 249, 253, 254, 257, 258, 260, 261, 262, 264, 269, 271, 275, 278, 291, 292
toxicity, xii, 40, 146, 181, 206, 231
tropical forests, 4
tumor(s), 39, 47, 48, 50, 53, 232, 234, 240
tumor growth, 47

U

UV radiation, 139, 178, 231

V

varieties, 22, 30, 37, 73, 88, 243
vegetable oil, 148
vegetables, 48, 232
vegetative stages, 27

vitamin B2, 286
vitamin E, 111
vitamins, 184
work-related stress, 51
World Health Organization (WHO), 212
wound healing, 127, 136, 160, 166, 175

W

water, viii, x, xii, 76, 96, 97, 99, 101, 105, 106, 107, 109, 110, 112, 114, 115, 123, 135, 137, 145, 153, 165, 174, 176, 185, 188, 189, 211, 213, 214, 215, 216, 223, 229, 230, 234

Z

zinc, 184, 187
ZnO, 210